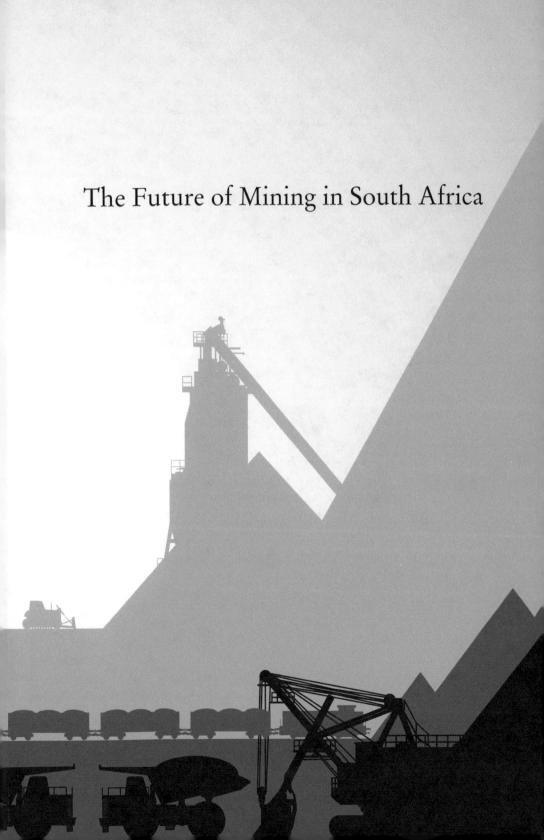

The Future of Mining in South Africa

The Future of Mining in South Africa:
Sunset or Sunrise?

EDITED BY

Salimah Valiani

THIS RESEARCH PROJECT WAS SUPPORTED BY:

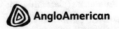

First published by the Mapungubwe Institute for Strategic Reflection (MISTRA) in 2018

142 Western Service Rd
Woodmead
Johannesburg, 2191

ISBN 978-0-6399238-2-6

© MISTRA, 2018

Production and design by Jacana Media, 2018
Text editors: Terry Shakinovsky, Wandile Ngcaweni, Susan Booysen
Copy editor: Linda Da Nova
Proofreader: Christopher Merrett
Designer: Alexandra Turner

Set in Stempel Garamond 10.5/15pt
Printed and bound by Creda Communications
Job no. 003413

When citing this publication, please list the publisher as MISTRA.

Contents

Preface

HOW CAN THINGS BE DIFFERENT in South Africa's mining industry? How can a shift away from the low-skilled, labour-intensive model change the work and workings of the industry? How can mining-affected communities, and the nation more broadly, benefit equitably from the country's mineral endowments? What are the links between mineral extraction and industrialisation? What are the ideal roles of the state, private corporations and unions in a 21st century mining industry?

This research project was born of these big questions. It was inspired not only by an appreciation of South Africa's extensive mineral endowments; but also by a realisation that, while the South African mining industry performs relatively well on many technical indicators, its management of the broader social issues leaves much to be desired. Critically, the principle needs to be debated whether, going forward, the mining industry can play as critical a role as it did in the evolution of the South African economy.

Collectively, the authors pursue these questions with a transdisciplinary lens, in line with the defining principles of the research endeavours of the Mapungubwe Institute for Strategic Reflection (MISTRA). Questions about the interplay of mining with the environment and climate change come into the mix. So do issues around the experiences of women – arguably the 'lesser gender' in the mining industry more than in any other major industry.

At the same time, the way in which the historical roots of dispossession take shape in mining industry dynamics today are examined, whether in the desperation and 'illegality' of informal mining or the trials and tribulations of mining-affected communities. Beyond this is the critical challenge brought out sharply in the sprawling mining dumps across the face of South Africa: what happens after mines are long gone?

This volume brings together political economists, sociologists, policy experts, environmental experts, community activists, historians and labour activists. The process of producing the volume was highly collaborative. The authors first gathered to conceptualise the chapters collectively and then assembled to hear one another's findings and to refine these, both collectively and individually. The editor played the role of interlocuter, prodding the contributors to consider relevant insights from other contributors in the multiple rounds of chapter drafting.

In the course of these interactions, new issues came to the fore. For instance, while the history of South African mining is treated against the backdrop of gold mining, the decline of the latter has coincided with the emergence of platinum group metals (PGMs), of which South Africa is estimated to have some three-quarters of world reserves. Currently, though, this sub-sector is facing serious headwinds of low prices, high costs, community unrest and uncertain trajectories of global demand. However, viewed against the emergent hydrogen economy and growing demand for jewellery, PGM mining does seem to have a long-term future barely appreciated.

More often than not, workers in the mining industry, as in most other economic sectors, are treated as an appendage, a cost to be managed down. In appreciation of their central role, the issue of their involvement at all levels of the industry – from the bowels of the earth to the skyscrapers where the product of their labour is apportioned –

has to form a critical part of the examination of the political economy of South African mining. All matters to do with raising the mining sector onto a higher trajectory have to place workers at the centre, including skills development, the opportunities that should come with the development of a mature mining cluster and the organisational adaptations required of a labour union of the 21st century.

As this book goes to press, the debate on the Mining Charter, meant to improve the sector's inclusivity, has reached a crescendo. This is understandable, given the weaknesses identified in several chapters in this volume. At the same time, the spate of accidents in South African mines has intensified; there is concern that, after a steady decline in the past 10 years, worker fatalities may be trending upwards.

It is the hope of the authors that the issues raised in this volume will encourage broader discussion on the role of mining in a South Africa that is striving to improve its growth and development. Nay more, if the ideas contained in this volume were to serve as catalyst for action, MISTRA would have more than achieved its ambition.

MISTRA wishes to express its profound appreciation to the team that put this volume together, from the authors to the peer reviewers, editors and project coordinators. Our gratitude is also extended to the Department of Science and Technology, Anglo American and AngloGold Ashanti who funded this project, as well as the many other contributors to MISTRA's operations and sustainability.

– Professor Sibusiso Vil-Nkomo
Chairperson: MISTRA Board of Governors

Acknowledgements

THE MAPUNGUBWE INSTITUTE for Strategic Reflection (MISTRA) would like to express its gratitude to the coordinator and project leader, Salimah Valiani, who edited this volume. Thanks go to Wandile Ngcaweni, who provided valuable support and assistance, and to the former MISTRA researchers who were part of this project from its inception: Mcebisi Ndletyana, David Maimela, Patience Kelebogile Salane and Temoso Mashile.

Thank you to the MISTRA staff who contributed to the successful outcome of this project: the project management directorate led by Xolelwa Kashe-Katiya, supported by Dzunisani Mathonsi and Towela Ng'ambi; Lorraine Pillay for fundraising and financial management activities, with support from Magati Nindi-Galenge; Terry Shakinovsky for editing the manuscript and Susan Booysen and Barry Gilder for their valuable contributions in exercising oversight and safeguarding the integrity and quality of the publication.

We also wish to express our thanks to the authors who contributed to this volume. MISTRA is honoured to have worked with them. Our gratitude also goes to the peer reviewers for their valuable assessments of the draft chapters.

MISTRA extends its appreciation to Jacana Media which was responsible for the design, layout and production of this publication.

PROJECT FUNDERS

Intellectual endeavours of this magnitude are not possible without financial resources. Special thanks go to **the Department of Science and Technology (DST), Anglo American Platinum and AngloGold Ashanti** without whom this project would not have been possible.

MISTRA would also like to acknowledge the donors who were not directly involved with this particular research project but who support the Institute and make its work possible. They include:
- ABSA
- Airports Company of South Africa Limited (ACSA)
- Albertinah Kekana
- Anglo American
- Anglo Coal
- Aspen Pharmacare
- Batho Batho Trust
- Belelani Group
- Brimstone
- Chancellor House
- Discovery
- First Rand Foundation
- Friedrich-Ebert-Stiftung (FES)
- Goldman Sachs
- Harith General Partners
- Jackie Mphafudi
- Kumba Iron Ore
- Mitochondria
- National Institute for the Humanities and Social Sciences (NIHSS)
- National Lotteries Commission (NLC)
- Oppenheimer Memorial Trust (OMT)
- OSISA (Open Society Initiative South Africa)
- Phembani Group

- Power Lumens Africa
- Royal Bafokeng Holdings
- Safika
- Shell South Africa
- Simeka
- South Africa Breweries
- Standard Bank
- Vhonani Mufamadi
- Yellowwoods

Contributors

Crispen Chinguno is Senior Lecturer (Sociology) at Sol Plaatje University. He is currently working on a book looking at worker struggles and shifting trade union organising strategies, drawing on case studies from South Africa's platinum belt.

Duma Gqubule is the founder and director of the Centre for Economic Development and Transformation (CEDT). He has spent the past two decades as a financial journalist, analyst, adviser and consultant on issues of economic development and transformation.

Edwin Ritchken is the coordinator of the Mining Phakisa, an industry-wide collaborative initiative between government, the mining industry, labour and other stakeholders, to galvanise growth and technology development in the mining sector. He specialises in industrial policy, strategic sourcing and supplier development, and state-owned enterprise governance and oversight.

Hameda Deedat is Acting Executive Director at Naledi, the research arm of the Congress of South African Trade Unions (COSATU). She is also a senior researcher in gender, trade, climate change, BRICS, the future of work and other labour research.

Hibist Kassa is an Executive Committee member of Development Alternatives with Women in a New Era (DAWN). She is a PhD candidate, attached to the Centre for Social Change and to the Department of Sociology at the University of Johannesburg.

Joel Netshitenzhe is Executive Director and Vice-Chairperson of the Board of Governors of the Mapungubwe Institute for Strategic Reflection (MISTRA). He has been a member of the National Executive Committee of the ANC since 1991.

Khwezi Mabasa is a researcher in the Faculty of Political Economy at MISTRA and a part-time lecturer in the Department of Political Sciences at the University of Pretoria.

Lorenzo Fioramonti is Professor of Political Economy at the University of Pretoria, where he directs the Centre for the Study of Governance Innovation, and is Deputy Project Leader of the Future Africa initiative. He is also a member of Parliament in Italy, since March 2018.

Nester Ndebele is the chair of Women Affected by Mining United in Action (WAMUA), a South African organisation that aims to mobilise women in grassroots communities. She has been engaged in mining and environmental activism globally for the past 10 years.

Ross Harvey leads the extractive industries research work of the Governance of Africa's Resources Programme at the South African Institute of International Affairs (SAIIA). His research focuses on how to transform natural resource endowments into inclusive, sustainable development.

Salimah Valiani is Senior Researcher, Faculty of Political Economy, at MISTRA. She has been working as an advocate and researcher in mining and development issues since the mid-1990s, with particular focus on Indonesia, the Philippines, India, Canada, Libya, Uganda and South Africa. She is also Visiting Professor at the Centre for Researching Education and Labour at Wits University.

Shingirirai S. Mutanga is a senior research specialist under the Science and Technology programme of the Africa Institute of South Africa, a division of the Human Science Research Council (HSRC), and is also a Senior Research Associate at the University of Johannesburg. He co-edited a publication titled *Africa in a Changing Global Environment: Perspectives of Climate Change Adaptation and Mitigation Strategies in Africa.*

Sonwabile Mnwana is Associate Professor and Head of Department of Sociology at the University of Fort Hare, East London. He is the former deputy director of the Society, Work and Development Institute (SWOP), University of the Witwatersrand. He is also a Research Associate at SWOP and project leader for the investigation *Mineral Wealth and Politics of Distribution on the Platinum Belt.*

ONE

Introducing the debates

SALIMAH VALIANI

AT THE TIME OF WRITING, a number of developments in the world of South African mining were unfolding. The agitation caused by Mining Charter 2017 had largely subsided, particularly for mining companies wanting far less responsibility in sharing the wealth of mining than that assigned to them in the Charter. The Draft Mining Charter 2018 – softening the black economic empowerment requirements on companies and giving them more time to achieve them – had been released for public comment after satisfying companies and other consulted stakeholders. The Minister of Minerals and Resources had just completed visits to mining-affected communities across South Africa – a cursory nod to a North Gauteng High Court ruling recognising mining-affected communities as relevant parties in the (later withdrawn) legal review of Mining Charter 2017. And, after nearly two decades of litigation, 29 mining houses had settled out of court to compensate some 100,000 former workers still alive who had contracted silicosis and tuberculosis in gold mines from 1965 onwards.

With these developments, little has actually changed. Remaining, as ever, are the issues of long-term investment in mineral-based development; agreement on appropriate roles and responsibilities for capital and the state; mine worker health and safety; mining-induced societal and environmental damage; post-mine clean-up; the intersection of mining and climate change; mine worker retrenchment and union renewal; and the most socially effective distribution of mineral revenues and use of land. Though not always discussed in the same quarters, these issues give rise to a host of perspectives and positions. Many of them clash. To put the question of the future of mining in South Africa, as MISTRA does in this edited volume, is an attempt to frame debates in a unifying manner and bring them under a single roof – that of 'the future'.

But the debates must be had – views presented, contested, rebutted and expanded – to allow for the possibility of reaching consensus, if not unity. Given mounting socio-political tensions and intensifying environmental consequences of mining in South Africa, one of the oldest and largest extractive economies of the African continent, the time to debate with the aim of consensus, and ultimately change, is no later than now. The words of Somadoda Fikeni (2018), from the launch of the Indlulamithi South Africa Scenarios 2030, are instructive: 'Time belongs to all of us, and time belongs to no one.'

Many of the same debates around mineral extraction are unfolding in other countries and continents, with varying 'meetings of minds' resulting. Elements of these debates and meetings of minds are pertinent to discussions around the future of mining in South Africa, while the reverse is also true. The discussion in this introduction thus highlights key points of chapters in this volume while simultaneously linking them to mining-related concepts and developments emanating from beyond South Africa. The volume is offered as a tool to position debates on national, subnational, as well as international scales – particularly given the oneness of the world mineral market and the increasingly acknowledged ecological interconnectedness of all parts of the world. In other words, the positions, empirical evidence and questions presented in this volume are argued to be vital both to debating the future of mining in South Africa and to larger debates on mining and mineral dependence globally.

In one of the most recent essays from Africa about the natural resource curse in Africa – a notable piece in that the bulk of current scholarly work on African economies is by non-Africans based outside of Africa (Chelwa, 2017) – Takavafira Masarira Zhou (2017: 280) argues that 'the curse' is not natural resources, but rather 'bad stewardship of resources'. Zhou is responding to longstanding discussions about the tendency of resource-rich countries to experience low economic growth rates, overvalued or/and fluctuating currencies, unstable domestic demand leading to retarded investment, poor price prediction and the pauperisation of labour (Baran, 1957; Lewis, 1984; Gelb, 1988; Auty, 1993).

Similar to Zhou, Joel Netshitenzhe, in chapter 2 of this volume, takes the position that the resource curse is a product of social agency and hence strategic planning and focused interventions can amount to the inverse of the experiences associated theoretically with the phenomenon. What is required, according to Netshitenzhe, is that the collective of partners in mining – private companies, workers, mining communities and the state – come together to develop a vision and programme that aligns with the objectives of South Africa's National Development Plan (NDP). Netshitenzhe offers a comprehensive organising framework for the vision, highlighting key issues and the most current related innovations. The following are the components of the framework: extraction, infrastructure, modernisation, backward linkages, forward linkages, research and development, ownership, jobs and human resource development, social and labour plans, exploration regime, land utilisation, post-mining activities and informal mining. These components, and the systematic way in which they are presented and linked to the NDP, provide a sound springboard for envisioning and discussion, both in this volume and beyond.

Netshitenzhe is optimistic that a coming together of collective partners will lead to intellectual engagement, envisioning and programming, while Zhou (2017: 280) underlines that it is 'people', particularly those with power, who abuse natural resources. Nevertheless, both agree that judicious management of resources is the solution. Taking a more systemic view, Taft (2017) argues that 'institutional capture' is the process whereby energies set up to serve the public interest – regulators, government departments and so on – serve private interests instead. Taft, a former opposition party leader of the oil-rich Canadian province, Alberta, examines how the oil industry has grown into a state within the Canadian state over the past 25 years. The 'deep state of oil' thus makes for seemingly contradictory commitments of the Canadian federal government, for instance, assuring both the construction of a pipeline connecting Alberta's tar sand-derived oil[1] to the Pacific and the cutting of greenhouse emissions by 30 per cent by 2030. Taft's study shows that such institutional capture can be a phenomenon of not only African countries, but resource-rich countries broadly.

The task of social agency, however, remains. Edwin Ritchken, in chapter 3, elaborates on the status quo of the platinum group metals (PGMs) in South Africa, which is the near opposite of Taft's 'deep state'. Ritchken argues that what exists in South Africa is a fragmented, incapacitated state – uninformed about the potential national value of PGMs – combined with two companies in control of the supply of PGMs but not investing in them adequately. Concretising Netshitenzhe's emphasis on social agency through a focus on PGMs – one of South Africa's greatest, underdeveloped mineral endowments – Ritchken argues for a 'development coalition', or collaboration among a 'critical mass of key role players' aiming for mutually desirable development outcomes and the long-term future of the metals. Some of the actions he suggests can be taken up by such a coalition: supporting agricultural development by making land, water and other enabling assets (large amounts of which are held by mining companies) available to rural communities; making waste dumps and streams available for processing by emerging miners; leveraging procurement to drive local production of mining equipment; and providing long-term security of platinum supply to the associated export development zone.

The optimism around PGM-based industrial, export and green development potential in South Africa is problematised by Sonwabile Mnwana in chapter 7. Stressing the 'intricate dynamics' of rural land

---

1   An unconventional petroleum deposit found in sand and sandstone. One of the world's largest sources of tar sands is in north-east Alberta, Canada.

holding in South Africa, Mnwana argues that these are yet more pronounced in South Africa's platinum belt, where African families access land through customary rights. Through collusion with local chiefs and the state, mining companies enjoy easy entry into these lands. Using archival and interview data from the Bakgatla area of today's North West province, Mnwana shows how these processes have not only led to yet another round of dispossession in rural communities on the platinum belt, but also how they tend to privilege mining capital when struggles over rural land ensue.

On a deeper level of abstraction, Mnwana shows how the very meaning of land has changed for African peasants struggling through various rounds of dispossession. As a resource to be worked up into crops and food, land was seen as something to be shared widely, with expanding numbers of people when necessary. As the same land increasingly became a source of minerals, African peasants moved to seeing land as something to be possessed, in ever-smaller numbers. In a sense this reveals a lesser weighting given to minerals and money, and a greater weighting given to land and fruits of the land.

Put slightly differently, the value of land is infinite when used for cultivation and finite when used to extract minerals. Beyond Mnwana's study, this notion is reflected in current thinking around the valuing of natural capital. Natural capital can be defined as the world's stock of natural assets including geology, soil, air, water and all living things. 'Ecosystem services' are the many use-values derived from natural capital by humans: food, water, plant materials forming the basis of fuel, medicines and construction materials, as well as the less tangible such as climate regulation and flood defence provided by forests (World Forum on Natural Capital, 2018).

Former World Bank economist Herman Daly was one of the first to introduce these notions to development policy discussions as far back as the mid-1990s (Daly, 2007). Some of Daly's major contributions to policy and measurement issues are: to not count natural capital as income, to tax resource throughput more and labour and income less and to maximise the productivity of natural capital in the short run while investing in increasing its supply in the long run (Daly, 2007). Subsequent calculations in *The Little Green Data Book* of the

World Bank show that in sub-Saharan Africa, adjusted net savings (as percentage of gross national income)[2] in the first decade of the 21st century, thus including the years of relatively high growth, not only declined but dipped below zero to -1.3 per cent. According to the *Little Green Data Book*, this suggests 'unsustainable development and declining wealth' (World Bank, 2013: vii). For South Africa, adjusted net savings in 2014 was 2.1 per cent. This compares with a 2014 average of 5.5 per cent for the sub-Saharan Africa group and 23.7 per cent for the upper middle-income country group (World Bank, 2016: 189).

An example of the logic of valuing natural capital extended fully is El Salvador's 2017 law banning all metal mining. Political parties from across the spectrum and even the Roman Catholic Church united in the effort to turn a 10-year moratorium into national law. The shift was driven by the united will to save dwindling clean water supply in the country (Palumbo and Malkin, 2017).

Looking at the issue of valuing natural capital from the opposite end of the kaleidoscope, as it were, Shingirirai Mutanga tackles the question of post-mine clean-up in South Africa in chapter 8. The magnitude of the question is considerable given the more than 6,000 estimated derelict and ownerless mines in the country (Winde, 2018: 7). Mutanga underlines the major challenges posed by disused or resource-depleted mines: acidification of water bodies, degraded soil quality, biodiversity loss, obliteration of natural landscapes and the multiple ripple effects on human wellbeing. Beginning with a discussion of the 2015 United Nations Sustainable Development Goals (SDGs), Mutanga highlights the links between SDG 6 and mining. SDG 6 aims to ensure access to clean water and sanitation for all. As indicated by the United Nations Development Plan (2016), more than 40 per cent of the world's population is affected by water scarcity. Within this global conundrum,

---

2   The 2013 equation for adjusted net savings was gross savings minus consumption of fixed capital, plus education expenditures, minus energy depletion, mineral depletion, net forest depletion, particulate emissions and carbon dioxide damage. The 2016 equation was: gross savings minus consumption of fixed capital, plus education expenditures, minus energy depletion, mineral depletion, net forest depletion and carbon dioxide and air pollution damage.

South Africa has been declared a water scarce country (Mujuru and Mutanga, 2016). The juxtaposition of the water scarcity challenge with the mining industry underlines the tremendous threat posed by abandoned mines, which are emitting acid mine drainage to dwindling fresh water bodies.

As Mutanga demonstrates, acid mine drainage (AMD) arising from gold mining has been the most commonly documented challenge in South Africa due to the volumes involved. While the source of AMD is largely abandoned mines and their associated waste dumps, the problem of acid water spreads far beyond. Mutanga, like others in the field, highlights the West Rand in Gauteng province, where AMD has exposed residents to numerous health hazards leading to displacement, for instance the displacement of 10,000 households in Khutsong. Mutanga points out that the environmental health impacts of AMD in South Africa have not been systematically surveyed, but long-term exposure to AMD has been shown to result in increased rates of cancer, decreased cognitive function, skin lesions, health problems in pregnant women, neural problems and possible mental retardation (Claassen, 2006). Illustrating the connections between water and land contamination, as well as those between rural and urban impacts, Mutanga gives the example of farms rendered unproductive by salts emanating from AMD which cannot be sold for urban expansion purposes because the land cannot support urban properties due to remaining underground mine tunnels.

As a passage out of this ensemble of problems, Mutanga offers 'systems thinking', a formal, abstract and structured endeavour to think about systems holistically. Mutanga argues that systems thinking makes explicit causal-effect assumptions between related variables in a system, enabling independent assessment and improvement of mental models behind particular thinking. In addition to understanding the causes, effects and feedback loops related to disused and abandoned mines, systems thinking can be applied to understanding the linkages between different agencies – both private and public – proposed to intervene in post-mine rehabilitation. The methodology is useful to deepen understanding of observed phenomena and to establish consequences of different options available at a decision point. Mutanga concludes that comprehensive cost and benefit analysis of mining prior to extraction is the route to a full shift to sustainability so that unsolvable destruction is prevented from the onset.

Focusing on the Fourth Industrial Revolution, Ross Harvey, in chapter 5, addresses the challenge of preventing mining-related environmental destruction in a different way. Harvey showcases emerging digital technologies which can be used at the rock face to selectively mine and pre-concentrate material for subsequent metal extraction and avoid many of the negative environmental impacts associated with mining. These technologies can be built into mining equipment and pre-programmed for specific mines in the South African context. They include automated rock-face mapping, material characterisation and fragmentation analysis, and rock preconditioning. The machines that cut hard rock are also increasingly able to identify and exploit natural rock cleavages to make cutting more efficient.

Similarly, Harvey shows how crushing technology is becoming more effective, phasing out the big crushers typically required at a processing plant. In the case of copper mining, for instance, crushing is one of the largest components of a mine's energy consumption and greenhouse gas emissions. These can be reduced by in-pit mobile crushing and a shovel feeding the run-of-mine ore directly onto a belt conveyor handling system, thereby eliminating the use of trucks.

Connecting a number of new technologies together means that mining operations could become less energy intensive, argues Harvey, rendering the option of being solar or wind powered both more financially attractive and more operationally viable. But the technologies have to be adopted by mining companies and adapted to differing ore bodies. Harvey thus challenges the mining industry in South Africa to transform itself structurally by taking up new technologies and by creating vertical and horizontal linkages to drive changes in the economy as a whole. This would be not unlike the industrially innovative path chosen by the mining industry in the early 20th century, as demonstrated by Gqubule in chapter 4. For the 21st century, Harvey proposes the industry become a driver in South Africa's move towards a low-carbon growth trajectory, realising the potential of the Fourth Industrial Revolution to enable systems of production and consumption that renew rather than destroy the earth's ecological systems (Harvey, 2016).

While Harvey acknowledges the need for a 'just transition' for the increasing numbers of mine workers facing the job losses implied in automated technology and the phasing out of coal, Hameda Deedat, in chapter 11, addresses the 'messy' details of just transition in the South African context. From Deedat's standpoint as a labour educator, she revisits the struggle to define and defend just transition against the backdrop of high unemployment and mounting levels of extreme poverty.[3] This process has consisted of several rounds of definition, involving unions as well as other players in the climate change and energy justice movements. From this account, it can be concluded that the sheer scale of job losses, and corporate as well as state inertia around adopting new technologies that would benefit the majority of South Africa's electricity consumers, have made it difficult for unions to convince their members that the transition to a low-carbon economy can be a just one.

Almost 30 years ago, Jazairy et al. (1992) were among the first to suggest that poverty alleviation requires the full participation of those affected and must emerge from respect for human life, dignity and democratic values. This notion can be useful as a guide to evaluate models of resource use, job creation, poverty alleviation and even economic growth. Lorenzo Fioramonti takes on Jazairy et al.'s (1992) melding of material and less material aspirations at the macro level in chapter 12. Fioramonti argues for a 'balanced economy' which reconnects human beings with each other as well as with the natural ecosystems underpinning human existence. Building on his notion of 'wellbeing economy' (Fioramonti, 2017), which values both natural and social systems of value creation and breaks from traditionally assumed, limitless consumption, Fioramonti discusses the transition to a wellbeing economy in South Africa, and the continent at large, with a focus on mining. In the 'circular value chain' outlined – from 're-mining' to product use and back – there is a role to be played by a range of economic players, from large companies to artisanal miners to public corporations.

---

3   For instance, between 2000 and 2012, more than 200,000 jobs were lost in mining in South Africa (Bhorat et al., 2014: 4).

In many ways, the debates in this volume are about the past versus the future. This juxtaposition features particularly in the chapters of Kassa, Gqubule, Valiani and Ndebele, and Mabasa and Chinguno. Tracing, for instance, the rise of large claim owners, financiers and British company representatives in the colony of Griqualand West, Hibist Kassa, in chapter 6, shows how indigenous, artisanal miners were squeezed out of diamond mining in the second half of the 19th century. Kassa argues that within the current context of deindustrialisation, artisanal and small-scale mining represents an opportunity to right historical wrongs effected during the colonial and apartheid periods. Following from this, the question that can be put is: are mining corporations and the South African state capable of taking this opportunity – as the June 2018 awarding to artisanal miners of mining permits and access to a Kimberley mine dump may be taken to suggest?

In chapter 4, Duma Gqubule draws out, in detail, the history of centralisation in the South African mining industry, whereby a minority of mining houses have controlled mineral supply, extraction and production – from gold to diamonds, iron ore, coal and PGMs – from colonial times to the present. Elaborating on the narrow scope of successive mining charters in South Africa and moving far beyond, Gqubule recommends 25 per cent public ownership of the mining industry, in addition to the existing black ownership target of 26 per cent. Two vehicles for this are proposed: a sovereign wealth fund and a public mining company. A question for readers to consider: does this policy prescription have potential to begin undoing the historic monopoly structure of mining and move South Africa into a genuinely new future?

Also employing a historical survey to evaluate the industry in chapter 9, Salimah Valiani and Nester Ndebele argue that from a feminist perspective, mining, as experienced thus far by key groups of women in South Africa, has amounted to stunted social reproduction, the suboptimal use of female labour and the destruction of community wealth. The sub-argument they make, drawing from fieldwork they undertook with female artisanal diamond miners in Kimberley, is that within a national context of high unemployment and undervalued female work, female artisanal diamond miners fare better than underground female mine workers, though living conditions in artisanal mining are notably harsh. Would women fare better in a capital-intensive mining industry of the sorts envisioned by Netshitenzhe or Harvey, or as in the open-pit mining of Australia, where female machine operators and truck drivers are reported to be increasingly employed at close to the very high salaries of male workers (see Connell and Claughton, 2018)? Would this be desirable for the majority of women in and affected by mining in South Africa, or only those employed as mine workers?

Placing the question of mine worker organising front and centre in chapter 10, Khwezi Mabasa and Crispen Chinguno analyse how structural phenomena have shaped worker agency in the mining industry from colonial times to the present. They elaborate on two predominant labour regimes: 'non-hegemonic and coercive', and 'hegemonic' based on manufacturing consent of workers and their organisations. The non-hegemonic regime which reigned in colonial and apartheid South Africa had the following structural features: the intertwining of the mining industry with national political governance, controlled migration and geographic organisation of the labour force, and social differentiation of labour based on race and ethnicity. Mabasa and Chinguno argue that the antecedents of the post-apartheid, hegemonic labour regime based on institutionalised capital-labour relations can be traced to the formation of the National Union of Mineworkers in the early 1980s.

A significant piece of the mining union past brought out by these authors and important to union renewal debates today relates to the ability of mine workers to assess political economic structures and formulate an organising strategy accordingly. More specifically, in the 1940s, when black mine workers mobilised and formed their first union, labour activists had identified controlled labour migration largely driven by employers as a key structural feature to be combated via organising. Do mining unions today, increasingly wrapped up in cycles of violence resembling colonial and apartheid times, possess the collective ability to identify particularly problematic structural features of the mining political economy around which to mobilise and strategise anew?

The overarching question of the past versus the future is taken up again, at a macro-scale, in the conclusion of this volume. With this, let the debates begin.

REFERENCES

Auty, R. (1993). *Sustainable Development in Mineral Economies: The Resource Curse Thesis*. Routledge, London.

Baran, P. (1957). *The Political Economy of Growth*. Penguin, Harmondsworth.

Bhorat, H., Hirsch, A., Kanbur, R. and Ncube, M. (eds). (2014). *The Oxford Companion to the Economics of South Africa*. Oxford University Press, Oxford.

Chelwa, G. (2017). 'Does economics have an "Africa problem"? Some data and preliminary thoughts'. Available at *WISER* website: https://wiser. wits.ac.za/system/files/seminar/Chelwa2017.pdf. Accessed 27 July 2018.

Claassen, M. (21 November 2006). 'Water resources in support of socio-economic development'. In *VAALCO Supplement: Water for a Sustainable Future 21*. Shorten Publications [Vaal River Catchment Association], Johannesburg.

Connell, C. and Claughton, D. (23 May 2018). 'Women in mining: Dig the changing face of Australia's mining industry'. ABC Upper Hunter. Available at http://www.abc.net.au/news/2018-05-22/dig-the-changing-face-of-mining-as-women-make-inroads/9786020. Accessed 1 July 2018.

Daly, H.E. (2007). *Ecological Economics and Sustainable Development – Selected Essays*. Edward Elgar Publishing Limited, Cheltenham.

Fikeni, S. (21 June 2018). 'Opening remarks'. Launch of the Indlulamithi South Africa Scenarios 2030. Midrand.

Fioramonti, L. (2017). *The World after GDP: Economics, Politics and International Relations in the Post-growth Era*. Polity Press, Cambridge.

Gelb, A. (1988). *Oil Windfalls: Blessing or Curse?* Oxford University Press, New York.

Harvey, R. (2016). 'Book review essay: Envisioning a more equitable and sustainable future'. *South African Journal of International Affairs*, Vol. 24(3), pp. 541–550.

Jazairy, I., Mohiuddin, A. and Paluccio, T. (1992). *The State of World Rural Poverty*. International Fund for Agricultural Development, New York.

Lewis, S.R. (1984). 'Development problems of mineral rich countries'. In Syrquin, M., Taylor, L. and Westphal, L.E. (eds) *Economic Structure and Performance: Essays in Honour of Hollis B. Chenery*. pp. 157–177. Academic Press, Orlando.

Mujuru, M. and Mutanga, S. (2016). *Management and Mitigation of Acid Mine Drainage in South Africa*. Africa Institute of South Africa, Pretoria.

Palumbo, G. and Malkin, E. (29 March 2017). 'El Salvador, prizing water over gold, bans all metal mining'. *New York Times*. Available at: https://www.nytimes.com/2017/03/29/world/americas/el-salvador-prizing-water-over-gold-bans-all-metal-mining.html. Accessed 18 June 2018.

Taft, K. (2017). *Oil's Deep State*. Lorimer, Toronto.

United Nations Development Plan. (2016). 'Sustainable Development Goal 6: Ensure access to water and sanitation for all'. Available at Sustainable Development Goals website: http://www.un.org/sustainabledevelopment/water-and-sanitation/. Accessed 30 June 2018.

Winde, F. (2018). 'Science, business, society conference proceedings: Linking science, society, business and policy for the sustainable use of abandoned mines in the SADC region'. Keynote Address, 28–30 November 2017. Academy of Science of South Africa, Pretoria.

World Bank. (2013). *The Little Green Data Book 2013*. World Bank, Washington.

World Bank. (2016). *The Little Green Data Book 2016*. World Bank, Washington.

World Forum on Natural Capital. (2018). 'What is natural capital?' Available at: https://naturalcapitalforum.com/about/. Accessed 29 June 2018.

Zhou, T. M. (2017). 'Poverty, Natural Resources "Curse" and Underdevelopment in Africa'. In Mawere, M. (ed.). *Underdevelopment, Development and the Future of Africa*. Langaa Research and Publishing Common Initiative Group, Bamenda.

Section One

Transforming Mining for a More
Inclusive Future

TWO

Towards Mining Vision 2030

JOEL NETSHITENZHE

IN 2012, THE SOUTH AFRICAN GOVERNMENT adopted the National
Development Plan (NDP) with its overarching vision for 2030, which
sets out the macrosocial targets for a more equitable society. The major
outcomes contained in the plan include the elimination of extreme
income poverty (R419 per person per month) from 39 per cent; reduction
of unemployment to 6 per cent from around 25 per cent; reduction of
income inequality as measured by the Gini co-efficient from 0.69 to
0.6, more than doubling the per capita income to R120,000; increasing
'the share of national income of the bottom 40% from 6% to 10', and
ensuring 'household food and nutrition security' (NPC, 2012: 34).

While some of these objectives may be modest in terms of the ideals
of an equitable society as envisaged in the country's constitution, their
attainment would constitute a major advancement from current levels
of social inequity. The plan identifies a variety of actions required in the
economic and other areas of social endeavour to realise these outcomes.
It calls for a social compact of joint and varied actions by various social
partners – government, business, labour and broader civil society.

This chapter deals with the role that the mining sector can play in attaining Vision 2030, proceeding from the understanding that the sector has a critical contribution to make, given its historical role in the evolution of South African society, the endowments that the country commands, and the position the sector occupies in the socio-economic dynamics of South African society. The core argument is that the collective of partners in mining – private companies, workers, mining communities and the state – need together to develop a vision and programme that aligns with the objectives of the NDP.

The central hypothesis of the chapter is that the sector contains massive potential to make such a contribution. This is informed by two considerations. Firstly, the resource curse that has afflicted most mineral-exporting economies – pertaining to such challenges as diversifying the economy, major swings in economic growth and value of the currency, limited diversification of the economy and difficulties in dealing with inequality and corruption – is not the natural order of things. Proceeding from the understanding that a resource curse is a product of social agency, this chapter argues that strategic planning and focused interventions can produce the inverse of this phenomenon. Mining can serve as a catalyst for an industrialisation drive, a skills and technological revolution and, broadly, as a bedrock of societal efforts to deal with poverty and inequality.

Secondly, pursuing a developmental path that includes mining as part of its core strategies is unavoidable for South Africa. The mineral endowments the country commands dictate that it should find ways of utilising them to the benefit of society. Beyond this, the global dynamics of industrialisation, urbanisation and a growing 'middle class'; the pursuit of new energy sources such as hydrogen and fuel cells; and technological applications that require a variety of minerals – all these and more speak to a sector with major potential. Viewed beyond mere extraction and export, in the context of an industrial cluster, the impact that mining can have on the entire economy beggars the pessimistic belief that mining is a 'sunset industry'. South Africa needs consciously to exploit this comparative and competitive advantage.

Post-1994, the mining industry has had its ebbs and flows in the context of a changing society and contradictory dynamics in the global economy. These range from rates of investment to contribution to the gross domestic product (GDP); racial and gender dynamics in terms of ownership, management and board composition; as well as labour and community relations. As other chapters in this book show, insufficient progress has been made across all of these indicators. What is even starker is the fact that, while the country is one of the most highly endowed in terms of mineral reserves, the mining sector features at fifth position globally with regard to value added to GDP (Global Insight as cited by DMR, 2013), and 'has failed to match the global growth trend in mineral exports' (NDP, 2012: 42). This has to do not merely with the complex geological location of the endowments, nor with anything related to deliberate diversification of the economy as such. Rather, it is a consequence of the lack of an overall societal strategy and divergent interests among the major role-players, with such issues as poor infrastructure and policy weaknesses also playing a major role. This chapter provides a framework for how these deficits can be addressed, so the mining industry can realise its full potential as a critical part of the country's development trajectory.

The next chapter of this book focuses on the application of the visioning approach to the platinum group metals (PGM) sub-sector. Besides providing a concrete illustration of how a vision can be crafted at a sub-sectoral level, the PGM case study is informed by the reality of PGM deposits that are estimated as constituting over 80 per cent of world reserves (SAMI 2009/2010; Wilson and Anhaeusser, 1998, as cited in the 2012 African National Congress (ANC) State Intervention in the Mining Sector (SIMS) Summary Report: 5). The variety of usages of PGMs in relation to reduction in greenhouse gas emissions in combustion engines, the nascent hydrogen economy, medical and surgical instruments and jewellery, among others, speaks to the utility of these minerals well into the future. Further, the sub-sector has been experiencing difficulties in terms of its cost-price ratio, labour relations and community partnerships – all of which threaten the sustenance of many operations. Indeed, it is in this intersection of danger and opportunity that strategic thinking and social compacting can stand South Africa in good stead.

GENERIC ATTRIBUTES OF THE MINING SECTOR

As reflected in other chapters of this book, the mining sector has been at the core of the evolution of the South African political economy.

Over the past 100 years, the South African economy evolved on the basis of two pillars: mining and agriculture. In the early years, the manufacturing sector grew in the main to service these two industries and provide for a growing domestic consumer base as well as markets in developed countries (NPC, 2011: 7–8).

From the turn of the 20th century, with the discovery of diamonds around Kimberley in the (now) Northern Cape and gold around Johannesburg in the (now) Gauteng province, the mining industry can be said to have profoundly influenced South Africa's spatial patterns of economic development, human settlement and infrastructure networks. It was the driver of the evolution of South Africa's manufacturing sector as well as energy sourcing and intensity. While its proportion of GDP has declined, especially since 1994, mining's combined contribution – taking into account direct mining activity, forward and backward linkages and the induced effect – stood at about 18.7 per cent in 2012 (Creamer, 2013). In 2005, it contributed 50 per cent of primary and beneficiated merchandise exports, 50 per cent of Transnet's rail and ports volume, 16 per cent of electricity demand, 30 per cent of liquid fuels from Sasol's coal-to-liquid process and 93 per cent of electricity generation, which is from coal power (Swanepoel, 2006).

According to the Manufacturing Circle,

Manufacturing is also still tied to the mining industry, depending to a large extent on the health of this sector for its own wellbeing. The impact then of the precipitous decline in the mining sector's contribution to GDP from above 20% in 1980 to 8% in 2016 is clear, alongside a global sourcing strategy by mining conglomerates as they became global players post 1994; as is the uncertainty created by the modern-day Mining Charter (Manufacturing Circle, 2017: 10).

Mining's contribution to the evolution of the country's skills base is reflected in the fact that one of the country's premier universities (Witwatersrand) started off in 1896 in Kimberley as the South African School of Mining (University of the Witwatersrand, undated). Similarly, it influenced much of the evolution of South African trade unionism and even the configuration of political parties and political discourse. The white colonial political establishment reflected dynamics of alliances and conflict among the agricultural and mining moguls, in the earlier years configured around the Afrikaner and English establishments. Policies on land tenure were developed at the turn of the 20th century to meet the needs mainly of the emergent mining sector.

As resistance to the apartheid colonial system reached its peak in the 1980s, the captains of the mining industry were among the first sections of the white ruling class to initiate interactions with the banned ANC, seeking accommodation in a negotiated settlement.

It is logical that, today, contestation around inclusion of black people in the mainstream of the economy plays out most intensely in the mining sector. This is reflective of a deep sense of grievance around sharing of the sheer wealth of mineral endowments that South Africa commands, estimated by Citibank at about US$2.5 trillion, the largest in the world (I-Net Bridge, 2012). As detailed in chapter 4 by Duma Gqubule, there is much contestation around the issue of ownership, and the Chamber and government have been locked in battle in the courts around a new Mining Charter. Whatever the detail of the issues under debate, the fact of the matter is that, across all measures of economic empowerment, the sector is still far from reflecting the demographics of the country.

As shown in Table 1, most of these endowments are estimated to have lifespans that amount to hundreds of years into the future. Yet, as mentioned above, South Africa ranked only fifth in terms of value added to GDP in US dollars and fourth in terms of mining employment. Its investment trends compared to Australia (see Figure 1) were counter-cyclical to the mining super-cycle of the past decade, a trend that seems to persist in terms of levels of investment against the backdrop of resurgent mineral prices. The Minerals Council of South Africa (previously the Chamber of Mines) estimates that the mining

Table 1: South Africa's mineral reserves, world ranking, 2009 production and nominal life (assuming no further reserves) at 2009 extraction rates

Mineral	Reserves				Production 2009			Life
		Mass	% World	Rank	Mass	% World	Rank	Years
Alumino-silicates	Mt	51	*	*	0.265	60.2	1	192
Antimony	kt	350	16.7	3	3	1.6	3	117
Chromium ore	Mt	5500	72.4	1	6.762	*	1	813
Coal	Mt	30408	7.4	6	250.6	3.6	7	121
Copper	Mt	13	2.4	6	0.089	*	*	146
Fluorspar	Mt	80	17	2	0.18	3.5	•	444
Gold	t	6000	12.7	1	197	7.8	5	30
Iron ore	Mt	1500	0.8	13	55.4	3.5	6	27
Iron ore - incl. BC	Mt	25000	~10	*	55.4	3.5	6	451
Lead	kt	3000	2.1	6	49	1.2	10	61
Manganese ore	Mt	4000	80	1	4.576	17.1	2	874
Nickel	Mt	3.7	5.2	8	0.0346	2.4	12	107
PGMs	t	70000	87.7	1	271	58.7	1	258
Phosphate rock	Mt	2500	5.3	4	2.237	1.4	11	1118
Titanium minerals	Mt	71	9.8	2	1.1	19.2	2	65
Titanium- incl. BC	Mt	400	65	1	1.1	19.2	2	364
Uranium	kt	435	8	4	0.623	1.3	10	698
Vanadium	kt	12000	32	2	11.6	25.4	1	1034
Vermiculite	Mt	80	40	2	0.1943	35	1	412
Zinc	Mt	15	3.3	8	0.029	0.2	25	517
Zirconium	Mt	14	25	2	0.395	32	2	35

Source: Wilson and Anhaeusser, 1998

sector expanded by 3.7 per cent and its employment by 1.6 per cent in 2017 (Breytenbach, 2018). However, according to the Chamber's 2017 survey, in a 'more certain and conducive' environment, capital spending stretching over four years could be 84 per cent higher. 'The impact on employment creation, according to the survey results, would be nearly 48,000 people' (Chamber of Mines, 2017: 3).

A number of factors, pertinent to the crafting of a long-term vision, account for this; not least the short-termism in the outlook of most mining companies in terms of generating shareholder value. Other factors include policy uncertainty in a polity that is transforming from a colonial past, poor relations with workers and communities and the persistence of old production as well as management and labour-sourcing methods.

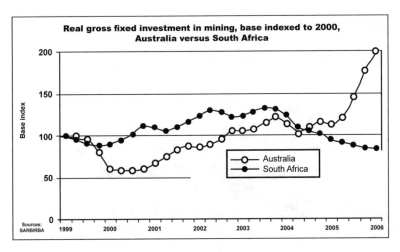

Figure 1: Mining fixed investments (South Africa and Australia)
Source: Bernard Swanepoel, Vice-President of the Chamber of Mines: Presentation to Mining Summit, September 2006

MOTIVATION FOR VISIONING:
A SUNRISE INDUSTRY

Given South Africa's mineral endowments, the country cannot avoid using mining as a critical platform in defining its growth and development trajectory, if only on the basis of self-interest. But there is more to this argument than sheer self-interest.

The first reality that informs this argument pertains to global demand for minerals, most of which South Africa has in abundance. According to Cynthia Carroll, former Chief Executive Officer of Anglo American plc, it is estimated that some three billion more people will live in urban areas by 2050; and as early as 2025, global cities will have to construct 'the equivalent of the entire land area of Australia ... in residential and commercial floor space, which would require US$80-trillion worth of investments' (Creamer, 2012). These trends, which will be even more manifest in developing countries, including those in Asia and Africa, will also influence, and in turn be impacted on, by the growth of the middle strata and consumption patterns. For instance, 'the global car fleet ... [is] projected to double to 1.7-billion by 2030' (Creamer, 2012). This is further illustrated in Figure 2.

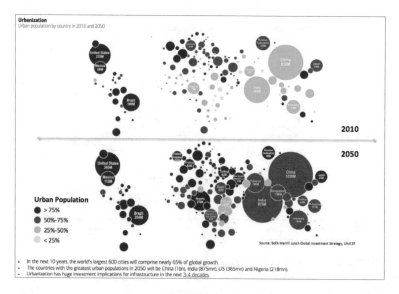

Figure 2: Trends in urbanisation to 2050
Source: truehuenews.com/2015/09/06/infographic-urbanization-urban-population-by-country-in-2010-and-2050/

In addition, as China climbs up the manufacturing sophistication ladder, low-end manufacturing is shifting to countries that seek to industrialise, such as Vietnam, Bangladesh, Ethiopia, Rwanda and Kenya. While the so-called Fourth Industrial Revolution may disrupt these trends, it will take a few decades before its impact affects all areas of manufacturing. At the same time, India has embarked on a new trajectory of high economic growth, comparable to China in the past three decades. Thus, it is expected that demand for minerals will remain high well into the future, albeit on a more muted scale. It is also worth noting that a combination of urbanisation, large infrastructure projects and an emergent 'middle class' should, steadily, transform African countries from being largely exporters to significant consumers of industrial minerals. Already, according to Kaizer Nyatsumba of the Steel and Engineering Industries Federation of Southern Africa, sub-Saharan Africa has become the most important market for metal-product exports from South Africa (Creamer, 2018).

The second reason – which also emphasises the need to differentiate between cyclical factors and long-term structural ones in relation to demand for commodities – is the fact that many of the minerals are needed for emergent technologies and to improve performance of existing applications. This applies to high-tech materials and new energy sources, chemical as well as water and food sectors, logistics and life sciences. Two examples in this regard are PGMs, which are not only used in auto-catalysts but also in proton exchange membrane fuel cells in the emergent hydrogen economy, and the titanium-aluminium-vanadium alloy which finds application in high-speed aircraft and jet engines. On the extreme, it can also be argued that South Africa's coal deposits, with a lifespan of some 121 years (see Table 1), can also be utilised in a manner that reduces the country's current carbon footprint: for instance, through more advanced coal-fired power stations, underground coal gasification and as sources for hydrogen.

The third reality about the central role mining can play in South Africa's development trajectory going forward pertains to all-round capabilities that the country has mustered over the years. As pointed out earlier, one of the premier universities in South Africa (Wits) emerged towards the end of the 19th century as a school of mining. With the endowments the country commands, the complex location of some of the minerals and the depths at which they have to be accessed, South Africa has over the years developed related manufacturing, engineering and other skills that are globally renowned. This is reflected in export of mining equipment, skills pertaining to turnkey services 'in new mine design and operations', as well as the number and quality of patents (Kaplan et al., 2012 as reported in Benjamin, 2013).

With regard to sophisticated beneficiation efforts, the government has sought to bring together research centres in universities and the mining sector in a partnership on Hydrogen South Africa for the fuel cell industry; and promoted focus on the part of the Council for Scientific and Industrial Research (CSIR) on the application of such materials as titanium. In other words, building on historical learning and focusing on current and future possibilities, the country has the potential to develop unique capabilities and the necessary human resources in a mining cluster. Many of the skills and equipment also have applications beyond the mining sector as such.

The fourth consideration pertains to the role of mining in infrastructure development within South Africa and the rest of sub-Saharan Africa. At one level is the issue of supplies for such infrastructure programmes. It is estimated, for instance, that infrastructure spending in sub-Saharan Africa would grow 'from $70-billion in 2013 to $180-billion by 2025' (Venter, 2014). This presents a major opportunity for supplies, most of which are processed from the very minerals that the continent produces. At another level is the infrastructure required by the mining industry such as energy and water, the transport value chain and other services. Given the location of these endowments (see Figure 3), these mining operations and related activities can also have a major impact on the country's spatial development, both in terms of export routes, provision of these minerals to locations of industrial activity and the possibility for such industrial and other activities being systematically developed closer to the mining operations and possibly even leading to the emergence of new cities and much-needed development in these areas.

In summary, South Africa's mineral endowments present a unique opportunity for a new industrialisation drive and advancement in the economy as a whole. This applies across the value chain: from mining equipment and services, to extraction, infrastructure development, beneficiation, skills development as well as research and development. Along with this, there are many vistas for more

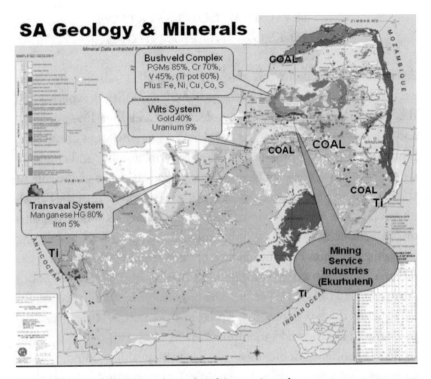

Figure 3: Spatial geology of South African minerals
Source: ANC SIMS Report, slide presentation, 2012

profound empowerment of the previously disadvantaged, including workers, women, communities and entrepreneurs. This requires a deliberate strategy to develop a mature mining cluster that touches virtually every aspect of the South African economy and political economy, including adaptation to new realities in terms of application of technology and labour sourcing. In other words, mining in South Africa can, in earnest, become a sunrise industry. The responsibility to take advantage of this opportunity lies as much with the state as it does with mining companies, workers, mining communities and indeed society at large.

Before turning to the building blocks of Mining Vision 2030, we briefly reflect on the paradigms, contestations and policy interventions that have been at play around the mining industry. This is covered at a high level, as most of these issues are canvassed in some detail in other chapters of this book.

OVERVIEW OF CONTESTATION
AND EVOLVING PARADIGMS

As indicated earlier in this chapter, the mining industry has been central to the evolution of the South African political economy. In this context, during the years of colonialism and after 1994, political and economic contestation has revolved in large measure around the mining sector; and efforts to abolish the colonial polity and eliminate its socio-economic impact have always placed the issue of the role of this sector at the top of the agenda. Mining was at the core of the British invasion of the provinces controlled by the Dutch settlers at the turn of the 20th century and the subsequent compromises struck among the whites premised on the alliance of 'wealthy Boer farmers and the imperialist mine-owners' during the formation of the Union of South Africa in 1910 (SACP, 1962: 19). The colonial political alliance sought to 'squeeze the last drop of cheap labour out of the African people' primarily in mining and agriculture (SACP, 1962: 19).

The economic dominance of English-speaking mine owners spawned a movement among the politically ascendant Afrikaner elites campaigning for their own pound of gold. Given the leading role of some Jewish families in the sector, this campaign opportunistically also contained tinges of anti-Semitism. As Shain (2016) observes: 'Hoggenheimer appeared precisely at a time the Witwatersrand mining magnates appeared to dominate politics.' In the cartoon sketches, 'the overweight and diamond-studded Randlord was depicted as having inordinate power' (Shain, 2016). In time, the extant mining magnates sought to ameliorate this by assisting in the growth and expansion of Afrikaner-led mining companies such as Federale Mynbou, which later became Gencor on the back of major acquisitions of the 1960s (Gencor Ltd Company Profile, undated).[1]

The contestation was and remains even more intense between mine owners and the workers. In the Rand Revolt of 1922, white workers fought against the weakening of the colour bar in the

---

1   It is not the purpose of this chapter to interrogate the seeming parallels between the historical Hoggenheimer discourse and the current debates within the ANC on 'white monopoly capital', intellectually intriguing as this may be.

mining industry and perpetrated violence against blacks, demanding job reservation; yet with tinges of a socialist outlook, reflected in the slogan: 'Workers of the world, unite and fight for a white South Africa' (Byrnes, 1996). On a larger scale, while the major strikes of black mine workers such as in 1946 – and later in 1987 and 2012 – had as immediate demands issues pertaining to wages and working conditions, they were largely undergirded by a deeper sense of grievance around issues of ownership of mineral wealth and the character of the socio-economic system.

It is against this background that the 1955 Freedom Charter, adopted by the ANC and its allies, had as one of its core economic propositions (which resonates to this day) the demand that 'the mineral wealth beneath the soil, the Banks and monopoly industry shall be transferred to the ownership of the people as a whole' (Freedom Charter, 1955).

Over the years, the issue of ownership has been at the centre of contestation around the mining industry. Broader strategic reflection within the liberation movement, beyond ownership, started to find expression in the early 1990s as the ANC prepared for government:

> *The ANC will, in consultation with unions and employers, introduce a mining strategy which will involve the introduction of a new system of taxation, financing, mineral rights and leasing. The strategy will require the normalisation of miners' living and working conditions, with full trade union rights and an end to private security forces on the mines. In addition, the strategy will, where appropriate, involve public ownership and joint ventures.*
>
> *Policies will be developed to integrate the mining industry with other sectors of the economy by encouraging mineral beneficiation and the creation of a world class mining and mineral processing capital goods industry* (ANC, 1992).

This is further elaborated in the Reconstruction and Development Programme (RDP) adopted by the democratic government in 1994, in which the development of 'South Africa's mineral wealth to its full potential and to the maximum benefit of the entire population' is outlined. The RDP included such issues as facilitating mineral development, assisting small-scale mining ventures, value addition through technology development, affirmative action and development of human resources (Reconstruction and Development Programme, 1994).

The Mineral and Petroleum Resources Development Act (MPRDA) No. 22 of 2002 sought to codify all these varied objectives proceeding from the premise that 'South Africa's mineral and petroleum resources belong to the nation and ... the state is the custodian thereof'. It emphasises issues of equitable access, expanding opportunities for those historically disadvantaged, the role of the sector in promoting economic growth and employment, sustainability and local development (South African Government, 2002). The attendant Mining Charter promulgated in 2004 elaborated on incentives and disincentives to ensure that the established companies promote the empowerment of black people and women in terms of ownership, governance and management structures, professions and other skilled jobs, social and labour plans, as well as procurement and enterprise development. Critically, ownership was defined as embracing employee share-ownership schemes and beneficiation. Pursuant to the principle of custodianship, mineral royalties were introduced through the Mineral and Petroleum Resources Royalty Act No. 28, finalised in 2008.

In chapter 4 of this book, Gqubule asserts that, while there has been some progress in terms of ownership and other measures of empowerment as defined in the Act and the 2004 Mining Charter, the basic structure of the industry still reflects the situation during the apartheid era. Critically, in terms of evolving paradigms, Gqubule further argues that from the mine owners' resistance of the early 2000s, there is now acknowledgement of the importance of empowerment and some level of compliance. But this is largely a reluctant tick-box exercise. Further, since 1994, a number of major mining companies have transferred their primary listings to foreign shores. At the same time, an attempt to address these deficits and speed up the empowerment project has resulted in a careless approach on the part of the Department of Mineral Resources (DMR), as reflected in contestation around the amended Mineral and Petroleum Resources Development Bill and 2017 draft Mining Charter. Poor consultation with other stakeholders, including business and mining communities, and a dogged determination to proceed in spite of protests, resulted in court challenges and, by early 2018, a change in the leadership of the Ministry and the initiation of meaningful stakeholder interactions.

More comprehensive paradigmatic reflections on mining – across the value chain – emerged with the SIMS Report of 2012, instructively arising from a debate within the ANC on ownership and calls for nationalisation. The report proposes the forging of a mature industrial cluster that takes into account linkages pertaining to fiscal policies, backward and forward nexuses as well as knowledge and spatial considerations:

> *International experience indicates that the growth, development and employment potential of our mineral assets can only be realised through the maximisation of the mineral economic linkages (e.g. Sweden, Finland, Brazil, China, etc.) as proposed by the Africa Mining Vision. The mineral linkage industries can survive beyond the resource exhaustion and provide the nurseries for more generalised industrialisation and job creation* (ANC SIMS Summary Report, 2012a: 36).

In the context of the debate on the draft of the new MPRDA, the ANC and government have sought to designate some minerals as strategic, so as to ensure their availability at reasonable cost for beneficiation (ANC, 2012c). Following up on the SIMS Report, the South African Department of Trade and Industry in 2013 initiated the Mineral Value Chain Study. This has led to a more systematic approach to the role a mining cluster can play in economic growth and development. Arising from this, recent iterations of the Industrial Policy Action Plans allude to the importance of the mining value chain in the economy as a whole.

Parallel to, and somewhat influenced by, the SIMS Report, the conceptualisation of the role of mining at a more strategic level also found expression in the 2012 National Development Plan. It argues:

> *If the [pitfalls of a resource curse] are consciously avoided, and if the mineral endowments are used to facilitate long-term*

capabilities, these resources can serve as a springboard for a new wave of industrialisation and services for domestic use and exports (NDP, 2012: 112).

Most probably inspired by a discourse that had started to take shape among other sectors of society, the Chamber of Mines in 2012 commissioned a research report on what the content of, and process towards, Mining Vision 2030 may look like. This includes such issues as how mining can support national development, how the profile of operations is bound to change with modernisation, spatial considerations and how the sector can come to terms with its past. It also proposes that the Mining Industry Growth Development and Employment Task Team (MIGDETT) made up of representatives of government, business and labour, appropriately expanded, should be utilised as a platform of engagement among the partners in developing such a vision (CSMI, 2012).

Mining Phakisa, an offshoot of the NDP, provided a platform in 2016 for comprehensive interaction between government departments, workers' unions, mining houses, research institutes and non-governmental organisations. Out of this a joint institution, the Mandela Mining Precinct, has been set up, seeking to position the sector as a 'centrepiece within a new, cohesive mining cluster that requires across-the-board competitiveness'. According to Edwin Ritchken, this initiative also seeks to confront the downsizing of mining research and development facilities that occurred when the capability in the (then) Chamber of Mines was 'transferred to the public-sector Council for Scientific and Industrial Research (CSIR), where it had been badly under-funded' (Creamer, 2016). In addition to this are initiatives that have been carried out through the Department of Science and Technology, pertaining to the hydrogen economy and fuel cell technology, as well as such minerals as titanium, a versatile alloy, and lithium which is used in batteries for energy storage.

It is instructive, though, that when the MIGDETT leadership met in 2015 to review progress in the implementation of the Mining Charter and to plan for Mining Phakisa, the issues covered in their report focused on ownership; housing and living conditions; procurement and enterprise development in terms of capital goods, services and consumable goods; employment equity; human resources development; mine community development; and sustainable development (Ramatlhodi, 2015). The draft Mining Charter gazetted in 2017 identifies as one of its primary objectives the promotion of 'beneficiation of South Africa's mineral commodities by South African-based companies' and identifies incentives towards that end (DMR, 2017). However, it does not at all refer to the need to forge a mining industrial cluster.

What the foregoing in this section speaks to is an acknowledgement among all social partners of the importance of mining in the history, present and future of the South African economy. This has evolved with time beyond ownership and labour relations to encompass broader strategic questions about the totality of the value chain and, as such, the role of the sector in the economy as a whole. It is subject to debate whether this has developed fully to exercise the minds of all government departments and indeed of all social partners. The next section explores the building blocks of Mining Vision 2030. It is informed by ideas from all the initiatives briefly outlined above; and it seeks to use them as the foundation and pillars of the vision.

BUILDING BLOCKS OF MINING VISION 2030

In the remaining sections of this chapter, the building blocks of Mining Vision 2030 are outlined. The starting point in this regard is the motivation outlined earlier regarding the opportunities that present themselves at domestic, regional and global levels. In the same manner that mining was central to South Africa's industrialisation drive and broader political economy over the past century, it can play a central role in the country's growth and development trajectory going forward – in a new context, much different from the mineral-based industrialisation of the previous era.

Though organised differently in terms of areas of focus, the building blocks identified hereunder are in part inspired by work done in a number of studies such as the 2012 proposals developed by the CSMI, the Africa Mining Vision (AMV) adopted by the African Union in February 2009 (African Union, 2009), the 2012 SIMS Report, the

2012 NDP and the 2014 World Economic Forum (WEF) Scoping Paper on Mining and Metals in a Sustainable World. What follows is a brief outline of relevant highlights from these studies.

Based on interviews with stakeholders, the 2012 CSMI report argues for an appreciation of the role mining can play in supporting national development, the changing profile of the sector and mining operations, the need for 'regional-scale plans' and the imperative for the sector to come to terms with its past. It also identifies gaps in terms of human resource development programmes, re-investment of resource rents, environmental considerations and 'deeper consideration of natural resource constraints' (CSMI, 2012: 4).

The NDP (2012) identifies constraints and opportunities in the development of the mining sector such as lack of certainty on property rights; the need to develop linkages with sectors other than mining, research and development; regional partnerships and, as with CSMI, MIGDETT-related process issues. '[I]t should be possible,' the NDP asserts, 'to create about 300,000 jobs in the mining cluster, including indirect jobs' by 2030 (NDP, 2012: 151).

The AMV identifies continent-wide interventions on resource-potential data, state contracting and negotiating capacity, resource development and governance, societal capacity to manage mineral wealth, attending to infrastructure constraints as well as dealing with artisanal and small-scale mining (AMV, 2012).

In addition to identifying matters pertaining to ownership and governance, the SIMS Report argues for a mature mining cluster connected in a mutually beneficial manner to other sectors of the economy. It identifies economic linkages through which such a cluster can be developed: fiscal, backward, forward, knowledge and spatial. It also addresses the regional dimension (SIMS, 2012).

Asserting a sustainable approach in which investors 'value and trade metals and mineral resources based on a shared understanding of and commitment to economic and social development', the WEF Scoping Paper promotes mineral-intensity reduction in economic activity, as well as reuse and recycling of minerals. It also notes the rise of automation technologies, 3D printing and other elements of the so-called Fourth Industrial Revolution which would change skills required in, and employment profile of, the mining sector (WEF, 2014: 8).

These observations are of much relevance and do serve as an important backdrop to this outline of the South African Mining Vision 2030. For purposes of presentation, the outline of the Mining Vision is categorised into the following issues: extraction, infrastructure, modernisation, backward linkages, forward linkages, research and development and issues of ownership. Generic issues that intersect with all these, such as human resources, social and labour plans, informal mining, post-mining dispensations and utilisation of land are briefly dealt with in one section. These matters are dealt with briefly and at a high level, given that each of the issues requires detailed plans that elaborate concrete steps. Some of the matters are canvassed in the other chapters of this book; and such are the varied properties, uses and dynamics for each class of minerals that it is not possible to reflect on each one of them in detail in a single chapter. The proposals should therefore be understood as an organising framework for Mining Vision 2030, rather than the vision as such.

JUDICIOUS EXTRACTION AND EXPORTS

Extraction and export constitute a critical part of South African mining and they should form part of the sector's long-term vision. As argued above, various factors in the global economy will continue to drive demand for natural resources, particularly for minerals that South Africa has in abundance.

As illustrated in Figure 4, South Africa ranks quite high in terms of many globally strategic minerals, with over 30 per cent of global reserves of PGMs, manganese, chromium, alumino-silicates, vermiculite and vanadium.

On the one hand, while the super-cycle created mostly by China is ebbing, other countries such as India, Indonesia, Vietnam, Ethiopia and Rwanda – with large economies and populations, or in the midst of rapid industrialisation as part of global value chains – are taking the baton. Indeed, according to Lennart Evrell, 'almost half of the world's population have not yet started the journey toward a higher standard of living', which is a major driver of the mining industry (Bloomberg,

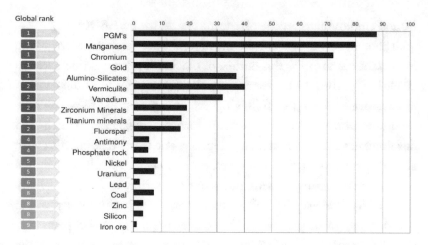

Figure 4: South African reserves of key minerals, based on 2008 data (horizontal axis shows % of global reserves)
Source: Baxter, 2011, cited in Makhuvela and Msimang, 2013

2017). On the other hand, many of these strategic minerals are critical for new technologies, including the green economy, as discussed in chapter 5 in this volume, which deals with matters related to the Fourth Industrial Revolution.

Mineral extraction, the NDP estimates, can create 100,000 additional jobs in around eight years if the sector were to grow at about 3 to 4 per cent (NDP, 2012: 146). This, similarly, would increase the foreign exchange and tax revenue that the country needs. South Africa needs to ensure that it takes advantage of global demand for these commodities, whilst taking into account the trends and principles (reduce, reuse, recycle) identified by the WEF. A judicious approach to extraction should embrace these principles – as an opportunity rather than a threat – and out of such considerations, extraction rates should be aligned to the desired nominal life of the reserves that the country has (see Table 1 for nominal life estimates).

A deliberate approach also needs to be adopted with regard to raw materials that are designated as strategic by South Africa's trading partners. In other words, South Africa's mineral endowments place on the country the responsibility to ensure global security of supply. For instance, in a report by the European Union's (EU, 2010) Ad Hoc Working Group on Defining Critical Raw Materials, the

European Union (EU) ranks raw materials that are economically important and/or face supply risk. As shown in Figure 5, many of these are found in significant quantities in South Africa (EU, 2010). While the subjective nature of the assessments can be contested, the fact that, in an exercise of this nature, the bloc ranks source countries on the basis of governance and environmental protection speaks to the extent of such dependence and a sense of insecurity – with South Africa in 2006 featuring as an EU import source that accounts for some 79 per cent of chromium, 60 per cent of PGMs and 33 per cent of manganese. This kind of dependence and the strategic calculations that derive from it apply to many of the major global powers, East and West. Viewed negatively, this can present a danger for the South African body politic. But approached strategically and positively, it is a major opportunity for the crafting of mutually beneficial terms of engagement.

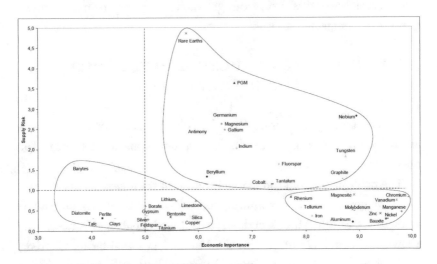

Figure 5: Ranking of critical raw materials for the EU
Source: EU, Report of Ad Hoc Working Group on Defining Critical Raw Materials, 2010

A visionary approach to extraction and exports has to include the creation of an environment conducive for investments, proceeding from the understanding that mining investments are long-term in nature and take many years before projects can break even. On the one hand, the private sector needs to embrace the variety of measures introduced by government to ensure inclusivity in ownership, management and the professions, worker rights, community development, environmental responsibility and other indicators (embracing both race and gender) contained in empowerment charters, some of which are dealt with later in this chapter. Aside from considerations of ethics, sense and sensibility, these measures should be seen as a necessary condition for sustainable social cohesion, which is in the long-term interest of investors. On the other hand, policy-makers have to consciously relate to private operators as partners. Government needs to understand the logic of investor imperatives and mining operations, and act as a strategic facilitator of investment and a professional regulator and enforcer of statutes. In this way, disasters such as the unseemly contestation around the 2017 Mining Charter would be averted. Such a professional and strategic approach on the part of the public sector should apply across the chain of the extraction sub-sector, from geo-survey capacity to management of trading arrangements.

MINING INFRASTRUCTURE DEVELOPMENT AND OPPORTUNITIES

There are three dimensions from which the issue of mining and infrastructure needs to feature in developing a long-term strategy for the sector.

The first one is about the spatial location of the endowments and the infrastructure demands for mining operations and trade. As illustrated in Figure 3, South Africa's endowments are concentrated variously in a few parts of the country, with the Mpumalanga coal basin, Gauteng gold reef and the Northern Cape ('Transvaal') system historically being the most prominent. Over the past few years, the rise of the Bushveld PGM complex from the western Rustenburg node and the eastern edge in Burgersfort, the Waterberg coal basin and more intense activities in the Northern Cape has reconfigured spatial concentration of operations. This reality has defined the configuration of the country's infrastructure.

Quite often, however, this has not been addressed in a proactive manner; the country does not have a framework to respond to the changing spatial intensity of mining operations. A long-term strategy for mineral extraction and its other linkages demand all-round infrastructure programmes to provide for the operations (electricity, water, human settlements and other needs), to ensure basic processing capacity for the raw materials, to manage a variety of community development issues and to avail efficient transportation of the minerals to the industries that need them and for export. As the spatial location of the activities shifts, there should be deliberate infrastructure plans (a turnkey framework), defining the role of each partner in the realisation of these plans. This demands dexterity and creativity on the part of the state-owned enterprises charged with providing such infrastructure – including forward planning, partnerships in sourcing capital, concessioning arrangements that ensure mutual benefit between the infrastructure companies and the mining operators and an 'open access' or 'shared use' infrastructure model to benefit the economy and society at large.

The second, related dimension is at a macro-policy level. It pertains to the manner in which, historically, mining carved up South Africa's economic geography. A visionary approach would ease the upheavals that come with new discoveries and the end of the extraction life-cycle. This should be informed by basic principles of spatial planning. Heightened economic (mining) activity naturally results in the migration of large numbers of people to areas experiencing such activity and the demographic exuberance can hardly be controlled. In a country without 'influx control' or a Chinese-type *hukou* or *huji* system[2] of managing migration, any plans need to take account of this reality. Integrated approaches among government departments and state entities (e.g., public enterprises, energy, water, transport, education and housing) in terms of planning and project implementation are fundamental to this.

Proceeding from the premise that mine operators also have a responsibility within and beyond the prescripts of social and labour plans, individual mining companies should not only allocate resources for these responsibilities, but also pool these resources and partner with all spheres of government in providing such services. At a macro-level, national spatial frameworks and plans pertaining to infrastructure needs

2   A system of registering households in terms of areas in which they are meant to reside, work and get social benefits; also variously prevalent in other East Asian countries.

should be developed in rolling pre-emptive cycles taking into account the shifting centres of mining operations. Such plans should be linked to the designation of relevant transport/development corridors linked to these operations. At a provincial and local level, mine operators, along with other sectors, should take an active part in the development of provincial economic strategies and municipal integrated development plans. They should also take it as their responsibility to provide varied capacities where such are required, especially at local level.

Related to this (and discussed later) is the issue of bold decisions on the location of upstream and downstream operations, as well as creative ways of preparing for, and managing, post-mining dispensations. Critically, South Africa needs to envision the emergence of new cities that combine various activities in the mining value chain and other sectors of the economy. In this regard, dynamics in the Waterberg and Burgersfort areas need to be interrogated – against the backdrop of the rise and decline of the Rustenburg belt – to ensure that spontaneous private-sector-based activities are harnessed through deliberate long-term planning led by government.

The third dimension pertains to how the infrastructure programme for mineral extraction and trade can spawn focused industrialisation. South Africa spends some R1 trillion every five years in infrastructure programmes and a large part of this supports mining operations. As indicated earlier, larger amounts are increasingly being allocated to such programmes in sub-Saharan Africa. From rail to roads, ports, housing, water and other infrastructure linked to mining operations, so massive have these programmes become that simple commercial logic dictates that supplies for the infrastructure projects should as far as possible be sourced from the region. For instance, given the rail networks being developed, the manufacturing of whole or parts of locomotives, carriages, signalling and braking systems, and railway tracks can be undertaken in South Africa and other parts of the region.

Attached to this should be focused economic diplomacy that includes, beyond agreements on trade, a drive towards standardisation in the infrastructure products such as railway gauges. An approach of this kind should extend from mining operations to issues of post-mining rehabilitation and technologies attached to this, such as the productive use of mine dumps and the processing of acid mine water.

South Africa and other parts of sub-Saharan Africa need to undertake these activities with a keen eye on the necessary division of labour. Critically, this should also be spurred on by the realisable ambition to attain such efficiency and excellence that these supplies for the infrastructure programmes not only feed the region but also position it as a necessary part of the global value chains of the original equipment manufacturers. For instance, the American General Electric group, which is supplying Transnet's locomotives, has many operations across the globe. Aside from the negative reports regarding whether the China South Rail (CSR) is meeting its localisation commitments, the partnership with CSR against the backdrop of the supplier value chain of China's mammoth 'One-Belt One-Road'[3] initiative presents opportunities that would redound to the benefit of South Africa and the region for many decades to come. Needless to say, localisation programmes are easier said than done. Monitoring capacity is crucial to ensure that objectives set are actually realised.

THE INCESSANT BUGLE OF MODERNISATION

The bugle of modernisation has sounded, and none can escape its incessant call. The form that modernisation takes, ranging from automation to management practices, and the consequences it unleashes, will depend on whether there is deliberate planning. Such planning should proceed from the premise that, fully embraced, technological change can be phased in and therefore optimally managed. But there should also be appreciation that the trajectory of modernisation cannot be fully predicted. The sector should therefore also be geared to respond to technological disruptions.

South African mining is renowned for the complexity of its geological formations, especially with regard to endowments that have been mined for over a century such as gold, which are now located in narrower tabular reefs and lie deeper in the bowels of the earth. While

---

3   China's programme to build major multiple road, rail and sea links across Asia, the Middle East, Europe and Africa.

this presents an objective condition which in itself demands more creative methods of extraction, subjectively the old mode of operation is generally slowly disappearing as new technologies come on stream. Productivity and competitiveness dictate that South African mining generally should follow global trends. But beyond this, South Africa's old methods of extraction and labour-sourcing still dominate, while the character of the labour force is changing, with younger and more educated workers joining it. Their expectations of wages, as well as working and living conditions are very different from the previous era. Without modernisation, these expectations increasingly collide with cost curves, thus rendering many operations unsustainable.

At the same time, on a global scale, shareholders and customers expect responsible mining that takes into account environmental concerns including energy and water intensity, in addition to safety and other working conditions. Research by the then Chamber of Mines predicts that, without a change in mining methods, South African gold and platinum mining will become unsustainable within the coming two decades (Chamber of Mines of South Africa as cited by Solomons, 2016a).

What then are some of the emergent operational technologies? We categorise the instances quoted below for illustrative purposes in terms of their primary impact. This is not meant to detract from the fact that virtually all of the categories do affect productivity and safety. Issues pertaining to research and patenting are addressed in a separate section. Various editions of the *Mining Weekly* journal (especially Solomons, 2106b) are the main source of the examples cited in this section.

Productivity: Technologies that are showing huge potential to improve productivity range from select blast mining to raise-boring technology which is fully automated and operates at great depths, platinum-fuel-cell-powered ultralow-profile bulldozers, a remotely navigated stope-cutting system operated from a surface control room and so on. At a more advanced level, high-precision swarm robotic mining which combines automation and digitisation, drones utilised above and below ground, driverless remotely controlled haul trucks and loaders, three-dimensional printing and medical imaging equipment are also being introduced. All these and other technologies create possibilities massively to improve productivity by extending mining hours, ensuring better mineral recovery and reducing costs in a manner that could also lead to the resuscitation of mining operations that have become uneconomical.

Safety: In addition to the technologies and systems identified above, many initiatives are being undertaken to reduce the dangers associated with mineral extraction. Non-explosive mining, which uses foam to break rocks in an inert way and does not require hand-drilling, is one example of this. There is also an increasing adoption of slot borer machines which can drill reefs as small as 100 centimetres compared to the heights required for humans to operate, let alone conditions of heat, dust and tremors which negatively impact morbidity and fatality rates in mining operations. The same applies to other ultralow-profile machinery, attached to some of which are sweepers and dozers operated by batteries in narrow areas.

Environment: Beyond the 'reduce, reuse and recycle' principles propagated by the WEF, some of the technologies being introduced to enhance environmentally friendly mining include underground coal-to-gas initiatives which would not only reduce the carbon footprint but also allow for the exploitation of un-minable coal. The usage of hydrogen-fuel-cell-powered bulldozers can also be linked to this system, as would on-site electricity generation. In the recent period, there have been attempts at agro-mining or phytomining where plants which have properties to accumulate high concentrations of minerals are used as metal crops. They can be used in sub-economic zones or to harvest tailings in mine dumps.

Research also points to a process of coarse-particle recovery which requires less energy in crushing and grinding and allows for water to be extracted and recycled. The product can be dry-stacked, and, as such, tailing dams would not be required. Other technologies are now being used to treat acid mine water; some propose that the potassium nitrate and ammonium sulphate from such waste water can then be used to produce agricultural fertiliser. The same applies to how zinc mines can be linked to the production of sulphate fertiliser in deliberate cluster arrangements. This is in line with systems thinking as outlined by Mutanga (2018) in chapter 8 of this volume.

These illustrations speak to the multiple positive outputs and outcomes that a systematic adoption of technology can yield. The process needs to be planned for and adopted as an integrated system, allowing for advancement across the whole front of mineral extraction. This implies a combination of smart approaches in a future 'digital mine':

> *smart surveying and mapping visualisation systems, smart climate control systems and energy savings, smart rock engineering systems, smart data processing and smart mine design, mining planning and decision-making ... [A] mine of the future [would] include underground communication systems that allow for real-time intervention to manage all risks, lamp room camera systems – environmental, health and safety monitoring for security, preventing illegal access, monitoring worker health and face recognition and underground drone technology – a floating technology that sees, maps and collects data* (Kilian, 2015).

Modernisation should also encompass procurement and supply chain management to improve efficiency and to eliminate nepotism and corruption.

Overall, all these initiatives would clearly have serious implications for the labour intensity of the extractive leg of the mining industry. As elaborated later, the mining cluster needs to be examined ultimately as an integrated system to determine the net employment effects.

ENHANCING BACKWARD LINKAGES

Upstream mining-supplier industries and services present the most critical opportunity for South Africa to benefit from its mineral endowments. A number of factors have seen to the development of capacities in this area: the knowledge that South Africa has accumulated over the years of the evolution of its mining industry; the complex nature of its geology in terms of rock formation and depth of mining operations; and the relative self-sufficiency pursued during years of anti-apartheid sanctions.

In an important intervention on this issue, Kaplan (2012) demonstrates that South Africa ranks among the best globally with regard to mining equipment and specialist services. Many of the companies operating in this area are South African, and in some areas they compete quite well with transnational corporations. Some of them are medium- and small-sized companies. Given the specificity of demands in the country's mining sector, they have developed what Kaplan refers to as 'economies of scope' as distinct from 'economies of scale'. South Africa features quite high in terms of patent numbers and quality – alongside the United States and Australia (Kaplan, 2012). What is also critical is that technologies developed for the mining industry such as hydraulic and haulage equipment lend themselves to usage in other economic sectors.

The Manufacturing Circle (2017: 8), citing the World Bank, similarly asserts that manufacturing 'provides the scope to develop world-class mining equipment and capabilities, as well as the best means of processing South Africa's own natural resources and driving innovative activity'.

What then hampers the development of this sub-sector, and what needs to be done to deal with these challenges? There are immediate factors related to the global slowdown in demand and challenges of mining policy uncertainty. However, there are deeper issues which require attention in the context of Mining Vision 2030. At the core of this is industrial policy. Kaplan asserts that mining 'equipment and specialist services, despite being by far the most competitive and export-oriented part of the capital goods sector, receives no particular or special attention' in government's industrial policy (Kaplan, 2012: 7).

The need to address this issue is now recognised across the board, including in the NDP which calls for 'substantially more attention' to be 'devoted to stimulating backward linkages or supplier industries (such as capital equipment, chemicals and engineering services)' (NDP, 2012: 147). What is required, as the Department of Trade and Industry (DTI) has in the recent period argued, is the incorporation of these critical imperatives in government mining policy and praxis. This includes using the Mining Charter as a key tool for mining-related local procurement and as support for the supplier industry in terms of skills, research and development, capital and operational expenditure and exports. A supplier cluster that addresses these challenges would be critical to ensure implementation. A narrow focus on black economic empowerment that can be exploited for fronting and import purposes should be discouraged (DTI, 2014).

Arising out of Mining Phakisa,[4] one of the initiatives has been the establishment of Mining Equipment Manufacturing of South Africa where government and original equipment manufacturers operate from one centre to develop trackless, rail-bound, pump-driven, battery-driven and other equipment. Attached to this is the need more deliberately to insert South Africa into global value chains, including through co-operation with transnational original equipment manufacturers. Critically, the further growth of the mining industry in sub-Saharan Africa offers great opportunity. It therefore stands to reason that, besides promoting the value-addition that would derive from the sub-sector's inherent strengths, the country's economic diplomacy should be geared towards fostering an understanding of the competitive and comparative advantages the various countries have, and the standardisation that would help expand the economies of scale across the region.

While the current focus in the contestation around the Mining Charter is on issues of ownership, the discourse needs to be broadened to address the major opportunities attached to upstream mining supplies. Related to this, in an organic process, black industrialists would also emerge. Critical in this regard is the urgent challenge to deal with poor alignment and co-ordination within government.

ENHANCING FORWARD LINKAGES

Forward linkages or beneficiation is, for good reason, a prerogative passionately propagated by most nations, particularly developing countries, as it speaks to the issue of societal benefit from natural endowments. Beyond this, there are also debates about the costs. It is in this context that the NDP argues:

---

4   An initiative where government departments, mining companies, unions and NGOs worked together over a number of weeks to position mining as a catalyst for economic growth.

> *Beneficiation or downstream production can raise the unit value of South African exports. In this regard, resource-cluster development, including the identification of sophisticated resource-based products that South Africa can manufacture, will be critical … In general, beneficiation is not a panacea because it is also usually capital intensive, contributing little to overall job creation* (NDP, 2012: 146).

The NDP qualification is also influenced by factors such as the energy intensity of most beneficiation industries, against the backdrop, during its adoption in 2012, of the electricity challenges South Africa was experiencing.

Unlike with backward linkages, South Africa has not developed much capacity in relation to most of these opportunities. While its steel industry, agricultural 'minerals', electricity generation and coal/gas-to-liquids technologies had been among the most advanced in the world, these were driven by need in the context of the limited industrialisation philosophies of the 20th century and the imperative in later years to circumvent economic sanctions. Otherwise, historically, the country sought more to earn foreign exchange from the export of minimally beneficiated minerals and pursued backward linkages to improve extraction efficiency. Attempts in the post-1994 period to change this trajectory have hit the wall of import parity pricing which effectively reflects abuse of market power, such that domestic steel prices, for instance, are 'in the highest quartile' in the world (DTI, 2014).

The question is whether it is possible, in the medium-term, to develop an approach that benefits the country, the economy and the various players in the sector and related industries. At the level of principle, there is agreement that the country should diversify 'away from resource extraction and reliance on commodity exports towards … manufacturing, value-adding and more labour-intensive growth' (DTI, 2014: 4).

Opportunities for beneficiation are identified in four categories of minerals: manufacturing (e.g., steel and iron ore, nickel, copper and zinc); infrastructure (e.g., those for manufacturing plus cement which would also incorporate gypsum, limestone and coal); energy (e.g., coal, uranium and gas); and for agriculture (e.g., phosphates, potassium and sulphur). To these can be added jewellery for such precious metals as gold, diamonds and PGMs. According to the DTI, downstream potential exists in a variety of areas (see Table 2).

Table 2: Assessment of beneficiation possibilities

Quantec HS – 4 digit, 2013	Export value, rands	Downstream potential	Downstream jobs
H7110: Platinum, unwrought, semi-manufactured or powder form	R81 319 519 856	High to medium	Medium (catalysts)
H2601: Iron ores and concentrates, roasted iron pyrites	R73 998 564 487	Very high	Very high (manufacturing-autos, construction)
H7108: Gold, unwrought, semi-manufactured, powder form	R63 571 314 217	Medium to low	Low (jewellery)
H2701: Coal, briquettes, ovoids etc, made from coal	R55 855 559 400	High to medium	Medium (polymers)
H7202: Ferro-alloys	R34 821 623 292	Medium to low	Low (SS)
H2602: Manganese ores, concentrates, iron ores >20% Manganese	R15 029 874 497	Medium to low	Low (Fe-alloys and stainless steel)
H2610: Chromium ores and concentrates	R13 131 271 379	Medium to low	Low (Fe-alloys and stainless steel)
H7102: Diamonds, not mounted or set	R12 162 947 876	Medium to low	Medium (jewellery)
H7601: Unwrought aluminium	R11 064 742 022	High to medium	Medium (parts)
H2614: Titanium ores and concentrates	R5 999 252 818	High to medium	Low (pigment, metal)
H7204: Ferrous waste or scrap, ingots or iron or steel	R4 802 413 299	Very high	Very high (manufacturing)
H7606: Aluminium plates, sheets and strip, thickness > 0.2 mm	R4 584 915 868	High to medium	Medium
H2603: Copper ores and concentrates	R4 541 138 458	High to medium	Medium (wire, brass)
H7404: Copper, copper alloy, waste or scrap	R4 439 989 630	High to medium	Medium (wire, brass)

H2615: Niobium, tantalum, vanadium zirconium ores, concentrates	R4 331 548 972	Low	Low
H2618: Granulated slag (slag sand) from iron and steel industry	R3 420 317 958	High	High

Source: DTI, 2014.

Table 2 shows that few of the categories with high to medium beneficiation potential translate into equivalent potential with regard to job creation. Many of the downstream opportunities such as aluminium smelters are very energy intensive. South Africa thus needs to determine whether it should revive the preferential electricity pricing approach in order to attract and service such investments – and thus the surpluses it should command in terms of electricity generation. Further, as argued earlier in this chapter, to the opportunities outlined can be added PGM and fuel cell technology, vanadium and jet engines and other minerals attached to emergent technologies.

How then can the constraints to forward linkages be addressed? These pertain to the infancy of the beneficiation industry generally and matters related to monopoly pricing. Quite logically, if a comprehensive approach is to be adopted on this matter, some of the endowments would need to be classified as 'strategic minerals', as resolved by the ANC at its 2012 National Conference (ANC, 2012c). This, it can be argued, is not contested among the partners in the industry. The question is how this is done and with what implications!

Firstly, it would be better that (rather than confer powers to a minister, willy-nilly, to declare such 'strategic minerals') this should be the product of consultation linked to a clear industrial strategy. Secondly, rather than relying on blunt instruments such as legal provisions that impose actual 'developmental prices', there should be negotiation around a cost-plus approach that takes into account the gate price and transportation costs. For precious minerals such as PGM, gold and diamonds, as opposed to the 'bulkies' (e.g., coal and iron) other relevant formulae can be devised. Such approaches would need to be combined with firm measures to deal with market distortions including cheap (often dumped) imported inputs. This should be handled in a manner that does not encourage operational and management practices that undermine domestic productivity and competitiveness. In addition, a comprehensive cost-benefit analysis needs to be done on the trade in scrap metals including its impact on prices and such economic crimes as cable theft. Indeed, calls have been made that, at the very least, there should be a tariff placed on the export of scrap metal.[5]

Given the infancy of the South African beneficiation sector, consideration should also be given to systems of vertical integration, with mining companies, at least in the early stages of the development of a beneficiation industry, holding shares in linked entities and assisting with funding, human resource development, research and, where applicable, off-take. The PGM sub-sector seems to have adopted this approach in relation to the nascent fuel cell industry.[6] Added to this is the intervention by government to create dedicated PGM special economic zones (SEZ) that offer a variety of incentives. As with backward linkages, development of domestic capacity should focus on South African companies; but consideration should also be given to the forging of partnerships with transnational corporations, in a manner that helps to insert relevant beneficiation activities into global value chains.

While the discourse on beneficiation has focused on manufacturing or physical industrial activities, South Africa needs creatively to look beyond this. For instance, in relation to the PGM industry: platinum could be designated as a foreign exchange reserve and the minting and promotion of platinum coins can be expanded. Further, the broader question of 'financial beneficiation' needs to be interrogated in terms of metal exchanges for endowments that South Africa has in abundance.[7]

---

5   This is one of the proposals emerging from the ANC National Executive Committee Lekgotla held in January 2018.
6   Anglo American Platinum has been engaged in research on electricity generation in rural areas working in partnership with fuel cell producers and, along with Implats and Sibanye, these companies have started using fuel cell technology in mining equipment and electricity generation.
7   Among other things, metal exchanges would assist with price discovery and help smooth out market volatility. The Mapungubwe Institute (MISTRA), working with Pan African Investments and the Johannesburg Stock Exchange has initiated research into the desirability and practicability of establishing a PGM Exchange.

RESEARCH AND DEVELOPMENT

In the section on modernisation, various emergent technologies were identified, and it is quite evident that there is preparedness and even enthusiasm on the part of mining companies operating in South Africa to embrace new technologies. Research on, and production of, equipment in the country has mostly undergirded extraction, spurred on by the desire to improve productivity and safety and to minimise the negative environmental impact that mining operations exact. There are also some commendable research activities in beneficiation, as in PGMs and fuel cell technology, vanadium and lithium, for the manufacturing of strong, light materials and energy storage, respectively.

With regard to forward linkages, citing work by Marin et al., Kaplan (2012) identifies the many positive attributes of the South African mining sector – driven by market requirements, advances in science and technology, as well as market context and market volume. South Africa has developed a technologically sophisticated and globally competitive mining equipment and specialist service sector (Kaplan, 2012: 1), as reflected in the number and quality of patents and the fact that South Africa is a 'world leader and first to market in a host of mining equipment and especially services', in large measure provided by South African companies (Kaplan, 2012: 2). This capacity developed during the pre-1994 era with government support and tapered off after the attainment of democracy. Weaknesses pertain to poor alignment between the departments of Science and Technology (DST) and Trade and Industry – between blue-sky research and technical, product-oriented research – as well as the lack of focus, poor government support and the degrading of institutions (such as CSIR Miningtek). As such, 'Australia is increasingly becoming the location of choice for research and development on the part of South African firms and indeed of South Africans themselves' (Kaplan, 2012: 6).

There have been conscious attempts, through Mining Phakisa to reverse this situation. However, though R100 million was earmarked for mining research and development for 2018 – from R5 million in 2014 – this amount pales into insignificance compared to the estimated R4 billion that other developing mining countries spend (Breytenbach, 2017).

Mining Vision 2030, therefore, should include the allocation of more resources for research and development and better co-ordination between blue-sky research and industrial policy. The initiatives around PGM-based fuel cell technology – ranging from research support to operationalisation of technology and establishment of SEZs – demonstrate the potential in this regard. However, there is also the need to cut the time between conceptualisation of industrial strategies and implementation, setting clear guidelines on localisation and corresponding human resource development strategies that undergird the paths chosen.

Success would largely be dependent on the conscious adoption of what Ferreira and Perot refer to as a 'triple helix' approach. Citing Leydesdorff and Etzkowitz (1998), they argue that in a triple helix arrangement, 'the institutional domains of government, university, and industry, in addition to executing their customary objectives, assume the roles of the other actors', with deliberate knowledge and operational flows among these partners (Ferrira and Perrot, 2013: 107).

A critical element in enhancing the research and development initiative is what the SIMS Report refers to as knowledge linkages. It argues for the reinvestment of a large part of taxes from mining into the training of engineers, artisans and technicians, and for deliberate strategies to discourage the migration of research and technical skills. While estimating that mining could create some 200,000 additional jobs in about 10 years, the NDP argues that, potentially, a further 100,000 jobs could be stimulated through linkages (NDP, 2012).

Research and development, in the context of a long-term vision, should address issues across the value chain of the industry and its environment. This includes the interrogation of global market demand and tailoring activities to extract maximum benefit, as shown in the the Richards Bay Coal Terminal case study cited by Pogue and Ferraz (2016). It should also extend to geo-survey expertise and how society can collectively benefit from its endowments. Critically, research should also focus on the political economy of the sector, including its history; changing demographics of the workforce and labour relations; incentivisation of workers rather than just the managers; issues of ownership and general inclusivity and relations with mining-affected communities and government across the spheres. A network of research partners, straddling the value chain, should form part of the triple helix.

EXPANDING INCLUSIVITY IN OWNERSHIP

Issues pertaining to ownership and progress in advancing black and women empowerment, as a critical part of policy imperatives in the post-1994 period, are canvassed by Gqubule (2012) in chapter 4 of this volume. The fundamental observation in this regard is that the mining sector is still largely white- and foreign-owned.

How should these matters be approached in the context of a long-term vision? In the first instance, the step taken in the 2002 MPRDA to declare mineral resources as the property of the nation as a whole with the state as custodian was a logical starting point. Licensees would be afforded the right to explore, prospect, mine and trade and would pay a royalty fee to the state. Secondly, the logic behind the empowerment imperative, as it applies in this instance to ownership, needs to be embraced by all; proceeding from the premise that correction of a historical injustice of deliberate exclusion is a logical, ethical and constitutional imperative. Thirdly, in macrosocial terms, demographic balances in ownership of the country's mineral endowments are in the long-term interest of all South Africans. Failure to address this issue can only breed social anomie.

It should also be underlined that the state itself acts not only as custodian and regulator; but it also has the right to own and/or operate mines through a state-owned company. Many years after this was asserted in ANC and government policies, little has come of efforts to merge the various state entities operating in the sector and to expand the ownership and operations of state-owned entities.

The weaknesses identified by MIGDETT include the fact that there has been little focus on share ownership by workers, the so-called employee stock ownership plans (ESOPs). There has also been inadequate attention to meaningful participation by communities. Mining Vision 2030 should address these matters also as an important contribution to dealing with inequality and ensuring that all stakeholders become a full part of the mining sector at all levels. This should be undergirded by systematic approaches to deal with matters of worker representation at decision-making level, agreed processes of identifying the most appropriate arrangement (or combination of arrangements) in each instance in terms of ESOPs and profit-sharing and financial education of the beneficiaries.

Reckless as some of the decisions by the previous Minister of Mineral Resources (2015–2018) appeared to be as reflected in both the content of, and process around, the 2017 draft Mining Charter, they are impelled by palpable impatience among various strata within the black population. Perceptions of opportunism and unethical behaviour on the part of specific government actors should not detract from the factual reality that breeds such impatience on the issue of ownership.

But how can some of the detail under contestation be addressed in the context of developing Mining Vision 2030? Firstly, the contestation around these issues should preferably take place in frank interactions among the partners rather than in the courtrooms. Indeed, the majority judgment of the High Court (Gauteng local division) in April 2018 in favour of the 'once-empowered, always-empowered' principle has confounded, rather than decisively resolved, this and related complex issues (Seccombe, 2018).

Secondly, it may be better to avoid retrospectivity in dealing with the ownership empowerment principle, accepting that the beneficiaries have indeed benefited, even if they may have exited specific entities or the sector as such. Thirdly, whatever ownership percentages are pursued, they should be within magnitudes and time ranges that decisively enhance empowerment and are also reasonable and not destructive of the investment environment. Fourthly, going forward, the experiences of other empowerment schemes such as those of Sasol and MTN – which confine black economic empowerment share disposal to previously disadvantaged persons – should be considered. Fifthly, the state needs to use its regulatory powers and procurement muscle more strategically (through state-owned enterprises such as Transnet and Eskom) as leverage to encourage the direct involvement of black people in the ownership and operation of mines. Lastly, when approached as a cluster with a variety of linkages, across the entire mining value chain, this sector can also play a critical role in the emergence of black industrialists, and this can include vertical integration with both the private and state mining entities.

SOME GENERIC ISSUES

In this section, brief assertions are made on other issues that are pertinent to the realisation of the core matters outlined above. This treatment is not meant to detract from the importance of these issues in relation to Mining Vision 2030. Rather, some of them are either dealt with in other chapters of this book, or would require such specialised treatment as this chapter is unable to provide.

Jobs and human resource development: Modernisation will definitely have a negative impact on the labour intensity of extraction in the mining industry. As outlined earlier, this has started to play out in a number of operations, and the trend itself cannot be resisted for long. Nor can the stakeholders in the industry and broader society adopt a Luddite[8] approach to the bugle of modernisation. Critical factors that need to be considered in such modelling include: the quality of jobs that would emerge from modernisation, job opportunities from operations in areas either ignored or abandoned as un-minable, improvement in safety standards, sustainable post-mining activities and the employment opportunities that the establishment of a mature mining cluster would generate in backward, forward and other linkages. To determine the net effect of these and other factors requires appropriate modelling, as well as clear plans on how to phase in modernisation in a manner that preserves employment as much as possible. A critical stream of the visioning should be a systematic approach to knowledge linkages and the allocation of resources and expertise to the development of human resources. This should include an assessment of opportunities across the board, along the lines attempted by the DTI in Table 2.

Social and labour plans: The Mining Charter argues for systematic

---

8   Named after Ned Ludd, an apprentice who is said to have destroyed a textile machine during the late 18th century, the term Luddite is used to refer to workers who protested the introduction of machines that rendered their jobs obsolete in Britain during the Industrial Revolution.

approaches to the development of human resources, labour relations and community development around mining settlements. One of the most critical lessons from the 2012 Marikana tragedy, the 2014 platinum strike as well as social instability in mining areas generally, pertains to community infrastructure development, the living-out allowance and related practices.

While many mine operators have allocated resources for community development, a major weakness relates to the extent of co-operation between the mining houses and various spheres of government. The major deficit in this regard is the level of social compacting, with each partner identifying and playing the role that is required of it. A long-term vision should oblige mining companies to be more actively involved in the conceptualisation and implementation of municipal and provincial development strategies and plans – an approach that demands a major mindset shift on the part of all the stakeholders. Systems and structures of accountability to communities need to be improved and should involve the highest levels of the companies. The capacities and reputational capital of each of the partners need to be mobilised for a common purpose and for common benefit.

Exploration regime: A variety of interventions are required to improve exploration activities, including during periods of low mineral prices. This should be backed up by a well-funded and capacitated national effort to enhance the country's geo-knowledge. National data on geographic incidence of mineral endowments would stand the country in good stead also through promoting small and mid-level mine operators. At the same time, there should be strict regulations to discourage mineral rights speculators and squatters – companies and individuals who access prospecting rights with little or no intention to engage in actual mining activities, but who seek to extract maximum financial benefit by sitting on the rights and/or opportunistically selling them. In this regard, effective implementation of the 'use-it or lose-it' principle would be required.

Utilisation of land: Mining houses own large tracts of land, most of which lies fallow. Especially in the context of Mining Phakisa, as described by Ritchken in chapter 3, creative initiatives are starting to emerge on how such land can be used for agricultural and other purposes. Mining Vision 2030 needs to encompass a systematic approach to this issue on a national scale, with possibilities that would help address not only the challenge of land hunger, but also contribute to community development, job creation and enterprise development.

Post-mining activities: Mutanga, in chapter 8, addresses the matter of cleaning up after disused mines. The chapter is instructive in outlining systems thinking, a theoretical model that should inform extractive activities from the start, so as to ensure sustainability in the long term. In addition to this, an indication was given earlier in this chapter and more extensively in Ross Harvey's chapter 5, on how new technologies can reduce such environmental disadvantages as water usage, tunnelling and extraction of large quantities of rock and soil and energy intensity, among others. All these and other initiatives, including what Ferraz (2013) in the MISTRA publication on *South Africa and the Global Hydrogen Economy: The Strategic Role of Platinum Group Metals* refers to as secondary beneficiation, need to form part of new approaches to mining.

Informal mining: The issue of informal mining is dealt with extensively by Kassa in chapter 6. Historically, South African policy-makers and mining operators seem largely to have buried their heads in the sand on this issue. The belief that law enforcement on its own can resolve this matter is shattered every time scores of illegal operators are 'discovered', piles of dead bodies are lifted from the bowels of the earth and inconsequential evocative statements are uttered by leaders. A long-term vision for mining needs to take on board a comprehensive understanding of the factors that drive the scourge of illegal mining, from the macrosocial to the detailed operational issues. In this regard, as Kassa argues, artisanal and small-scale mining needs to form part of economic inclusivity and poverty alleviation. Experiences across the continent and further afield would stand South Africa in good stead in addressing this issue, taking into account both intended and unintended consequences of such programmes.

CONCLUSION: THE FRAMING PARADIGM AND PRAXIS

An acknowledgement of the role the mining sector has played in the evolution of South Africa's macrosocial dynamics and polity is fundamental to the development of a mining sector vision that can and will be embraced by all stakeholders. The South African mining industry has played a central role in the social exclusion that has historically defined South African society, the legacy of which the country continues to experience. Land dispossession, rural underdevelopment, the migratory labour system, the energy intensity of South Africa's economy and the country's subordinate relationship with developed economies across the globe are some of the blights that the mining sector has historically wrought on South African society.

The inverse of this is the positive force that mining has been to the development of the South African economy. The industrialisation that the country experienced over the past century owes much of its dynamism and relatively advanced character to the mining sector. Though much of the direct contribution to industrialisation may have been skewed towards backward linkages, the indirect ripple effects in the manufacturing sector and the contribution of mining to the development of South Africa's economic infrastructure speak not only to a history that should be acknowledged, but also to a dynamic potential that should be harnessed for the country's future development trajectory.

An appreciation of this reality is a critical starting point in defining the collective mindset that should inform the development of Mining Vision 2030. Acknowledging the negative consequences of past injustice and embracing the potential of a future mutually beneficial trajectory – with a mining industry that catalyses South Africa's development going forward – constitute important first steps towards that common future.

Mining Vision 2030 should have this paradigm as its point of navigational reference. Arising out of this should then be general plans that address the variety of building blocks outlined in this chapter: judicious extraction and exports that also ensure global security of supply; the role of mining in infrastructure development; industry modernisation, backward and forward linkages; research and development; and issues of equitable ownership and community development. Added to these are factors such as quality of jobs, safety standards, environmental responsibility and the attendant skills revolution.

Quite clearly, all of these issues need to be addressed at a generic level. However, the generic approach will need to be complemented by plans that are relevant and applicable to specific sub-sectors such as the PGMs, iron and steel, coal, gold and other minerals. At each level, policy-makers, operators, workers, communities and other partners should see themselves as catalysts for a new industrialisation drive across the economy, and as core participants in addressing the marginalisation that afflicts society as a whole. Social compacting – characterised by the identification of common objectives and a commitment collectively and variously to contribute to their realisation – should define the relationship among the partners. The triple helix approach that brings on board policy-makers, operators and researchers is also fundamental. Vertical integration should be used to help forge and support emergent sub-industries.

The state has a central role to play as custodian of the country's mineral endowments: as a regulator, owner-operator, leading force in skills development and as a facilitator of the necessary macro-economic environment conducive for investment in and growth of the mining sector.

The private sector has, as a starting point, to view itself as an integral part of societal efforts to attain national objectives at the same time as it pursues the reasonable returns that are integral to its very existence. This should include a deliberate focus on the emergence of the multifaceted mining industrial cluster, and on imperatives outlined in the empowerment charter such as skills development, equitable representation in professional and management positions, social and labour plans, as well as ownership and profit-sharing arrangements that include meaningful worker and community participation. In the context of an emergent mining industrial cluster, the organic rise of black industrialists will also find practical expression.

The union movement and mine workers generally have to help define and at the same time adapt to the changing environment in the mining sector. This applies to the opportunities and challenges pertaining to modernisation and the emergence of a mature mining industrial cluster. Attached to this is the need for creativity on the part of unionists in taking advantage of opportunities that come with constitutional freedoms, without succumbing to the allure of social distance between worker-leaders and their constituents. The unions are also called upon to tailor their organisational approaches to the changing environment, including the transformation of the demographics of the workforce. It is also critical that, in their crafting of organisational strategies and tactics, unions keenly factor in the concrete circumstances of the sector and how to respond to realities of 'fat and lean years'.

The community sector, particularly mining-affected communities, requires the level of organisation and strategy development which ensures that mining operations in their localities redound to the benefit of all. In this regard, consistent community democracy, equitability in terms of traditional hierarchies and gender, active involvement in the crafting and implementation of social infrastructure programmes, skills development in management of resources and long-term approaches that include extra- and post-mining activities should be encouraged to ensure maximum community benefit.

Appropriate organisational forms are critical for all the social partners, particularly in the negotiations around, as well as crafting and implementation of, the long-term vision. Required of government are integrated approaches that bring together all relevant departmental functions (primarily mining, trade and industry, science and technology, housing, transport, water and environment) and the sub-national spheres. The Chamber of Mines and related sectoral organisations have an important role to play in this regard – a function that they can only adequately fulfil if they enjoy respect among their constituents and other social partners alike.

For purposes of crafting the vision and monitoring its implementation, MIGDETT, which is made up of representatives of government, business and labour should be expanded and appropriately capacitated. Such a platform, also referred to as MIGDETT+, would serve as the organisational mechanism for social compacting, with requisite strategic foresight and technical capacity.

Mining Phakisa – arising from efforts to implement elements of the NDP, which has provided a platform for interaction among government departments, labour unions, mining houses and non-governmental organisations, seems to have embraced the MIGDETT+ approach, and has the potential to serve as the technical infrastructure for these discussions and for monitoring the implementation of the vision and plans. It should also be emphasised that the formal platforms of interaction should not preclude informal interfaces (and mini-compacting at sub-sector or geographic levels), as long as these ultimately feed into the larger formal processes.

REFERENCES

African National Congress (ANC). (1992). 'Ready to govern: ANC policy guidelines for a democratic South Africa adopted at the National Conference'. Available at: http://www.anc.org.za/docs/pol/1992/readyto. html. Accessed 24 November 2017.

ANC. (2012a). 'State intervention in the minerals sector (SIMS)'. Summary Report. Available at: http://www.anc.org.za/docs/discus/2012/sims.pdf. Accessed 4 December 2017.

ANC. (2012b). 'State intervention in the minerals sector (SIMS)'. Slide presentation to the SACP, Johannesburg.

ANC. (2012c). '53rd National Conference resolutions'. Available at: http:// www.anc.org.za/docs/res/2013/resolutions53r.pdf. Accessed 4 December 2017.

African Union. (2009). 'Africa Mining Vision'. Available at: http://www. africaminingvision.org/amv_resources/AMV/Africa_Mining_Vision_ English.pdf. Accessed 15 November 2017.

Benjamin, C. (8 March 2013). 'SA enjoys spin-offs of mining boom'. *Mail & Guardian.* Available at: https://mg.co.za/article/2013-03-08-00-sa-enjoys-spin-offs-of-mining-boom. Accessed 7 December 2017.

Bloomberg. (28 March 2017). 'Swedish miner says Donald Trump doesn't hold key to metal prices'. *Mining Weekly.* Available at: http://www. engineeringnews.co.za/article/swedish-miner-says-donald-trump-doesnt-hold-key-to-metal-prices-2017-03-28. Accessed 15 November 2017.

Breytenbach, M. (6 October 2017). 'South African mining industry still not spending enough on R&D'. *Mining Weekly.* Available at: http://www. miningweekly.com/article/south-african-mining-industry-still-not-spending-enough-on-rd-2017-10-06/rep_id:3650. Accessed 4 December 2017.

Breytenbach, M. (5 February 2018). 'CoM sees resurgence in hope, green shoots in South African mining industry'. *Mining Weekly*. Available at: http://www.miningweekly.com/article/com-sees-resurgence-in-hope-green-shoots-in-south-african-mining-industry-2018-02-05. Accessed 6 February 2018.

Byrnes, R.M. (ed.). (1996). *South Africa: A Country Study*. GPO for the Library of Congress, Washington. Available at: http://countrystudies.us/south-africa/. Accessed 23 November 2017.

Centre for Sustainability in Mining and Industry (CSMI). (2012). 'Mining Vision 2030'. CSMI, University of the Witwatersrand.

Chamber of Mines. (2017). 'What if? Mining investment in South Africa in an improved policy and regulatory environment'. Media statement. Chamber of Mines of South Africa.

Creamer, C. (10 August 2012). 'Positive projections'. *Mining Weekly*. Available at: http://www.miningweekly.com/login.php?url=/article/positive-projections-2012-08-10. Accessed 17 November 2017.

Creamer, M. (5 February 2013). 'Mining sustains quarter of South Africa's people'. *Polity*. Available at: http://www.polity.org.za/article/mining-sustains-quarter-of-south-africas-people-anglo-american-2013-02-05. Accessed 17 November 2017.

Creamer, M. (9 June 2016). 'Mining's rebirth under way'. *Engineering News*. Available at: http://www.engineeringnews.co.za/article/minings-rebirth-on-way-ritchken-2016-06-09. Accessed 16 November 2017.

Creamer, T. (6 March 2018). 'South African steel, aluminium forms watchful, but not overly anxious, as US tariff hikes loom'. *Engineering News*. Available at: http://www.engineeringnews.co.za/article/south-african-steel-aluminum-firms-watchful-but-not-overly-anxious-as-us-tariff-hikes-loom-2018-03-06. Accessed 7 March 2018.

Department of Mineral Resources (DMR). (2013). 'SA mining industry – Status, challenges and responses'. Presentation to ANC Economic Transformation Committee (ETC), Johannesburg, March 2013.

DMR. (2017). 'Reviewed broad based black economic empowerment charter for the South African mining and minerals industry'. *Government Gazette* 41062.

Department of Trade and Industry (DTI). (2014). 'Upstream mining and downstream mineral value chain issues and action plans'. Presentation to ANC ETC, Johannesburg.

European Union (EU). (2010). 'Report of the Ad Hoc Working Group on Defining Critical Raw Materials'. Available at: https://ec.europa.eu/growth/tools-databases/eip-raw-materials/en/system/files/ged/79%20report-b_en. Accessed 24 November 2017.

Ferraz, F. (2013). In MISTRA (ed.). *South Africa and the Global Hydrogen Economy: The Strategic Role of Platinum Group Metals*. Mapungubwe

Institute for Strategic Reflection (MISTRA), Johannesburg.

Ferreira, V. and Perrot, R. (2013). 'Emergence of new industries: Industry and knowledge factors'. In MISTRA (ed.) *South Africa and the Global Hydrogen Economy: The Strategic Role of Platinum Group Metals.* Mapungubwe Institute for Strategic Reflection (MISTRA), Johannesburg.

Freedom Charter. (1955). Available at: http://www.anc.org.za/sites/default/files/docs. Accessed 23 November 2017.

Gencor Ltd. (Undated). Gencor Ltd. Company profile, Information, Business description, History, Background information on. Available at: http://www.referenceforbusiness.com/history2/22/Gencor-Ltd.html. Accessed 23 November 2017.

I-Net Bridge. (7 February 2012). 'SA failing to capitalise on its mineral wealth'. *TimesLive.* Available at: https://www.timeslive.co.za/sunday-times/business/2012-02-06-sa-failing-to-capitalise-on-its-minerals-wealth. Accessed 23 November 2017.

Kaplan, D. (2012). 'South African mining equipment and specialist services: Technological capacity, export performance and policy'. *Resource Policy.* Vol. 34(4), pp. 425–433.

Kaplan, D., Morris, M. and Kaplinsky, R. (2012). 'One thing leads to another: Promoting industrialisation by making the most of the commodity boom in sub-Saharan Africa'. Available at: http://www.prism.uct.ac.za/Downloads/MMCP%20Book.pdf. Accessed 7 December 2017.

Kilian, A. (21 August 2015). 'Mine of the future will be digital'. *Mining Weekly.* Available at: http://www.miningweekly.com/article/mine-of-the-future-will-be-digital-2015-08-21. Accessed 11 November 2017.

Makhuvela, A. and Msimang, V. (2013). 'PGM and other strategic minerals in the hydrogen economy: South Africa's boom'. In MISTRA (ed.) *South Africa and the Global Hydrogen Economy: The Strategic Role of Platinum Group Metals.* Mapungubwe Institute for Strategic Reflection (MISTRA), Johannesburg, p.44.

Manufacturing Circle. (2017). 'Map to a million new jobs in a decade'. Available at: https://docs.wixstatic.com/ugd/d9d043_c5b0339de6ff4a3ca45ba9d2320690b0.pdf. Accessed 23 November 2017.

National Planning Commission (NPC). (2011). *Diagnostic Document.* The Presidency, Republic of South Africa.

NPC. (2012). *National Development Plan (NDP).* The Presidency, Republic of South Africa.

Pogue, T. and Ferraz, F. (2016). 'Mining and the South African national systems of innovation.' In Scerri, M. (ed.). *The Emergence of Systems of Innovation in South(ern) Africa: Long Histories and Contemporary Debates.* Mapungubwe Institute for Strategic Reflection (MISTRA), Johannesburg.

Ramatlhodi, N. (2015). 'Outcomes of the MIGDETT meeting'. Statement by Minister of Mineral Resources, Pretoria. Available at: https://www.gov.za/speeches/minister-ngoako-ramatlhodi-outcomes-migdett-meeting-14-may-2015-0000. Accessed 27 November 2017.

Seccombe, A. (4 April 2018). 'Court finds in favour of Chamber of Mines: Once empowered, always empowered'. *Business Day*. Available at: https://www.businesslive.co.za/bd/national/2018-04-04-high-court-finds-for-chamber-of-mines-regarding-mining-charter/. Accessed 7 December 2017.

Shain, M. (20 April 2016). 'From Hoggenheimer to the Guptas'. *Politicsweb*. Available at: http://www.politicsweb.co.za/opinion/from-hoggenheimer-to-the-guptas. Accessed 11 November 2017.

Solomons, I. (13 September 2016a). 'Platinum miner says mechanisation, modernisation is only way forward'. *Engineering News*. Available at: http://www.engineeringnews.co.za/article/platinum-miner-says-mechanisation-modernisation-is-only-way-forward-2016-09-13. Accessed 16 November 2017.

Solomons, I. (23 September 2016b). 'No future for conventional mining methods in South Africa – Amplats'. *Mining Weekly*. Available at: http://www.miningweekly.com/article/substantial-change-required-to-s-african-mining-methods-to-ensure-sustainability-2016-09-23. Accessed 7 December 2017.

South African Communist Party (SACP). (1962). *The Road to South African Freedom*. Inkululeko Publications, London.

South African Government. (2002). Mineral and Petroleum Resources Development Act (MPRDA). Preamble and Chapter 2. *Government Gazette*.

Swanepoel, B. (2006). 'Presentation to Mining Summit'. Also reported in: Webb, M. 12 September 2006. 'Regulatory regime stunting foreign mining investment – Swanepoel'. *Engineering News*. Available at: http://www.engineeringnews.co.za/article/regulatory-regime-stunting-foreign-mining-investment-swanepoel-2006-09-12. Accessed 15 November 2017.

Truehuenews. (6 September 2015). 'Trends in urbanisation to 2050'. Table. Available at: http://truehuenews.com/2015/09/06/infographic-urbanization-urban-population-by-country-in-2010-and-2050. Accessed 11 November 2017.

University of the Witwatersrand. (Undated). 'History and heritage'. Available at: https://www.wits.ac.za/about-wits/history-and-heritage. Accessed 30 October 2017.

Venter, I. (2 December 2014). 'Sub-Saharan infrastructure spend to reach $180bn by 2025'. *Engineering News*. Available at: http://www.engineeringnews.co.za/article/sub-saharan-infrastructure-spend-to-reach-180bn-by-2025-pwc-2014-12-02. Accessed 23 November 2017.

Wilson, M.G.C. and Anhaeusser, C.R. (1998). *The Mineral Resources of South Africa*. Council of Geoscience, Silverton.

World Economic Forum. (2014). 'Scoping paper: Mining and metals in a sustainable world'. World Economic Forum, Switzerland.

THREE

The gold of the 21st century

A vision for South Africa's platinum group metals

EDWIN RITCHKEN

THERE ARE A NUMBER of developmental opportunities that are common across the mining industry and different commodities markets. These include opportunities for upstream and downstream industry development, technology enhancement in core mining operations, surrounding community development, as well as environmental impacts. It is consequently possible to construct an over-arching 2030 vision for mining, which is a challenge undertaken by Netshitenzhe in this book.

It is also important to recognise that different commodities and types of mining open up substantially different opportunities for development impacts. Some of the drivers of these differences include: above-ground versus deep underground mining, bulk versus precious metal product and the nature of the market for the commodity.

Following a brief discussion of these drivers, this chapter explores how they need to be taken into account in constructing a commodity-specific vision for the platinum group metals (PGMs). PGMs are the largest mineral resource in South Africa by value by some margin. Hence, given the decline of gold mining, if any resource has the potential to become the cornerstone for a future mining cluster, it is PGMs.

Given the relative inelasticity of supply and dynamic demand conditions, mining by its nature will always be exposed to a commodity cycle. The market for PGMs needs to be continuously developed as there is no 'natural' global demand for the metals – demand is derived through a continuous process of PGM-based technology development and commercialisation. This requires co-ordinated and ongoing global investment into mining, into new PGM applications and into the development of markets for those applications. Failure to make adequate investments in any portion of the value chain will result in either over or under mining capacity and will contribute to the formation and/ or aggravation of a commodity cycle. However, as the PGM mining and technology development industry is highly concentrated with four major mining companies and five major technology developers (who account for 85 per cent of the market), the sector lends itself to strategic partnerships across the value chain (IDC, 2013). Historically, mining companies recognised the importance of investing aggressively in growing demand for platinum and the development of the industry was characterised by high levels of collaboration along the value chain as technology and markets were developed. This acted as a mechanism to moderate the commodity cycle. However, in 2013, the high level of collaboration broke down and historically strategic relationships became more transactional; and emphasis was put on the development of platinum as an investment vehicle, which was not successful. This chapter argues that it is critically important for government together with the Public Investment Corporation (PIC) and the mining industry, to rebuild key strategic relationships across the value chain.

Core to the management of a mining company's balance sheet is the building up of reserves during the boom and the utilisation of these reserves during the down cycle. The reality is that a number of major mining companies paid out a large portion of their retained earnings to shareholders and are presently investing inadequately in sustaining capital and in market development programmes.

In addition, South Africa requires a nationally driven effort to generate sustainable demand for the metals so as to secure the sustainability of the industry. This, in turn, will require an enabling policy and regulatory environment. The problem is that there is a high level of distrust between stakeholders, which makes the building of a broad consensus around an industry strategy very difficult. Consequently, it is further argued that a focused development coalition needs to be built, which needs to implement a range of concrete interventions to lay a strong foundation for the building of trust between stakeholders and the unlocking of a truly national initiative. These interventions include initiatives to redress historical legacies, to promote impact-driven shareholding by the PIC, to drive industrialisation and innovation in the supply chain, to support beneficiation and global PGM-demand creation and to drive large-scale agricultural and rural development. In the absence of such initiatives, the sustainability of the industry at anything like its present scale will be at severe risk.

KEY DRIVERS OF AN INDUSTRY VISION

South Africa is a relatively small segment of the global market for above-ground mining equipment. This limits the opportunities to leverage procurement within South Africa to establish a world-leading, above-ground mining equipment design and manufacturing industry. In contrast, South Africa is a leading customer for deep underground mining equipment and systems. In addition, the national industry has the unique challenge of tackling narrow precious metal seams embedded in hard rock types that occur four kilometres underground. This gives South Africa the opportunity to leverage this position to take a leading role globally in the design and manufacture of these technologies.

In terms of bulk versus precious metal product, from a beneficiation point of view the comparative advantage of mining product in South Africa is ultimately determined by the cost of getting it to a beneficiation centre in relation to the value of the product. For example, the cost of logistics for very precious metals is extremely low compared to the value of the metal, whereas for iron ore the logistics costs are more significant.

Furthermore, the key drivers of country competitiveness in the beneficiation of any product relate to technology mastery, skills and investment in machinery, equipment and enabling infrastructure. Hence, a realistic assessment of the comparative advantage of beneficiating minerals in South Africa requires an analysis of location advantage as well as the quality of the beneficiation cluster.

Lastly, each commodity market has its own dynamics and cycles and all commodities require investment in marketing and new applications. However, certain markets such as iron ore and gold are extremely mature, have well-established applications and are relatively predictable over the medium term. The PGM market in contrast is very dependent on the continuous development of new technologies. Consequently, this market requires a very strategic approach to development for the sustainability of PGM mining.

DYNAMICS OF THE MARKET FOR PGMS

The six PGMs – platinum, palladium, rhodium, osmium, ruthenium and iridium – occur together in nature alongside nickel and copper. Platinum, palladium and rhodium, the most economically significant of the PGMs, are found in the largest quantities. The remaining PGMs are produced as co-products. South Africa is the world's leading platinum and rhodium producer and the second largest palladium producer after Russia. South Africa's production is sourced entirely from the Bushveld Complex, the largest known PGM resource in the world.

World platinum resources are estimated at over 100 million kilograms, of which over 70 per cent are located in South Africa (US Geological Survey, 2018). In terms of reserves, South Africa has over 90 per cent of the world's known platinum reserves of 69 million kilograms (valued at over US$2 trillion), with Russia, the USA, Canada and Zimbabwe making up the balance (Table 1). At present, South Africa supplies approximately 70 per cent of global demand (Figure 1). De facto, South Africa is the global supplier and guardian of PGM resources.

PGM and associated mining activity are strategic elements of the South African economy. The platinum mines constitute around

40 per cent of mining employment through providing direct employment to 136,000 people, and further support 325,000 indirect jobs in South Africa (Chamber of Mines, 2016). PGMs as a group are the second largest mining-related export revenue generator for South Africa.

Table 1: Mine production of platinum and palladium

Platinum and palladium mine production and reserves by country (2017)						
Country	Platinum (mine prod)	Percentage	Palladium (mine prod)	Percentage	PGM reserves	Percentage
United States	3,900	2%	13,000	6%	900,000	1%
Canada	12,000	6%	19,000	9%	310,000	0,4%
Russia	21,000	10,5%	81,000	40%	3,900,000	6%
South Africa	140,000	70%	76,000	38%	63,000,000	91%
Zimbabwe	15,000	7,5%	12,000	6%	1,200,000	2%
Other countries	4,000	2%	8,400	4%	N/A	
World	200,000		210,000		69,000,000	

Source: USGS, 2018

PGMs have unique properties that make them a key asset in existing and new technologies to combat climate change. According to Mathys-Graaff (2015), currently 41 per cent of platinum demand comes from the autocatalytic convertor industry, 35 per cent from jewellery manufacturing and 19 per cent from investment (Figure 2). PGMs presently play a critical role reducing poisonous gas emissions from vehicles, and will be the key element in the unlocking of a hydrogen economy.[1] Hence, PGM demand is tied closely into the robustness of environmental policy and regulation.

---

1   For a comprehensive discussion on the opportunities associated with the hydrogen economy, please see *South Africa and the Hydrogen Economy: The Strategic Role of Platinum Group Metals*, Mapungubwe Institute for Strategic Reflection, 2013. See in particular chapter 1: 'PGM and Other Strategic Minerals in the Hydrogen Economy'.

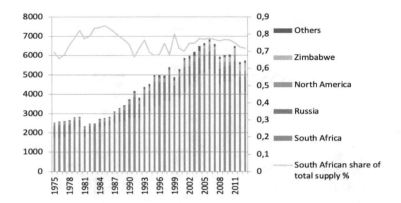

Figure 1: Global platinum supply
Source: Bowman, 2016

The PGM market needs to be continuously developed as there is no 'natural' global demand for the metals – demand is derived through a continuous process of PGM-based technology development and commercialisation. This requires co-ordinated and ongoing investment into mining systems, new PGM applications and into the development of markets for those applications. Put simply, the value of PGMs is intrinsically tied to the value of the intellectual property (IP) investment that turns the PGM into a useful metal (alongside its relative scarcity) and the commercialisation of this IP (see Figure 3). The development of PGM-based technology is dominated by five global original equipment manufacturers (OEMs) or 'fabricators' who are responsible for buying 85 per cent of platinum produced by refiners and who then develop the metal for different application (IDC, 2012). These fabricators also have the ability to generate the intellectual property to minimise PGM content in technology and consequently hold a very strategic position in the platinum value chain. Consequently, the PGM market is in a continuous state of flux as substitutes for PGM-based technologies are continuously developed, thrift methodologies[2] are advanced and an inevitable business cycle operates. These processes are exacerbated by perceptions of supply risk associated with instability in South Africa's mining policy and regulatory and social environment.

---

2   Thrift methodologies refer to explicitly designing technologies so as to decrease the quantity of platinum used in an application.

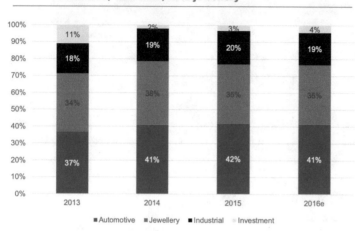

Figure 2: Platinum demand
Source: Mathys-Graaff, 2015

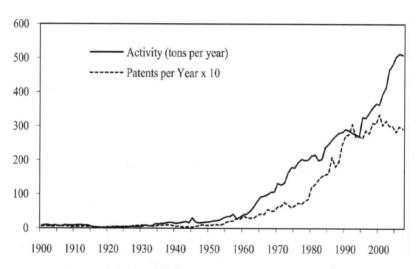

Figure 3: PGM production and patents
Source: Connelly, 2010

Co-ordinated investment in research and development (R&D) is particularly challenging because whilst the PGM resource is concentrated in South Africa, PGM technology development capabilities and markets are global (see Figure 4).

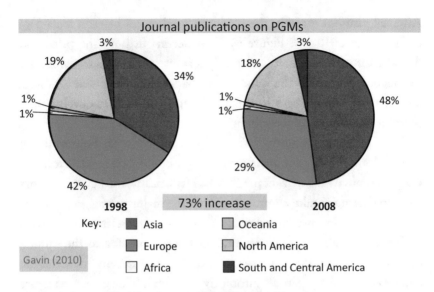

Figure 4: Journal publications on PGMs by world region
Source: Msimang, 2013

Given these dynamics, it is unlikely that the 'market' will optimally co-ordinate investments along the value chain to secure the long-term viability of the industry. In the absence of such co-ordination, demand will at best be volatile; at worst, it could enter into a process of continuous decline (see Figure 5).

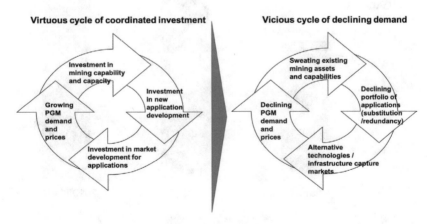

Figure 5: Vicious and virtuous cycles of PGM demand[3]

---

3   All figures in this chapter that do not identify a source are the personal work of the author.

After 20 years of growth, the platinum market has been in a state of decline since 2010 (see Figure 6). This decline in both the price and demand for platinum is at least partly a result of changes in the nature of collaboration across the value chain, changes in the demand for diesel autocatalysts and changes in the application and market development programmes driven by the mining industry.

The platinum industry is highly concentrated. Four mining companies at present dominate the mining sector whilst five fabricators/ OEMs dominate the market for PGMs. Historically, the key fabricators had long-term off-take contracts with all the major mining companies, although certain mining companies had special relationships with individual fabricators. Japanese trading houses invested in the mining companies in the 1980s to secure supply of platinum for catalytic converters for the Japanese automotive industry. This cemented more long-term relationships. It is notable that Japanese trading houses and state-owned entities continue to invest in mining exploration and mining and processing companies to secure supply for the nascent Japanese hydrogen economy (Speight, 2014). The metal is seen as strategic to Japan's industrial policy.

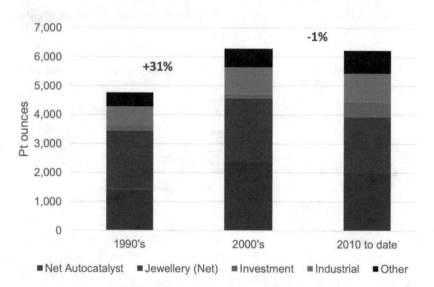

Figure 6: Platinum historical demand
Source: Mathys-Graaff, 2015

In 2013, Anglo Platinum (Anglo) was the dominant platinum miner responsible for 40 per cent of global primary platinum production and was the largest owner of platinum reserves in the world (Anglo American Platinum Limited, 2013). Anglo's open pit Mogalakwena Mine is at the lower end of the cost curve and extremely profitable, accounting for almost 70 per cent of 2017 operating profits (Anglo American Platinum Limited, 2017). While all four mining companies entered into long-term contracts with fabricators, Anglo's strategic partnership with Johnson Matthey (JM), a world leader in the development of chemical-based technologies, was special and was core to the development of the sector. The partnership started in the 1920s, to develop the platinum resource. JM developed the refining technology which was later localised in South Africa by Anglo. The strategic partnership evolved to incorporate downstream application and market development objectives (Bruce, 1996). Under the terms of the partnership, Anglo effectively sold its output in bulk to JM at a discounted price. In exchange for the discount, JM developed new applications, marketed the metal to smaller buyers and established lines of credit and logistics networks to support buyers and to provide market intelligence services to Anglo to help balance supply with projected market demand. In effect, the JM relationship created a strategic marketing organisation that was able to steadily increase demand in a number of different markets while the metal was finding its feet and supply was slowly growing. As important, but hard to measure, was a sense of joint interest in the management of the metal market by the major mining companies and the fabricators. In 2013, Anglo terminated the strategic market development partnership with JM and started to market the metal in-house on a transactional basis with the fabricators (IOL, 2014). The World Platinum Investment Council, a PGM mining driven organisation, supported the development of exchange traded funds (ETFs), particularly in South Africa, where the holdings of ETFs grew from nothing to well over a million ounces, to make up over half of global ETF holdings between 2012 and 2014 (World Platinum Investment Council, 2018). Although the investment funds stagnated thereafter and have not contributed to creating significant additional demand, a tightly managed market, characterised by high levels of co-ordination and collaboration along the value chain, became financialised and subject to speculative trading activity.

In the 1980s, platinum was the dominant catalyst for petrol engines. However, in the early 1990s, because palladium was cheaper than platinum, investment went into research and development aimed at substituting platinum with palladium in petrol catalysis. The result was that by the end of the 1990s, palladium had overwhelmingly substituted platinum in petrol catalysis (see Figure 7). Post 2010, with the diesel car emissions scandal, demand for diesel cars (and thus platinum) has dropped which has, together with strong growth in the demand for petrol cars in China, increased the demand for palladium, resulting in the price of palladium overtaking that of platinum for the first time in 17 years in 2018. The supply of palladium is inelastic as it is mined as a by-product of nickel in Russia, which accounts for just under 40 per cent of supply, and as a by-product of platinum in southern Africa which accounts for 43 per cent of supply. Given the current trends, this suggests that there will be a serious deficit of palladium for petrol catalysis in the coming years unless palladium is substituted by platinum. Hence, the outlook for platinum has, amongst other issues, a great deal to do with how quickly the IP can be developed to competitively utilise platinum in petrol catalysis.

The platinum market has historically been driven by a balance between industrial demand (which includes auto-catalysts and is price inelastic) and jewellery demand (which is price sensitive). In instances when industrial demand has dropped, resulting in a lowering of price, the jewellery market has served the function of absorbing the additional platinum ounces on the market. This role was enabled by investment in jewellery market development in both India and China, driven by an industry initiative, the Platinum Guild International (Anglo American Platinum, undated). This dynamic is illustrated in Figure 8, which shows how jewellery demand picked up, for a limited period only, in 2009 as a consequence of the lower platinum price. However, since 2012, the mining industry has cut the budget for jewellery market development. The result has been a structural change to the market where jewellery demand has shrunk from 2015 to 2018 despite a lowering in platinum price and an increase in platinum availability (see Figure 8). There was a particularly precipitous drop in jewellery demand by one million ounces in 2017. This was also influenced by a lower demand for gold jewellery, particularly in China, and increased competition from other sectors for millennials' disposable income. There is a challenge, therefore, to return investment in jewellery market development to at least 2012 levels, in order to make up for lost ground.

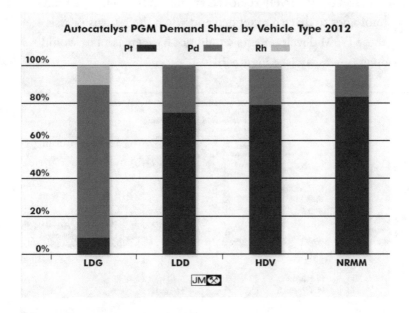

Figure 7: The use of platinum and palladium in auto-catalysis
Source: Matthey, 2013

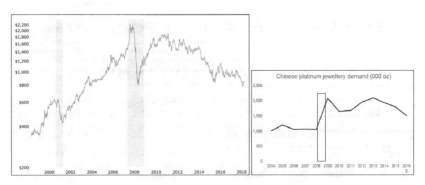

Figure 8: Declining platinum jewellery demand despite decrease in platinum price
Source: Jollie, 2016 and Macrotrends, undated (https://www.macrotrends net/2540/platinum-prices-historical-chart-data)

In addition, key mining companies decided to turn a range of their market development activities into venture capital activities which prioritise market-related financial return of a project over the project's impact on demand. As there already is a venture capital industry and large fabricators investing in PGM development, this structural change effectively removed an important source of funds for technologies to drive market demand when the returns or risk meant these PGM developments would not happen at all or would not happen timeously (see Figure 9).

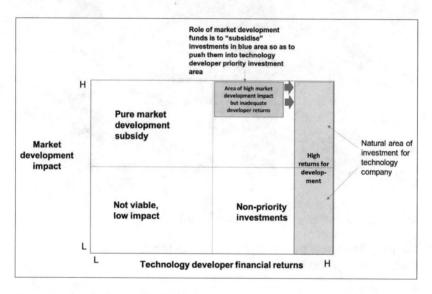

Figure 9: A coherent market development fund and a 'confused' fund

DECLINING INVESTMENT IN MINING

The Western Limb of the platinum belt, located in the North West province, is responsible for 55 per cent of South African mining production, 40 per cent of global mining production and 30 per cent of global platinum supply when taking recycling into account (see Figure 10; Mnguni, 2018). Two key platinum mining companies, Lonmin and Sibanye Platinum (particularly when the Sibanye assets were owned by Anglo Platinum until 2016), have not invested adequately in sustaining capital over the last five years (see Figure 11). If the trend is to continue, these cuts in sustaining capital, combined with depleted reserves, relatively low platinum prices and depleted company balance sheets are projected to result in a 51 per cent cut in platinum production over the next 10 years until 2028 (see Figure 12).

This is because even when demand for platinum recovers and the price increases, these companies will not be able to respond timeously to the opportunity and their production will continue to decline. The net result will be that, if the trend continues, according to Mnguni (2018), in 10 years' time, platinum output from the limb will be halved, which could create a serious deficit in the market, although Anglo's Mogalakwena Mine could ramp up production to meet some of the deficit. This will give the users of platinum additional incentive to substitute and thrift the metal because the resulting supply deficit will ultimately lead to a smaller industry.

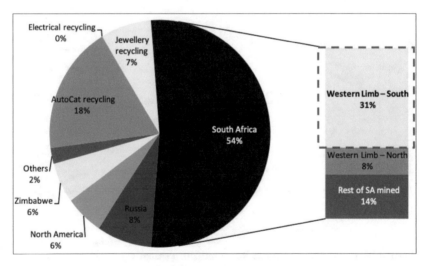

Figure 10: Western Limb contribution to global platinum production
Source: Mnguni, 2018

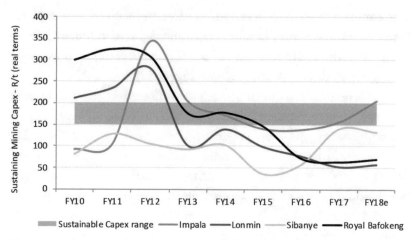

Figure 11: Shortage of investment in sustaining capital
Source: Mnguni, 2018

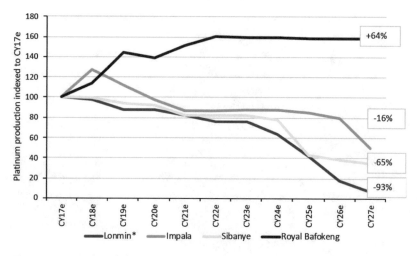

Figure 12: Projected decline in Western Limb output
Source: Mnguni, 2018

A further development that has undermined investment in PGM mines in the Western Limb was the Sibanye Gold acquisition of Stillwater combined with Sibanye taking over Lonmin in an all share buy-out. When Sibanye acquired Anglo American Rustenburg assets in 2015 (as discussed above), these shafts had already experienced under-investment. In 2017, Sibanye paid a price of US$2.2 billion cash for the purchase of Stillwater, an American palladium mining company. The deal was paid for through a US$1 billion rights offer and through taking out additional debt (Sibanye-Stillwater, 2017). The result of the acquisition was to heavily load Sibanye's balance sheet with debt and consequently to fundamentally change how the company invested in its different shafts and mines. In order to receive this debt, Sibanye had to enter into loan covenants with banks. An example of a loan covenant would be the proportion of cash that a mine generates in relation to the total amount of debt that is owed. If a covenant is broken, a bank can call in its loan which can result in the company going bankrupt. The meeting of loan covenants (i.e., agreements with banks that have provided debt to Sibanye) means that Sibanye has to optimise its short-term cash generation and preserve cash to decrease the debt owed to banks. However, the need to meet loan covenants puts immediate pressure on lesser-performing assets as the company has no space to cross-subsidise poorly performing shafts until prices recover, because the money is needed to pay debt. For example, by the end of the first quarter of 2018, the only growth projects in Sibanye's PGM assets relate to Stillwater (in the USA) with an expenditure of R335 million (including corporate costs). Growth capital of R300 million out of R350 million for PGM mine development in South Africa has been deferred (Sibanye-Stillwater, 2018). The problem is that today's growth capital is required to sustain production in three to five years – it is then that the impacts of the deferred growth capital expenditure will be felt. When the situation reaches a certain breaking point, it results in the closure or fire sale of shafts and the retrenchment of people; this would not have happened if there was a strong balance sheet (Figure 13). It is in this context that the takeover of Lonmin, which has experienced serious under-investment in sustaining infrastructure over a number of years, by a severely capital-constrained Sibanye-Stillwater should be analysed. The impact of investing offshore on local mining investment needs to be better understood, particularly when the funding is provided by the PIC.

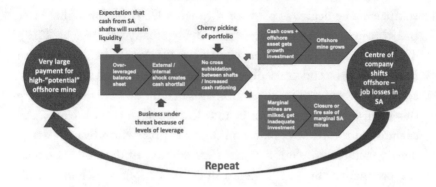

Figure 13: The process of mining companies moving offshore

POLITICAL ECONOMY ISSUES

Reversing this negative spiral in demand and investment will require a national effort involving a partnership between the mining cluster and a range of government departments and agencies in order to stimulate investment in mining, in PGM application development and in the commercialisation and market development of these applications. These partnerships need to be organised into a 'developmental coalition' that will have the ongoing objective of optimising the sustainability and developmental impact of the sector by undertaking a range of concrete initiatives. It will ultimately be the positive impact of these initiatives (rather than any rhetoric that may surround them), that will result in the rebuilding of trust between the industry and stakeholders. The key insight is that the extent to which there is an enabling policy and regulatory environment for the mining sector in general, and the PGM sector in particular, will ultimately be determined by the extent and quality of stakeholder support which, in turn, will be determined by the integrity and impact of industry behaviour (see Figure 14).

Globally, mining is a uniquely contested sector as mineral resources are seen as a non-renewable, national asset. Consequently, there is a great deal of sensitivity regarding the extent to which the value from the resource is being captured by a minority or being used to lay a broader foundation for prosperity for future generations. In South Africa, this contestation over the value flows from mining is particularly acute because of a history in which the benefits of mining were captured by a white minority, whilst the social costs were disproportionately borne by a black majority. Mining does not only have a history of externalising many social costs; the environmental costs of the industry can also be significant and amplify the social costs. This is particularly so where mining operations are concentrated, such as on the platinum belt in South Africa. These costs are most deeply felt by local communities who lose arable and grazing land and whose water resources are reduced or polluted. Health problems associated with air pollutants are also common in communities bordering mines.

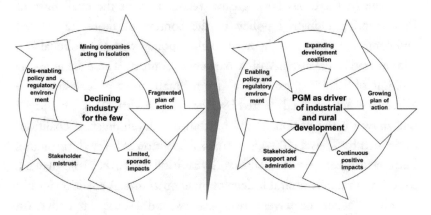

Figure 14: Cycles of industry decline and development

This complexity is exacerbated by the cyclic nature of commodities. Booms will invariably be followed by downturns. This creates an intrinsic tension between which profits should be distributed to shareholders and which should be used to strengthen the balance sheet, so that the company will be in a strong position to maintain investment during a down-cycle. The platinum mining industry went through a significant boom during which the industry achieved extraordinarily high profit margins (see Figure 15). The result is that if the industry does not retain cash, during downturns many shafts that are high up in the cost curve would be unsustainable as companies would not have capital on their balance sheet to invest in these shafts (let alone market development initiatives).

The 1994 South African political settlement was at too high a level to provide any resolution as to how the benefits of the mining sector should be shared amongst stakeholders. This has left the sector in flux and trapped in a process of continuous negotiation and contestation.

There is much literature on industrial policy in the context of a coherent, co-ordinated state that drives investment in areas of greatest medium-term economic return. These 'developmental states' are characterised by high levels of centralisation and an alignment (as a result of the use of appropriate incentives and penalties) between the state and business regarding the state's developmental agenda (Kohli, 2004 and Evans, 1995).

Recently, there has been greater recognition of the challenges of implementing industrial policy in the context of more fragmented and contested states and the need to align policy with the political and administrative constraints of a particular context. These include the building of 'islands of excellence' (from Brazil) and a process-intensive 'discovery' approach advocated by Rodrick (Lowitt, 2017). This paper will draw on the notion of the bottom-up 'developmental coalition' as defined by Meisel (2004),[4] whereby a coalition is formed, through a bottom-up agreement to collaborate by a critical mass of key role-players, to achieve mutually desirable development outcomes. A coalition is only meaningful when the players are capacitated and resourced to deliver on any commitments made to other participants. In other words, in highly fragmented political environments, a strategy is only as good as how much you can get a critical mass of stakeholders to deliver.

Behind every policy and regulatory environment is a coalition of stakeholders who have implemented a programme that has sufficient stakeholder support to substantively influence the nature of the policy regime. At present, the PGM mining sector is extremely politically isolated (as a result of a range of historical and contemporary issues) and does not have the trust or credibility to build bridges with key stakeholders through lobbying or rhetoric. Hence, the starting point for addressing the marginalisation of the sector is to build a development

---

4   Meisel does assume that the coalition will create a strategic 'focal monopoly' to transform the structure of the economy. This chapter uses the term 'development coalition' in a more evolutionary and tactical sense.

**Gross profit margins at Amplats (pre-1996, RPM), Implats and Lonmin, 1977–2015
(Source: Company annual reports and accounts)**

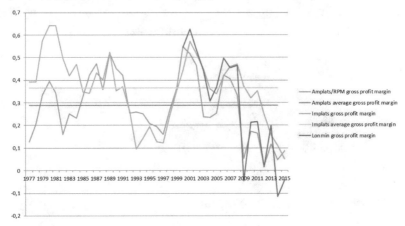

Figure 15: Gross profit margins at major South African PGM mines
Source: Bowman, 2016

coalition that has the resources and capability to start delivering on co-ordinated transformation, and on an industrial and rural development programme. A 'development coalition' is formed through a bottom-up agreement to collaborate by a critical mass of key role-players, to achieve a mutually desirable development outcome. A PGM market development intervention would be an integral component of this programme. This will require concrete commitments from all participating stakeholders and a co-ordinated programme of implementation.

THE CHALLENGE OF PATRIMONIALISM

There are inevitably three logics driving the trajectory of a company, namely the commercial, developmental and patrimonial logics (see Figure 16). Commercially based decisions are made to secure the financial sustainability of the business. Historically, these decisions were made without adequate regard for environmental and safety consequences, amongst other factors. A business will have an intrinsic developmental impact if it adequately invests in plant, skills and technology. However, the way in which a business relates to staff welfare, customers, suppliers and affected communities will also determine its development impact.

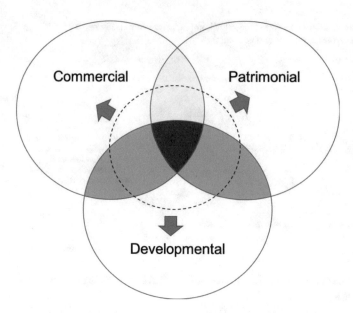

Figure 16: The three logics underlying organisational trajectory of a company

Historically, these impacts were greatly skewed towards white stakeholders. It is therefore inevitable that a certain level of patrimonialism exists in business, in that employment and procurement decisions will often be based on familiarity with an insider, rather than on a rigorous, external search. However, patrimonialism in South Africa is particularly politicised, given the racial nature of the management class in mining and the 'old boys clubs' that were established with a range of stakeholders during apartheid.

These networks form the thin edge of the wedge of racial exclusion, despite formal policies at both enterprise and government levels that seek to drive inclusion. This creates a great deal of cynicism about the industry as the lived experience of communities, black entrepreneurs and black management can contradict the company policy rhetoric voiced by out-of-touch board members who shy away from hands-on oversight.

For example, in an assessment of the efficacy of social and labour plans, the Centre for Applied Legal Studies (CALS) found that 'there is mounting evidence of a stark disjuncture between the rhetoric in

SLPs and the lived realities of mine-affected communities, who do not see the promised benefits of mining development materialising' (2016: 5). This poses questions about the relationship between practices of patrimonialism and how these may impact on policies designed to promote creative collaboration, in contrast to policies that result in creative compliance.

THE LOGIC OF COLLABORATION

It is fundamentally important to appreciate that the logic of collaboration within the context of a development coalition is completely different from that associated with practices of creative compliance. A developmental collaboration is about focusing on impact and operating in a transparent and accountable manner to achieve agreed goals. Coalitions depend on integrity and commitment to achieve an impact, rather than negative consequences for bad behaviour.

In contrast, whilst the logic of top-down policy compliance is intrinsically punitive in nature, it does not require the same level of transparency, and in many cases, effort (see Table 2). However, in situations of distrust between the regulator and the regulated, the regulator will invariably seek to establish a regulatory regime characterised by rigid rules and narrow reporting. In other words, the regime takes as its starting point an intrinsic assumption that the regulated have not substantively bought into the policy objectives that the regulator is tasked with implementing.

Top-down compliance regimes are almost universally associated with a 'creative' response from the regulated and can become 'comfortably uncomfortable' for all concerned (Hood, 2000). An example of this creative compliance response is the narrow view taken by many companies in response to the social and labour plan (SLP, 2016) framework which is discussed in greater depth later on in the chapter. In addition, cynical practices of creative compliance are extremely compatible with practices of patrimonialism where the rhetoric of compliance becomes a smokescreen for the reality of continuity and corruption.

Table 2: Collaboration versus compliance

Compliance practice	Collaborative practice
• Operates through threat of punitive consequence	• Operates through finding common cause
• Company focus	• Industry / cluster focus
• Objective is to comply	• Objective is to solve the problem / do the right thing
• Company acts in relative isolation	• Partnership and coalition building focus
• Focus on ticking boxes	
• Focus on individual, easy-to-measure projects of limited ambition	• Focus on innovation and impact
	• Focus on transformational programmes, regional and cluster development
• Ad-hoc, stand-alone interventions – no systematic integration with broader business processes and governance	• Systematic integration of programs into broader business processes and governance
• Funding through mining company balance sheet	
• Ritualised, superficial political positioning between stakeholders	• Funding based on which stakeholder is best equipped to manage risk and optimise value
• Creativity unlocked in finding loop-holes in compliance framework – does not encourage integrity or responsibility by any stakeholder	• Candid, honest, trust-building conversations between stakeholders
	• All stakeholders need to act with integrity, transparency and accountability – stakeholders need to take responsibility for behaviour.

POLICY, REGULATION AND
COMMUNITY DEVELOPMENT

The communities that are proximate to mining operations and labour sending areas are often characterised by high levels of internal division and an extremely complex political economy. This political economy includes conflicts between competing political factions, conflicts within and between traditional authorities and other stakeholders and gender struggles and generational struggles. It is very common for local governments to be, at best, suffering from a shortage of technical capacity, and at worst, be captured by local political factions to serve a very narrow (often economic) agenda. Given the high levels of acute poverty in these areas, conflicts around access to state and other external resources can be extremely volatile and intense. Consequently, local dynamics can vary considerably from time to time and place to place (McKinley, 2011).

SLPs are the legislative and regulatory mechanism that has been put in place to address this legacy. Under this system, mining rights applicants are required to draw up a plan that includes undertakings aimed at providing benefits to mine workers and communities. These plans need to be in place as part of a mining rights application and to become binding on the approval of mining rights. Technically, should the plan not be by the mining company, the company could lose its mining licence. The regulatory logic underlying SLPs is that a top-down, punitive-based regulatory system needs to be in place as mining companies cannot be trusted to embrace the spirit of the post-apartheid challenge.

A fundamental problem with the SLP approach is that it is based on the fallacy (or fantasy) that the solution to highly complex and dynamic problems is 'getting the plan right' and then enforcing implementation. However, this approach is understandable in an environment of acute distrust.[5] This fallacy is particularly problematic when applied

---

5   An example of an argument that falls into the trap of the 'planning fallacy' is that presented by CALS in their paper 'The Social and Labour Plan Series – Phase One: System Design'. The paper argues that the root of the problem lies in a lack of comprehensive planning and associated enforcement frameworks.

to community-related developmental processes in the mining sector. The problems with a rigid and narrow planning and accountability approach include:

- The operations of mines impact on local communities in very unpredictable and dynamic ways. The SLPs are developed as a precondition for getting mining licences, yet by the time the mine is in operation the local community will almost invariably have changed considerably from what existed prior to the establishment of the mine. This can radically impact upon what would constitute a priority developmental intervention. At worst, a rigid plan will force the implementation of a completely inappropriate project whilst preventing the implementation of a valuable intervention.

- The nature of local government itself is very unpredictable as local factions contest these institutions to access resources. Extensive feasibility studies, often performed well in advance of an intervention, that require local government co-operation could be a complete waste of time and very misleading.

- Local community politics is, as a rule, very fluid. This makes the engagement between mines and communities very complex. Attempting to 'freeze' agreements over a number of years, as per the SLP model, could lead to high degrees of misalignment.

- The framework discourages mining companies from developing robust governance systems that make these companies' management and boards accountable to communities for any social and development commitments made to the community.

- The framework discourages mining companies from systematically collaborating to achieve more comprehensive regional impacts.

At best, upfront planning will be speculative; at worst (like most predictions in a complex system) it will be entirely misleading. The setting of targets in this environment will be arbitrary. The underlying problem is that the focus on planning is part of a project to build a top-down compliance framework, rather than a dynamic framework for learning and implementation.

The irony of this is that the compliance framework is difficult to make effective. Firstly, the regulator will always struggle to have the capacity to keep up with the oversight required by the system. Secondly, the loss of a mining licence as the penalty for non-delivery creates the worst of all worlds. The regulator is put in a Catch-22 situation in that if a SLP is enforced to the letter, the regulator will need to close a mine and this action will have severe implications for the local economy. However, if the regulator does *not* close the mine, it makes the policy unenforceable. Furthermore, mining companies are also negatively affected by the situation. This is because the market will price in the risk of closure, seeing it as a systemic risk faced by the sector, while the mining companies will, in many cases, even with best endeavours, invariably fail to deliver on the letter of its SLP.

There is a wealth of literature, relating to both the public and private sectors, on the problems of planning and policy formation in highly complex environments (see in particular Ormerod, 2005). This literature emphasises the need to optimise bottom-up learning by engaging in processes focused on implementation and utilising experimentation on a small scale, with varied approaches, rather than jumping to large-scale solutions. According to Hartford (2012), Ries (2011) and Zhang and Marsh (2016), this involves providing focused incentives for innovation, creating learning and collaboration forums, focusing on adapting the implementation processes as learning takes place (rather than trying to plan for every contingency), entrenching a continuous improvement culture and creating key performance indicators that measure learning.

Sustainable regulation requires a complex balance of top-down defined incentives, penalties and processes to build bottom-up buy-in to the policy and regulatory objectives. The notion that a regulator can sustain a regime based on a predominantly 'command and control' mechanistic approach in a complex regulatory system is a myth. At best, the regulated will comply creatively, at worst the regulated (who are often well resourced) will enter into constant power struggles with the regulator which will wear down the regulator's resources, will and capacity. Furthermore, in the context where top-down regulation is based on penalties (such as the loss of a licence) that are both disproportionate and very difficult to implement (because of the consequences of closure) the regulator will become toothless. As a mechanism to avoid creative compliance and ongoing power struggles, a bottom-up approach seeks to build a principled consensus with all relevant stakeholders around regulatory objectives.

In a complex regulatory environment, a purely bottom-up approach without top-down incentives and penalties will not be effective in directing stakeholders' behaviour in any meaningful manner. A new balance between the two approaches needs to be found. However, this will require that the regulated truly come to the party in voluntarily implementing initiatives that positively impact on policy objectives. The building of a development coalition is a proposal to drive the rebalancing of the policy and regulatory environment through building trust through the delivery of concrete value. The next section provides an example of the construction of a development coalition, enabled by the mining industry, in the area of agricultural and rural development.

A DEVELOPMENT COALITION RESPONSE TO RURAL DEVELOPMENT

The ability for mining operations to take place is ultimately dependent on support from surrounding communities which are characterised by high levels of poverty and unemployment. Previous local economic development projects have been sub-scale, fragmented and have, on the whole, had negligible impact. Furthermore, in those areas where resources are being depleted, it is necessary to implement programmes to build a post-mining economy. The first issue that needs to be confronted is what the mining sector can do to tackle unemployment. The second issue is what can be achieved if the mining sector works in coalition with a range of stakeholders around programmes that are initiated in sectors, precisely because of their ability to impact employment. This section will explore these issues in relation to a large-scale agricultural programme.

Mining employment peaked in 1987, when the sector employed around 760,000 people and accounted for 12.6 per cent of total private sector employment (Fedderke and Pirouz, 1998). This has declined to around 6 per cent of total private non-agricultural employment today (Chamber of Mines, 2017). Given the safety and productivity challenges associated with labour-intensive mining, it is unlikely that mining employment is going to reverse this trend. Consequently, there needs to be realism about the impact of mining as a sector on rural unemployment and development.

An important structural explanation for the unemployment crisis in South Africa is the national failure to generate employment in the agricultural sector, particularly when compared to our middle-income peers. Agriculture accounts for around 39 per cent of employment in the average middle-income country, compared to around 5 to 6 per cent in South Africa (see Figure 17). It is notable that in 2001, agriculture accounted for 16 per cent of employment in the economy (see Figure 18).

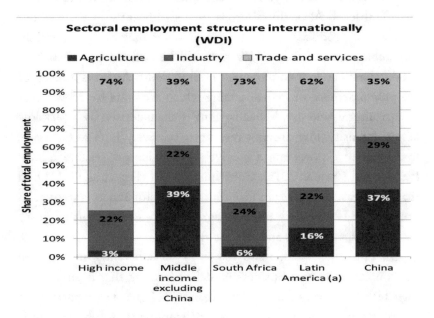

Figure 17: Benchmarks of employment in agriculture
Source: Makgetla, 2014

Agriculture is a high-risk enterprise, particularly in South Africa, where climatic vagaries, disease and sudden market price fluctuations can easily result in enterprise failure. These risks, combined with the inefficient use of water, inadequate investment in water infrastructure by the state and land tenure insecurity, have contributed to low levels of gross fixed capital investment in the sector (Sender and Cramer, 2015).

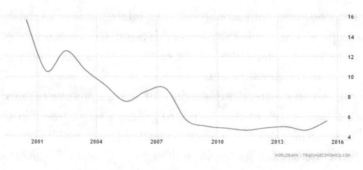

Figure 18: Agriculture as a percentage of total employment
Source: World Bank, 2015 and https://tradingeconomics.com

The low level of employment in agriculture is also driven at a more structural level. Agricultural crops vary significantly in their labour intensity of production, the amount of water they need, investment requirements and in the value that will be realised per hectare of production. The low employment level in South Africa is a result of the preponderance of maize farming which accounts for 58 per cent of agricultural land use. Maize farming is characterised by a very low labour intensity, low drought resistance, relatively high water usage, low levels of exports and a relatively low value per hectare farmed (Sender and Cramer, 2015). Consequently Greyling et al. (2011) find that to increase the value of the output of the sector and its employment potential, it is necessary to focus agricultural development on crops such as nuts, vegetables and selected fruits.

On the other hand, these risks can be coherently managed in the context of an agri-business model of farming that is large scale, high tech, high value and export focused. Higher-value agriculture is also associated with higher levels of employment. Job creation is also associated with the scale of these farms (enabled by better water usage), higher crop yields (creating more harvesting employment) and downstream employment created in the supply chain (Sender and Cramer, 2015). These farms can provide much higher wages and employment certainty than small-scale units. These core enterprises can (and in certain cases do) establish outgrower schemes where technology, extension services inputs and guaranteed markets can be provided so as to enable the growth of new agri-industrialists.

South Africa has some sophisticated agricultural champions at co-operative, farmer and agri-business level that have the capability to drive the expansion of next-generation agriculture in South Africa. At an agri-business level, certain South African companies are supplying the most demanding export markets in the world and have established the technology, supply chain, logistics and marketing networks and the operational methodologies to successfully implement large-scale agricultural projects. The key constraints on these companies are their limited access to land with an adequate water supply and secure property rights, and – to a lesser extent – financial capital (Sender and Cramer, 2015).

Mining companies have major landholdings running into the hundreds of thousands of hectares, much of which is being suboptimally utilised.[6] Furthermore, some leading mining companies are using what used to be termed 'buffer land' for agricultural development as a means of creating greater social stability – as unutilised land is a smack in the face for those who have been dispossessed. In addition, certain mining companies are pumping large quantities of water from underground, which can potentially be used for agriculture, into rivers.

Agriculture in South Africa, much like mining, has a brutal legacy of abuse and the exploitation of labour and communities living on farms. Hence, it is critical that any partnership with agri-business is sensitive to this legacy and explicitly addresses it. Consequently, it needs to be explored whether a deal can be done with large agri-businesses whereby these businesses get access to under-utilised mining land in exchange for guaranteeing conditions of employment, training investment, support for out-grower schemes, management development programmes, and labour and community equity participation. In addition, partnerships are required with specialised not-for-profit organisations to develop interfaces and promote dialogue with surrounding communities as well as to promote community gardens and food support networks.

---

6   For example, Sibanye-Stillwater and the Dolomitic Land Association effectively own and control 50,000 hectares on the West Rand.

Sibanye-Stillwater is in the process of implementing a comprehensive rural development process along the above lines. In a movement away from isolated company and compliance-based projects, Sibanye has made an initial allotment of 15,000 hectares of quality agricultural land with relatively plentiful supplies of water available for agricultural and rural development. Sibanye, in partnership with the Mining Phakisa, is moving towards the creation of a multi-stakeholder coalition involving mining companies, agri-business, small-scale farmers, provincial and local governments, development finance institutions and the PIC that can have a profoundly catalytic impact on the development of an agricultural-industrial cluster in the West Rand with associated job and value creation. The vision of the programme is to optimise agricultural and associated agricultural input and processing activities that can be unlocked through mining companies making quality land available on a negotiated basis to the right strategic partners so as to grow the local economy, create jobs for unemployed mine workers and members of the local community, and to facilitate wealth creation and development in surrounding communities. The strategic partners are identified through a structured discovery process.

In summary, the proposal is for mining companies to act as facilitators to support the building of a development coalition, rather than as the owners of the programmes. In certain areas it is necessary to focus immediately on the establishment of post-mining economies where shafts are projected to close in the next 10 years, rather than focusing on isolated social projects of limited impact, through the establishment of focused agri-industrial development coalitions. The following section explores the construction of an over-arching development coalition and vision for the PGM sector.

KEY ELEMENTS OF A PGM VISION: BUILDING A DEVELOPMENT COALITION

The previous sections argue that there are three core challenges facing the PGM industry. These relate to inadequate and uncoordinated investment along the value chain and a breakdown of trust between stakeholders at an enterprise and industry level (based on a perception that the value from PGM mining is distributed along patrimonial lines) combined with a very top-down, rules-based policy and regulatory environment. In order to turn this situation around it is necessary to build a development coalition to drive a comprehensive programme that will build bridges with stakeholders through implementing a range of strategic development programmes with associated governance innovations that will position the industry to move forward on a more coherent basis. These programmes include a shareholder management initiative to align shareholders, boards and management with the vision of a sustainable industry, a supplier development initiative to enhance the value of the PGM industry to the country through contributing to the national industrialisation effort, a set of initiatives around application and market development to promote demand for the metals, a rural and agricultural development programme to create employment and economic activity around the mines and a series of initiatives around addressing the historical legacies of mining (Figure 20).

The following is tentatively proposed as a guiding vision for the development coalition and for the PGM sector in South Africa: a growing and profitable PGM mining industry in South Africa that, through operating with the utmost integrity and driving an over-arching national industrial, rural development and transformation process, combined with global downstream application and market development partnerships, has the respect, support and admiration of all key stakeholders.

Building on this base the chapter offers an overview of key areas around which it is proposed that a development coalition needs to be constituted in response to this vision. The first focuses on the initiatives that are required to build stakeholder support for the sector, as without key stakeholder support it will be difficult to either bring production stability to the sector or additional investment support. It then turns to governance innovations required to drive sustainable investment in the value chain and finally identifies specific areas of the value chain that need to be developed to make an over-arching vision a reality.

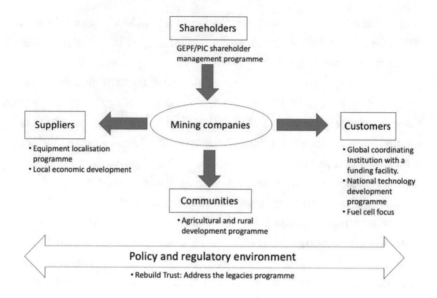

Figure 20: Key elements of a development coalition

KEY AREA: EARNING THE RESPECT AND ADMIRATION OF KEY STAKEHOLDERS

There are a number of legacies of the apartheid era of mining. In order to establish the integrity of mining in the post-apartheid era, the industry is required to recognise, address and show remorse for these legacies. They include the migrant labour system and associated rural impoverishment on which the apartheid system was based, environmental degradation and fatalities linked to unsafe working conditions. In addition, a racially based patrimonial culture was entrenched in the industry whereby black South Africans were excluded from participating in management, in training opportunities, in the supply chain and in downstream beneficiation processes. The result is an industry that is not trusted by many of its stakeholders. This trust deficit has been exacerbated, in recent years, by practices of transfer pricing, high levels of procurement of imported equipment, extremely weak supply-chain management practices, conflicts with affected communities and limited efforts to transform supply chains. Whilst these legacies cannot be addressed in a 'big bang', it is critical that the PGM industry makes significant, concrete efforts to rebuild trust with very sceptical stakeholders, and in doing so, acts as a role model for the rest of the mining industry.

The following initiatives[7] are consequently proposed to build bridges with stakeholders:

Ethical and value-based initiatives

At the most basic level the PGM industry needs to adopt a code of ethics and associated governance mechanisms relating to patrimonialism and corruption that will be binding on boards and management. This code needs to include mechanisms to make management accountable to communities for any commitments made to those communities, including but not limited to those regarding corporate social initiatives (CSI), supply chain and SLPs. Relevant board sub-committees need to be held accountable for overseeing the implementation of this code.

Community and rural development initiatives

Major PGM producers need to explore innovative arrangements to make any surplus land holdings, pumped underground water, sports and other useful infrastructure available for developmental purposes with the ultimate intention of placing the relevant assets into an appropriate corporate structure so as to enable large-scale rural development.

Major PGM producers need to mobilise resources through their CSI budgets, supply chains and key customers to implement a comprehensive food security programme in the communities in which they operate. (This can include government incentives where possible.)

Mutually accepted formal and independent grievance mechanisms should be established through social dialogue to hear and propose solutions to community concerns.

Major PGM producers need to explore how the land usually sterilised by mining and mine waste can be returned to a level of utility that at least covers the long-term costs of managing such land. This is allied with a new and growing field in which mining land can be developed or optimised for post-mining use to the benefit of communities affected

---

7   The initiatives that are suggested in this paper are a result of workshops and many discussions involving a range of industry stakeholders.

by mining and society as a whole.

PGM producers should urgently address particulate and sulphur-dioxide emissions from smelters and improve the cleaning and scrubbing infrastructure.

Strategic transformation initiatives

Individual companies and the industry as a whole need to comprehensively demonstrate how the transformation of top management will take place in the next five years. This should include search processes and programmes for on-the-job exposure, coaching and mentorship and leadership development.

Mining companies undertake that they will establish a task team with the PIC to identify shafts that have been effectively closed and are under minimal 'care and maintenance' (often to avoid the difficulties of getting a closure licence) that could be contracted out to a black-owned company on the basis of the established mining company procuring the ore from the contractor at an agreed-upon rate.

There are many dumps, tailings dams and waste streams that contain significant resources as they were established when processing technology was less effective, before the many innovative technologies recently developed to exploit these resources. The fact that these resources are above ground dramatically decreases the capital costs required to enter these niches. Mining companies should consequently undertake to reserve these resources for black-owned processing companies and to fast track the contracting of black-owned companies to process them.

Mining companies undertake to enter into a process of stakeholder consultation to establish a comprehensive, industry-wide code of ethics and conduct that will be published and monitored by the boards of the companies.

KEY AREA: SUSTAINABLE INVESTMENT IN THE MINING VALUE CHAIN

There has been a shift in the global financial services sector towards the management of savings by asset managers who are given incentives to seek short-term returns over long-term sustainability. This has created an environment in which management, in turn, has incentives to drive shareholder returns over long-term investments. This is particularly perilous in mining as it is a cyclical industry that requires continuous investment for productivity to be sustained. Indeed, it is almost impossible for management to coherently steer a mine through the commodity cycle with a focus on short-term returns and the payment of continuous cash dividends to shareholders.

During the PGM boom in the first decade of the millennium, certain platinum mining companies were focused on returning money to shareholders, rather than building war chests to see them through bad times (see Figure 21). This was exacerbated by certain mining companies who, during boom times, used leverage to increase shareholder returns which created additional fragility. With the downturn, between 2007 and 2016, 40 to 50 per cent of the top 50 shareholders (many of whom were foreign) in the big three fully exited their investment, whilst domestic shareholders, particularly the PIC, gained in importance (Bowman 2016).

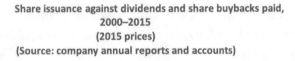

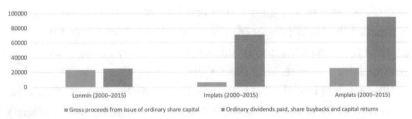

Figure 21: Capital raised for mining versus capital returned
Source: Bowman, 2016

Company law provides shareholders with considerable governance rights to oversee both the strategy and operations of commercial enterprises. This means that long-term shareholders in the PGM industry have significant leverage to play an anchor role in a PGM development coalition. The Government Employee Pension Fund (GEPF) and its asset manager, the PIC, have an overwhelming interest in building a sustainable PGM industry but they can use their positions as shareholders in a number of PGM mining companies far more effectively to nudge the industry in a more sustainable direction. The PIC is the biggest investor in PGM mining in the world and is the shareholder of reference[8] for Sibanye-Stillwater, Anglo Platinum (including Anglo American), Lonmin and Impala Platinum. In addition, as the trustee of pensioner funds invested throughout the Johannesburg Stock Exchange, that is as the universal shareholder, the GEPF through the PIC has an interest in securing the long-term, balanced growth of the economy. In other words, because GEPF's shareholding is so widespread, the decline of one sector – as a result of under investment – can have multiplier impacts on the broader portfolio. An additional concern for stakeholders regarding the sustainability of the PGM sector is the patrimonial nature of mining management composition, supply chain practices and community development programmes.

The problem is that the governance process along the investment chain is geared towards short-term returns (see Figure 22).[9] The workers who put their money into the GEPF (and the taxpayers who underwrite the payment of the defined benefits of the GEPF) have an interest in the long-term growth of the economy, as their pensions will only be paid out well into the future. However, the board of the GEPF is responsible for giving a mandate to the PIC.[10] Although the board has the mandate to establish specialised monitoring and impact-measuring capabilities, in practice the board, many of whose members are of limited financial capability, is overwhelmingly influenced by actuaries who have been groomed in a sector seeking short-term returns.

The result is that the PIC is given a mandate by the GEPF to seek short-term returns without any feedback mechanism (except financial) to measure the impact of PIC investment on the sustainability and growth of the economy. The PIC, in turn, does not have any systems, feedback mechanisms or shareholder management methodologies

---

8   The largest or very significant shareholder that companies turns to before making a serious decision.

9   Anything coherent in this section is a result of conversations with William Frater, a pension fund specialist.

10   Government Employees Pension Act 1996.

in place to recognise its position as asset manager for the economy's universal shareholder, but rather incentivises its investment managers around short-term returns.

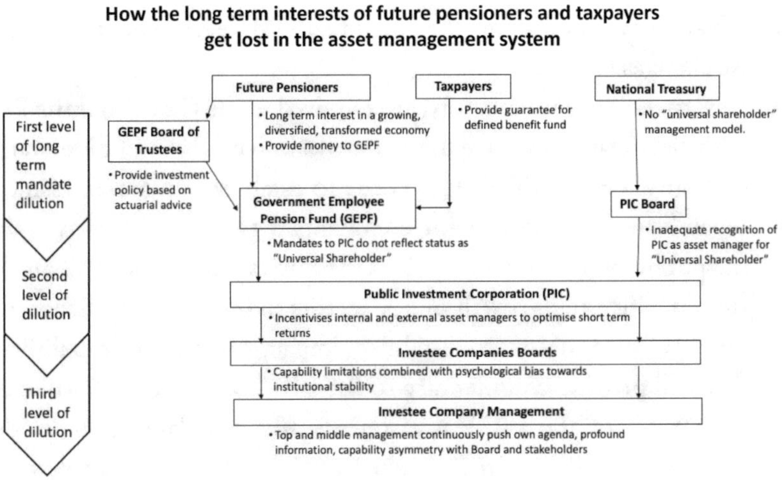

How the long term interests of future pensioners and taxpayers get lost in the asset management system

Figure 22: Loss of strategic impact – the GEPF asset management system

The long-term interests of workers and taxpayers are further compromised once investment is made in a company. Chief executive officers are given relatively short-term contracts with options linked to the share price and many will work to optimise their returns in this time period. Even a cursory examination of the composition of many mining company boards suggests that the capability and qualifications of most mining company board members to challenge management on strategic and operational matters is questionable – even if the board was not already at an overwhelming disadvantage due to information asymmetry. This is further exacerbated by a common, independent psychological bias of directors to protect 'colleagues on the Board from legal sanction' and avoid the 'collateral damage' that may follow from such a sanction (Cox and Munsinger, 1985). In a nutshell, no one wants to spoil the party. This leads to capture of the company strategy by management.

Hence, the way the GEPF-PIC governance system is presently operating will exacerbate the tendency of mining management to invest for the short term, will consequently exacerbate under-investment in key areas of the minerals value chain during the downturn of the commodity cycle and will undermine attempts to build a long-term vision for the PGM industry. However, with a number of changes, the GEPF and PIC can become the core of a PGM development coalition. These changes include:

- Key governance innovations are required in the GEPF and the PIC who, between them, have the status of a universal shareholder in the economy. Firstly, the GEPF needs to establish impact-assessment capabilities that will provide commentary to the GEPF board and the public on the impact of the GEPF mandate to the PIC on the sustainability and growth of key sectors in the economy. Secondly, the PIC also needs to establish impact-assessment capabilities that will provide commentary to the PIC board and the public on the impact of the PIC's investments and associated shareholder oversight initiatives on the economy. Thirdly, given the scale of the GEPF surplus and the fact that the taxpayer underwrites the defined benefit, it would make sense for a portion of this surplus to be used to make investments with a long-term time horizon focused on strategic impact on the economy (similar to a sovereign wealth fund). If the PIC continues acting to optimise short-term returns (based on their mandate from GEPF), it is unlikely that the sector will stabilise and reach its potential.

- All of the above suggests that the PIC should have a shareholder management model that recognises its unique position as the asset manager for the universal shareholder in the South African economy. Patient capital is required to shift from a quarterly reporting mentality, which forces institutions to put pressure on companies to deliver short-term financial returns at the expense of long-term business sustainability and socio-economic impact. The PIC needs to develop its own capability to more critically interrogate the investment and management incentive strategies of enterprises against a long-term goal. At present, the PIC's asset managers are overly rewarded for short-term returns, rather than for ensuring the businesses are run ethically and making the investments in plant, skills, technology, processes and markets to secure their sustainable medium- to long-term growth. Hence, the first step will require that the PIC investment managers are incentivised on a more complex matrix than short-term returns.

- The PIC should be producing a report that will demonstrate the linkages between different sectors of the economy and how investments in these sectors have a multiplier effect. A simple methodology can then be used to provide an indication of the broader impacts of investment. This alone will allow both investment managers and companies to understand how their decisions are impacting the overall growth of the economy.

- The PIC needs to develop the capability to design clear and public shareholder intent statements so that the boards, management and other stakeholders understand what the PIC is hoping to achieve from making a specific investment. In the case of PGM mining, these statements can provide a clear direction to the board to oversee collaboration when this is appropriate.

- The PIC's mandate needs to be extended to invest in metals when this will have leverage impacts on its equity portfolio. This will require increased governance requirements around price discovery and metal marketing.

- The PIC should explore strategies to 'crowd in' additional funding for PGM demand stimulation, such as being the anchor investor in niche funding facilities alongside the private sector.

- The accountability of boards to stakeholders remains significantly academic unless stakeholders assert their rights. The majority of stakeholders do not have the clout or resources to hold boards accountable for the ethical and patrimonial behaviour of a company. The PIC needs to start playing a leading role in this regard and needs to adequately resource this capability.

The PIC needs to establish a public PGM sustainability scorecard to keep track of the progress made by government, the PGM industry as a whole and individual mining companies in building a sustainable industry. The performance of companies against the scorecard could be included in the PIC investment manager's performance appraisals.

KEY AREA: PGM APPLICATION AND MARKET DEVELOPMENT

As discussed earlier in this chapter, the sustainability of the sector depends on co-ordinated investment across the value chain so that demand is generated to absorb mining production. In addition, in order to embed the PGM sector as strategic to the South African economy and to add to the sustainability of the sector, it is critical that technology and market development programmes are implemented in South Africa in a manner that recognises the global nature of the industry. This section will explore the initiatives that are required to achieve the goal of sustainable and strategically targeted application and market development.

GLOBAL MARKET DEVELOPMENT AND CO-ORDINATION INITIATIVES

- From a mining and national perspective, investment in new PGM applications and market development processes cannot be done on a strictly commercial basis to achieve an optimal outcome for the industry. There is also a strategic need to leverage off the capabilities and embedded intellectual property of the leading fabricators to drive development to grow the national market. There is a well-established global venture capital and private equity sector focused on investing into PGM applications, based on a strictly commercial assessment of returns. However, to develop a sustainable value chain, mining companies need their investment criteria to take into account the extent to which investment in an application can boost demand for a relevant metal, as well as the opportunities for technology development and beneficiation in South Africa. Consequently, it is proposed that a fund be established to specifically stimulate application development and commercialisation activities. This fund will make cheap finance available to downstream developers for ventures that would not naturally be prioritised or attract mainstream finance, on the back of their financial sustainability and their potential impact on demand. A portion of this fund should come from the South African government, to finance those proposals that will also enhance South Africa's application development capabilities (Figure 23).

- Although there is significant foreign investment in PGM mining in South Africa, national interest in the sustainability of the resource is overwhelmingly a South African concern. This creates a significant challenge because the bulk of technology development capabilities and the market lie outside of the country. It is consequently important to explore how government-driven country-to-country partnerships can be established that incorporate all three elements of the value chain.

- There is a need for an institutional mechanism to facilitate co-ordination between mining companies and, at present, there are a number of international bodies that are involved in the growing market. These include the International Platinum Group Metals Association, Platinum Guild International and the World Platinum Investment Council. It will be necessary to understand how individual, enterprise technology and market development initiatives relate to these associations. In addition, it will be necessary to perform a cost-benefit analysis of existing structures to assess where and how both the state and private sector can most effectively provide support for the process of co-ordinating investments across the value chain.

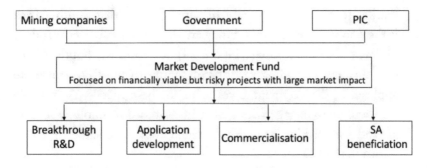

Figure 23: Structure of a fund to support market development

PGM applications development, beneficiation and marketing initiatives

- In order to secure the economic viability of mining operations as well as to optimise the national value that can be realised from the resource, it is important to create mechanisms in South Africa that will provide global partners with security of platinum supply, establish a physical space to locate a beneficiation cluster and allow these partners to view PGM as a portfolio of metals that require application development, beneficiation and marketing, rather than viewing platinum in isolation. The Precious Metals Act of 2005 empowers a regulator to promote access to precious metals to further the objectives of beneficiation in South Africa (*Government Gazette*, 2006). A platinum export development zone (EDZ) is in the process of being established in Ekurhuleni, adjacent to the Impala Platinum smelter. This zone creates the opportunity to attract global catalyst fabricators to invest in both R&D and export-orientated manufacturing facilities within the zone, with an agreement in place that these facilities will have absolute security of PGM supply, provided that PGMs continue to be mined in South Africa. This arrangement can be secured through a combination of an agreement between the regulator and the platinum mining industry each to provide security – pro rata – of PGM supply to the facility, and through inventories being held by the EDZ.

- The level of R&D into the 'minor' PGMs (namely iridium, rhodium, etc.) is suboptimal as a result of a combination of global market fears around security of supply, given that South Africa is the dominant supplier and they are only essentially a co-product of 'major' PGM mining (Figure 24). In contrast, South Africa can, through providing security of supply to the EDZ, promote application development investment in these metals. A possible start will be to engage with global PGM-based technology leaders around establishing a centre of technology excellence in South Africa in the EDZ.

- As a core component of the hydrogen economy, fuel cells are the most likely technology to make up for the downturn in PGM demand resulting from the projected increased use of electric vehicles (Matthey, 2018). The development of fuel cells for motor vehicles will be critically important in terms of demand for PGMs and although the consumption of platinum for motor vehicle applications is increasing, it is expected that a critical mass will only be reached in 2030 (see Figure 25). The competitiveness of stationary fuel cells is growing with the declining cost of equipment and the increased cost of alternatives. There are a number of fuel cell technologies that are either commercialised or on the cusp of being commercialised, which will be followed by the cost reductions associated with the manufacturing and design learning curve. It is important for South Africa to be positioned to support and catch the wave of projected increased fuel cell sales. Aside from providing finance to the above development fund, mining companies can act as fuel cell power off-takers and co-developers of underground mining equipment powered by fuel cells.

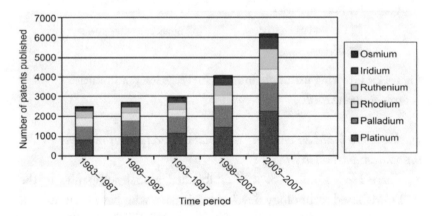

Figure 24: PGM patenting trends
Source: Msimang, 2013

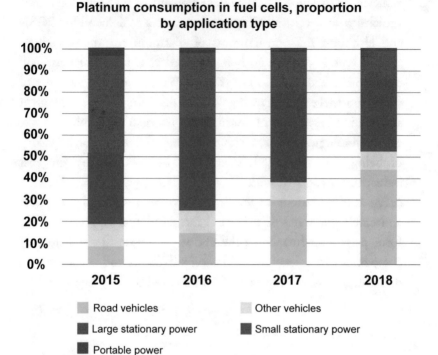

Figure 25: Growing use of platinum in fuel cells for road vehicles
Source: Matthey, 2018

Initiatives to build an enabling national environment for global PGM innovation partnerships

- There are a number of well-established global companies in the PGM-based technology development space who have accumulated deep levels of IP around both existing applications and how to design and implement PGM-based technology development programmes. These companies have established marketing networks and relationships with large customers and understand what value these customers are looking for in future applications. It is critical that South Africa provides an enabling environment for partnerships with these companies (see Ferreira and Perrot, 2013).

- In this context, Hydrogen South Africa (HySA) is seen as a South African government initiative to build a company to compete with these leading PGM fabricators (see Figure 26). Leaving aside the viability of this ambition in the context of a very mature and competitive global catalyst supplier market, whilst South Africa has the right to build PGM technology development capability, the positioning of HySA as the exclusive Department of Science and Technology agency that has a monopoly on government-supported PGM development is a profound mistake if South Africa wants to become part of global technology development networks. The status and strategy of HySA needs to be urgently reviewed by a combined public and private sector team. In particular, it is proposed that HySA should not have the status of the exclusive national initiative in the platinum catalysis space, nor should HySA have exclusivity in the national markets in which it has chosen to operate. In addition, mining companies (and other stakeholders) should have access to government support for fuel cells and catalysis initiatives, based on the competitiveness of the technology and how the process will increase national platinum-related technology and manufacturing capacity, regardless of whether this is in HySA or not.

- Government incentives associated with IP ownership conditions and South African Revenue Services regulations need to be reviewed to create an enabling environment for international partnerships.

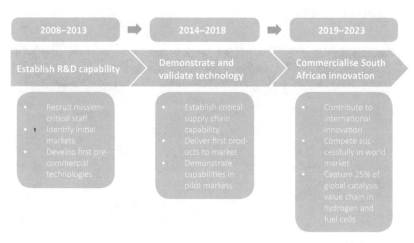

Figure 26: HySA programme to achieve global dominance by 2023
Source: North, B., 2015

KEY AREA: PGM MINING AS A DRIVER OF INDUSTRIALISATION

As easily accessible resources are mined, there is a global trend towards mining more waste in the ore body, using more energy and water and producing more carbon emissions in order to exploit less proximate reserves (see Figure 27). In the PGM (and gold) sectors in South Africa, there is also the reality that resources that are minable with present technologies, processes and skills are running out and/or are not economically viable. Figure 28 demonstrates the impact on the viability of the industry of staying with existing technologies versus developing new systems to tackle presently un-minable resources.

In order to modernise and then lead in the development and manufacture of new mining systems, it is necessary for the mining supply chain as a whole to be upgraded and more deeply embedded in the South African economy. The combined gold and platinum industry procures approximately R90 billion of goods and services a year (Smeiman, 2018). The PGM industry procures R32 billion per annum worth of equipment, components, supplies and services (Table 3). In addition, there is large expenditure in key technology areas that can support the development of a dynamic supplier industry (particularly when combined with gold demand).

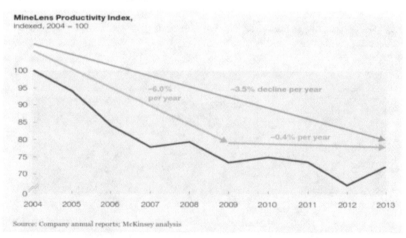

Figure 27: Declining mining productivity
Source: Anglo American, 2017

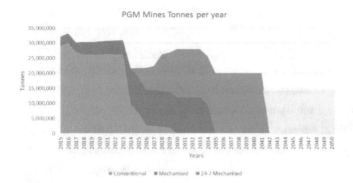

Figure 28: The outlook for PGM mining based on technology
Source: Singh, 2017

Table 3: Key PGM expenditure areas

UNSPSC family	Grand total	2015	2016
2010 – Mining and quarrying machinery and equipment	7,097M	3,502M	3,594M
2410 – Material-handling machinery and equipment	4,397M	2,262M	2,135M
3025 – Underground mining structures and materials	4,019M	2,064M	1,955M
2512 – Railway and tramway machinery and equipment	1,350M	537M	813M
4017 – Pipe piping and pipe fittings	630M	264M	366M
4618 – Personal safety and protection	624M	262M	362M
7611 – Cleaning and janitorial services	537M	280M	256M
3133 – Fabricated structural assemblies	429M	165M	264M
4619 – Fire protection	342M	156M	186M
3019 – Construction and maintenance support equipment	143M	59M	84M

Source: Smeiman, 2018

The problem is that the supply chain maturity of the mining industry is strikingly low. In manufacturing, the quality of supply chain management became a key driver of competitiveness, particularly at times when companies needed to source globally and produce globally.

Technologies were developed, processes designed and governance systems put in place to support the building of extremely sophisticated supply chains. In contrast, mining (and particularly underground mining) has never felt these pressures to the same extent, and supply chain capabilities remain relatively rudimentary. For example, there are no industry product naming, numbering and classification standards within the supply chain. This means that it is difficult to aggregate expenditure on products and services. Where companies have adopted their own classification system it is usually ignored, and a text description is used to define what is being procured (see Table 4). In short, many companies have no systematic way of knowing what they are buying, who they are buying it from and what they are spending. This makes strategic and tactical demand planning almost impossible, which results in many ad hoc procurements.

Table 4: Use of free text as product descriptor in procurement database

	2015		2016	
	% of transactions	% of value	% of transactions	% of value
Free text	39%	72%	40%	71%
Stock item	61%	28%	60%	29%

Source: Smeiman, 2018

Historically, there is a strong culture in the mining industry in South Africa that favours the decentralised management of mines. There are obviously good reasons to empower mine managers and engineers on the ground as there are numerous contingencies that can occur on a daily basis. This resulted in a very decentralised procurement system, whereby mine managers and engineers had a high degree of autonomy to make the procurements they deemed necessary. However, in order to reap the benefits of standardisation and scale, mining companies established centralised supply chain functions to more effectively plan and contract procurements, whilst engineers retained their right to define what needed to be procured. Policies that empowered the centralised supply chain function existed alongside (necessary) policies that allowed for deviations to cater for urgent requirements. The problem is that in a number of companies, the bulk of procurements are deviations that makes supplier chain planning and development (both at an industrial and community level) impossible

All the above has resulted in higher than necessary imports, higher costs, corruption and limited national innovation down the supply chain. In other words, building supply chain capabilities in the mining inWdustry and leveraging procurement to build South African industrial capability will be good for the mines. South Africa has the potential to become the global centre of excellence in the development, design, manufacture and export of cutting-edge mining systems, with a particular emphasis on deep-level, hard-rock capabilities. In driving this objective, a number of major programmes consequently need to be launched both within specific mining companies and in collaboration across mining companies and with the state to rapidly enhance the supply chain capability in the PGM industry. The following could serve as guidelines:

- A collaborative programme, involving mining companies and the CSIR Mining Tech, has been launched to enhance existing equipment and to develop and commercialise the next generation of deep-level, narrow-reef, hard-rock mining technologies. It is critical that this programme is built on developing a strategy that will make explicit the commitments from different stakeholders to drive the process.
- Mining companies need to undergo and publish the results of an independent supply chain maturity assessment, together with plans to improve their supply chain capabilities. In particular, mining companies need to publish the number and value of procurements that are 'deviations' from a proper contracting process.
- Mining companies need to individually and jointly develop supply chain localisation and transformation plans to increase the level of national and empowered content in their supply chains and to publicly report on progress made in implementing these plans.
- Mining companies need to establish board sub-committees with oversight of procurement that will have the responsibility of verifying for stakeholders that the integrity of procurement processes as per company policy has been respected. This should include a level of governance innovation that will allow communities to seek

accountability from the boards for commitments that were made by the mining company and are perceived as not being honoured.

CONCLUSION

The PGM mining sector is at a watershed moment. It is facing declining platinum demand at precisely the time when the sector has limited resources to invest in application and market development and commercialisation. Ultimately, the sustainability of the PGM cluster in South Africa will require a nationally driven effort. The problem is that not only is the strategic importance of PGMs to the South African economy one of the country's better kept secrets, but that national stakeholders will be reluctant to throw their support behind the sector while it is perceived to disproportionately benefit a few, rather than the full range of stakeholders impacted by the sector's activity. A national effort will only take place if there is a concrete commitment by mining company boards and management to reposition the sector behind a comprehensive development programme and to guard the integrity of this repositioning. The PGM cluster has as much potential to become significant to the industrialisation and development of the South African economy in the 21st century as gold had in the 20th century. However, the nature of the cluster's impacts would be very different to that of gold:

- The sector could incubate the development of major exporters of cutting edge mining services and equipment.
- The cluster could position South Africa as a world leader in the development and commercialisation of all PGM, and in many instances green, technologies.
- The cluster can catalyse the growth of a high-value agriculture and agro-processing export-orientated sector.

The key innovations that are required for the sector to play this role relate to governance and leadership. Firstly, at an industry co-ordination level, there is an urgent need for an institution to oversee a process of industry investment and strategic collaboration in the areas of technology and market development, particularly in relation to the large fabricators. At an investment level, while the PIC, based on the GEPF mandate, is structured as just another asset manager chasing short-term returns, without any sustainable industry vision, boards and management will always be incentivised to invest for quarterly results, rather than long-term sustainability. It is not possible for mining companies to adequately and coherently invest in the PGM value chain throughout the commodity cycle if they are going to be pressured by key shareholders to optimise short-term returns. At a stakeholder level, there are no mechanisms for communities to hold companies accountable for commitments made to the communities. Unless supply chain management capabilities in the PGM sector are radically enhanced, the industrialisation objectives will be impossible to achieve – corporate boards need to take responsibility for this process. While the PIC is not vigilant around issues of good corporate governance, the sector will continue to remain plagued by corruption and patrimonialism. While the government-industry interface is fragmented and incoherent, the sustainability of the industry will be threatened.

What is required is that the industry builds strong bridges with its stakeholders so that PGMs can become a symbol of developmental mining and shared value. This will require a new kind of leadership from the boards and chief executive officers. Board members, in particular, need to ask themselves what their motivations are for joining the board – if the motivations are to oversee shareholder interests without actively engaging other stakeholders, then these board members are not the leadership the PGM industry needs at this point in time. It will be impossible to regulate the industry in a manner that solves the problem of its legitimacy. In order to build effective developmental coalitions and win public support, the boards will have to fill in the gap that lies between regulation and industry culture and strategy. This will require governance innovations relating to community engagement and rural development, supply chain development and associated industrialisation and internal and supplier transformation. In the absence of such innovations, it is likely that there will be no option but for hard regulation to rule the sector.

REFERENCES

Anglo American. (2017). Unpublished presentation given to a delegation of government officials, Pretoria.

Anglo American Platinum Limited. (Undated). 'PGM marketing strategy'. Presentation to Barclays Thought Leadership Forum. Available at: http://www.angloamericanplatinum.com/~/media/Files/A/Anglo-American-Platinum/investor-presentation/barclays-pgm-leadership.pdf. Accessed 10 September 2018.

Anglo American Platinum Limited. (2013). 'Financial Report 2013'. Anglo American Platinum Limited, South Africa.

Anglo American Platinum Limited. (2017). 'Audited annual financial statements 2017'. Anglo American Platinum Limited, Johannesburg.

Bath, A. (2014). 'Fuel cell interim feasibility study for the DTI SEZ fund'. Unpublished.

Bowman, A. (2016). 'Dilemmas of distribution: Financialisation, boom and bust in the post apartheid platinum industry'. Society, Work and Development Institute, University of the Witwatersrand. Johannesburg.

Bruce, J.T. (1996). 'Rustenburg and Johnson Matthey, an enduring relationship – Sixty five years of continuous committed development'. *Platinum Metal Review*, Vol. 40(1), pp. 2–7.

Centre for Applied Legal Studies (CALS). (2016). 'The social and labour plan series – Phase one: System design'. CALS, University of the Witwatersrand, Johannesburg.

Chamber of Mines. (2016). Presentation delivered at the Mining Phakisa, Johannesburg.

Chamber of Mines. (2017). 'Fact and figures pocketbook'. Chamber of Mines, Johannesburg.

Connelly, M. (2010). 'An analysis of innovation in materials and energy'. University of Cincinnati, Cincinnati, Ohio.

Cox, J. and Munsinger, H. (1985). 'Bias in the boardroom: Psychological foundations and legal implications of corporate cohesion'. *Law and Contemporary Problems,* Vol. 48(3), pp. 84–135.

Evans, P. (1995). *Embedded Autonomy: States and Industrial Transformation.* Princeton University Press, Princeton.

Fedderke, J. and Pirouz, F. (1998). 'The Role of mining in the South African economy'. Economic Research South Africa.

Ferreira, V. and Perrot, R. (2013). 'Emergence of new industries: Industry and knowledge factors'. In MISTRA (ed.). *South Africa and the Global Hydrogen Economy – The Strategic Role of Platinum Group Metals.* Mapungubwe Institute for Strategic Reflection (MISTRA), Johannesburg.

Government Gazette. (2006). Precious Metal Act, 2005. Vol. 490.

Greyling, J., Meyer, F., Punt, C., Reynolds, S., Van Der Burgh, G. and Vink, N.

(2011). *The Contribution of the Agro-Industrial Complex to Employment in South Africa.* Bureau for Food and Agricultural Policy, Pretoria.

Hartford, T. (2012). *Adapt: Why Success Always Starts with Failure.* Picador, New York.

Hood, C. (2000). *The Art of the State: Culture, Rhetoric, and Public Management.* Oxford University Press, Oxford.

Industrial Development Corporation (IDC). (2012). *Sector Report: Opportunities for Downstream Beneficiation in the Platinum Value Chain: Fuel Cells.* Industrial Development Corporation, Sandown.

IOL. (1 October 2014). 'Anglo Platinum struggles to go solo'. Independent-On-Line. Available at: https://www.iol.co.za/business-report/markets/commodities/anglo-platinum-struggles-to-go-solo-1758677. Accessed 2 August 2018.

Jollie, D. (2016). 'Overview of PGM markets'. Presentation, Anglo American. Available at: http://www.angloamerican.com/~/media/Files/A/Anglo-American-PLC-V2/documents/platinum-marketing-david-jollie-241116.pd.

Kohli, A. (2004). *State-Directed Development: Political Power and Industrialisation in the Global Periphery.* Cambridge University Press, New York.

Lapping, A. (2018) 'Do you enjoy pain? How about the platinum industry?' Allan Grey, Cape Town. Available at: https://www.allangray.co.za/latest-insights/investment-insights/do-you-enjoy-pain-how-about-the-platinum-industry/. Accessed 10 September 2018.

Lowitt, S. (2017). 'Three new practical ideas in industrial policy thinking'. Tips Policy Brief: 4. Available at: http://www.tips.org.za/policy-briefs/item/3304-three-new-practical-ideas-in-heterodox-industrial-policy-thinking. Accessed 2 August 2018.

Macrotrends. (Undated). 'Platinum prices – Interactive historical chart'. Available at: https://www.macrotrends.net/2540/platinum-prices-historical-chart-data. Accessed 10 September 2018.

Makgetla, N. (2014). 'Industrial progress in South Africa, 1994 to 2004: Manufacturing employment and equality?' Presentation to TIPS conference, Johannesburg.

Mathys-Graaff, M. (2015). 'Fuel cell investment case critical success factors'. Presentation to PGM Workshop, Pretoria.

Matthey, J. (2013). Briefing note on auto-catalysis market. Johnson Matthey Plc, Royston.

Matthey, J. (2017). *PGM Market Report.* Johnson Matthey Plc, Royston.

Matthey, J. (2018). *PGM Market Report.* Johnson Matthey Plc, Royston.

McKinley, D. (2011). 'A state of deep crisis in South Africa's local government'. South Africa Civil Society Information Service. Available at: http://sacsis.org.za/site/article/635.1. Accessed 1 August 2018.

Meisel, N. (2004). *Governance Culture and Development: A Different Perspective on Corporate Governance.* OECD Development Centre, Paris.

Mnguni, L. (2018). *Platinum Group Metals: Western Limb Twilight Years.* SBG Securities, Johannesburg.

Msimang, V. (2013). 'The role of platinum in the hydrogen economy'. Presentation. MISTRA, Sandton.

North, Brian. (19 October 2015). 'Hydrogen South Africa and SA H2 safety programs'. Presentation to ICHS, Yokohama.

Ormerod, P. (2005). *Why Most Things Fail.* John Wiley, New Jersey.

Ries, E. (2011). *The Lean Start-up.* Penguin, London.

Sender, J and Cramer, C. (2015). 'Agro-processing, wage employment and export revenue: Opportunities for strategic intervention'. Working Paper for the DTI, Pretoria.

Sibanye-Stillwater. (2017). *Annual Report.* Johannesburg.

Sibanye-Stillwater. (April 2018). *Quarterly Report.* Johannesburg.

Singh, N. (2017). 'The CSIR's role in mining'. Presentation to CSIR Annual Conference, Pretoria.

Smeiman, M. (2018). 'Opportunities for mining industrialisation'. Presentation. Unpublished.

Speight, R. (2014). *Japanese Investment in the African Mining Industry.* Mayer Brown International, USA.

US Geological Survey. (2018). 'Mineral commodity summaries 2018: U.S. Geological Survey'. Available at: https://doi.org/10.3133/70194932. Accessed 30 July 2018.

World Bank. (2015). 'South Africa – Employment in agriculture (% of total employment)'. Available at: https://tradingeconomics.com/south-africa/employment-in-agriculture-percent-of-total-employment-wb-data.html. Accessed 10 September 2018.

World Platinum Investment Council. (March 2018). 'Will new ETFs drive investment demand?'. London.

Zhang, Y. and Marsh, D. (2016). 'Learning by doing: the case of administrative policy transfer in China'. *Policy Studies*, Vol. 37(2), pp. 35–52.

FOUR

Transformation in South Africa's mining industry

DUMA GQUBULE

IN OCTOBER 2002, industry stakeholders in South Africa reached agreement on a charter for the transformation of the country's mining sector. There were hopes that the Broad-Based Socio-Economic Empowerment Charter for the South African Mining Industry (Department of Minerals and Energy, 2002) would make a contribution towards reversing more than a century of exploitation of black workers and the pitiful contributions that the industry had made towards addressing the horrific living conditions in many mining host communities and labour sending areas. The Charter came into effect in May 2004, following the enactment of a new law, the Mineral and Petroleum Resources Development Act (MPRDA) of 2002.

However, hopes that the Charter and the MPRDA would become powerful policy and legislative tools for economic development and the empowerment of black people have turned to despair. The Department of Mineral Resources (DMR) has failed three times – in

2010, 2017 and 2018 when it introduced changes to the Charter – to address the original document's egregious shortcomings.

This chapter seeks to inform contemporary South African debates around issues such as public (or state) ownership, white monopoly capital and radical socio-economic transformation, the latter a political buzz phrase that is now the policy of the ruling African National Congress (ANC). In his 2017 state of the nation address, former South African President Jacob Zuma said:

> *What do we mean by radical socio-economic transformation? We mean fundamental change in the structure, systems, institutions and patterns of ownership, management and control of the economy in favour of all South Africans, especially the poor, the majority of whom are African and female* (Zuma, 2017).

The premise of this chapter is that the country's mineral resources belong to all its citizens. They do not belong to the companies that extract them. Therefore, there must be a public benefit from this extraction. The benefits for the state (as the custodian on behalf of the country's citizens) must be over and above tax revenues that apply to all industries, including those not publicly owned. There are two reasons why there must also be a public benefit accruing from mining, in addition to Broad-based Black Economic Empowerment (B-BBEE) requirements. First, B-BBEE applies to all industries. As is shown later, the current situation is that the benchmark for B-BBEE in mining is far below the one that applies in the rest of the economy. Second, B-BBEE is not a public benefit. No matter how it is structured, it will only benefit a few people.

The argument of this chapter is that there has been little structural change in the mining industry, despite the entry of new black shareholders. The Mining Charter did not change the basic structure of the industry and the four strategic sub-sectors of platinum group metals (PGMs), gold, coal and iron ore. All remain under the control of large apartheid-era companies. It is further argued that in order to begin to effect structural change in the industry, a strategy is required to achieve 25 per cent public ownership *in addition* to the existing black ownership target of 26 per cent. Two vehicles for this are proposed: a sovereign wealth fund (SWF) and a public mining company (PMC).

The chapter begins with a historical overview of ownership in the mining industry from the second half of the 19th century to the end of apartheid. The purpose is to set the scene for the discussion of the impact of the Mining Charter on the inherited industry structure. The chapter then provides an analysis of the Mining Charter. It focuses on ownership and not on elements of B-BBEE or issues relating to the multiple ways of maximising the developmental impact of mining, which are addressed in other chapters in this volume. The third section of the chapter presents statistics on the extent of black ownership in the industry. A short review of international best practices to harness mineral resources for economic development follows. It informs the recommendations at the end of the chapter.

HISTORICAL OVERVIEW OF OWNERSHIP

South Africa 'discovered' coal in the early 1860s, diamonds in 1866, gold in 1887 and platinum in 1924. Over the next few decades, the industry consolidated around seven mining houses. These were Union Corporation (formed in 1886), Goldfields (1887), Johannesburg Consolidated Investment Company or JCI (1889), Rand Mines (1893), General Mining and Finance Corporation or Genmin (1895), Anglo American (1917) and Anglo Transvaal Consolidated Investment Company or Anglovaal (1934). During the 1920s, Anglo expanded into diamonds in Namibia, Angola and Congo, and copper in Zambia. It started buying shares in De Beers Consolidated Mines (DBCM), the diamond company that was established in 1888 with controlling shares in the country's four leading mines. By 1929, Anglo controlled DBCM and its founder, Ernest Oppenheimer, became the company's chairman (Davenport, 2013: 264–276). Later, the two companies established a cross-shareholding structure – with each company owning about a third of the other – to prevent a hostile takeover. Anglo dismantled the structure in 2001.

During the first four decades, mining activities were based in the Central Rand, or present-day Johannesburg. With the gold price per ounce fixed – at US$20.67 between 1900 and 1933, and at $35 between 1934 and 1971 – there were long periods when there were severe cost pressures in the industry. But periodic devaluations of the South African pound in 1919, 1932 and 1949 alleviated the cost pressures. South Africa left the gold standard in December 1932, which resulted in a sharp depreciation of the South African pound. In local currency, the gold price more than doubled. Davenport (2013: 315) writes that the dramatic developments 'resulted in the most significant windfall in South Africa's economic history and facilitated the greatest mining boom the country has ever witnessed'.

The surplus from the windfall was used to develop new gold mines and finance unprecedented exploration activity. By 1936, the industry was developing 15 new mines, most of them in the Far East Rand of the current Gauteng province, the new centre of mining activities. Goldfields led the exploration activity that resulted in the discovery of the West Wits line in the Far West Rand. Anglo played a similar role in the discovery of the Klerksdorp gold fields, to the west of Gauteng. The company also led the charge into the Free State gold fields. Union Corporation discovered the Evander line on the north-eastern rim of the Witwatersrand Gold Basin. The commissioning of 13 new mines in the Free State, five mines along the West Wits line, four mines near Klerksdorp and three mines on the Evander gold fields during the 1940s, 1950s and 1960s, resulted in an exponential increase in gold production, which peaked at 1 million kg or 35.2 million ounces in 1970 (Davenport, 2013: 315–343).

By 1960, Anglo American was by far the largest South African mining house. According to Innes (1977: 145), Anglo had acquired more than 50 per cent of the shares in the Rand Mines Group and the JCI Group by 1960. Also, by that time, it had significant shareholdings in all the other mining groups: over 40 per cent in General Mining and about 10 per cent in the three remaining groups. For Innes, Anglo came to dominate South African mining for two key reasons. First, Anglo's competitors had imperialist concerns and channelled profits to London while Anglo chose to focus on southern Africa, involving an aggressive local reinvestment policy. Second, with its interests in diamonds and copper, Anglo was more diversified than other gold companies. The surplus from the Free State mines was used to add a new dimension to the Anglo empire after 1960: a growing international presence (Innes, 1977: 144).

The South African government set up state-owned companies that supported diversification from mining into industry. It established the Electricity Supply Commission (Eskom) in 1922, the Iron and Steel Corporation (Iscor) in 1928, the Industrial Development Corporation (IDC) in 1940 and the South African Coal, Oil and Gas Corporation (Sasol) in 1950. Fine and Rustomjee (1996) coined the term the minerals energy complex (MEC) to describe a unique system of accumulation based on a core set of industries and institutions that developed around mining. The critical actors within the MEC included the mining houses – or group producers – that fused with Afrikaner finance capital during the 1960s and diversified out of the minerals and energy core to spread an ever-widening net of conglomerate control across the national economy through the 1970s and 1980s (Capp, 2012: 65).

The state companies played an important role in 'lubricating both the growth of MEC core sectors and the ascendance of large-scale capital'. Afrikaner finance capital emerged during the 1950s under state tutelage prior to its integration with English mining capital during the next decade (Fine and Rustomjee, 1996: 97, 147). Sanlam, an insurance company that had been formed in 1918, established Federale Mynbou in 1953 to increase Afrikaner participation in mining. In 1964, it took over Genmin, after receiving assistance from Anglo. The transaction signalled a 'conscious accommodation of Afrikaner capital by English capital' (Fine and Rustomjee, 1996: 161).

From the 1970s, there was diversification within the mining industry as the expansion of coal, iron ore and PGMs reduced the country's dependence on gold. On the corporate front, Barlow, an industrial company, acquired Rand Mines in 1971 to form Barlow Rand, which eventually became South Africa's largest industrial company in the late 1980s (Fine and Rustomjee, 1996: 104). In 1980, Gencor was established following a merger between Genmin and Union Corporation (Fine and Rustomjee, 1996: 172–173). By the late apartheid years, there were 'six axes of capital', according to Fine and Rustomjee (1996: 98). These were Anglo, SA Mutual, Liberty/Standard, Anglovaal, Sanlam and Rembrandt/Volkskas. In 1990, they controlled 84 per cent of the shares on the Johannesburg Stock Exchange (JSE) by value, with Anglo accounting for 44.2 per cent (Chabane et al., 2003; McGregor and Zalk, 2017). In addition to the six axes of capital, there were six major mining houses, namely Anglo, Rand Mines (controlled by SA Mutual), Gencor (Sanlam), JCI (Anglo American), Anglovaal and Goldfields (Fine and Rustomjee 1996: 100, 108). During the early 1990s, Barlow Rand unbundled its industrial, mining and property interests into four companies.

The six conglomerates mined more than 70 per cent of all major minerals. At a sectoral level, they produced 91 per cent of the country's gold in 1988 (Fine and Rustomjee, 1996: 100). In platinum, due to a 1931 merger between two companies, JCI and Goldfields creating Rustenburg Platinum, the latter was the only producer for the next four decades. In the 1970s, two new producers – Impala Platinum (Implats) and Western Platinum (Westplats) – emerged as demand for the white metal increased. By the mid-1990s, there was a 'big three' of platinum producers that controlled 100 per cent of production: Rustenburg Platinum (51.6 per cent), Implats (39.1 per cent) and Westplats (9.3 per cent) (Capps, 2012).

In coal, three major private producers emerged, namely Anglo American Coal, Federale Mynbou (grown on the back of Eskom contracts) and Rand Mines's Randcoal (Fine and Rustomjee, 1996: 169). Within the state sector, Sasol and Iscor became major producers for their own operations. Finally, in iron ore, Iscor Mining became the largest producer. It established the Thabazimbi and Sishen mines in 1932 and 1954, respectively. In 1989 Iscor was privatised (Kumba Resources, 2006: 17).

In sum, by the late apartheid years, the South African economy had consolidated around 'six axes of capital' – led by Anglo – and six mining houses. According to Fine and Rustomjee (1996: 98), 'after the exhaustion of surface gold deposits, it was the enormous economies of scale required by deep mining conditions which led to early concentration of ownership.' Also, foreign exchange controls confined investment to the domestic economy. Later, as sanctions and the apartheid crisis intensified, subsidiaries of foreign companies became available for acquisition (Fine and Rustomjee, 1996: 11).

STRUCTURAL CHANGES OF THE MINING INDUSTRY AFTER 1994

At the end of apartheid, five of the original seven mining houses had remained. They were Anglo, JCI, Gencor, Goldfields and Anglovaal. During the post-apartheid era there was major restructuring in the mining industry as it navigated its way through volatile world commodity prices, the continued decline of gold and the rise of platinum and iron ore. Diversified conglomerates with their complex control structures were dismantled, operations and shareholdings were consolidated and simplified and the mining houses internationalised.

Following more than two decades of frenetic post-apartheid deal-making, Anglo remained in a strong position in the South African mining industry with interests in platinum, iron ore, coal, diamonds and manganese. It was the only mining house to survive of the five existing in 1994. JCI and Gencor unbundled themselves out of existence. Goldfields sold most of its assets to Sibanye Gold. Anglovaal's mining operations became part of black-owned African Rainbow Minerals (ARM). As is shown later, the restructuring created limited opportunities for black companies, but failed to decisively change the industry structure as conglomerates reasserted control within mining sub-sectors. According to McGregor and Zalk (2017), the unbundling and reconstitution of South Africa's apartheid-era conglomerates from the early 1990s is 'arguably the biggest restructuring of the South African economy since the discovery of diamonds in the 1880s, but it has received only limited attention'.

The restructuring was significant because it dismantled conglomerates that had survived for up to a century. Although it was structural change within mining, it had an impact on the rest of the economy as the links between the mining houses and finance and industry were also dismantled. Mohamed (2010) writes that the restructuring of Anglo was one of the most important changes in the South African economy. Its shedding of almost all manufacturing operations in particular was ultimately 'the withdrawal of the most important and influential conglomerate from downstream value-added production in South Africa' (Mohamed, 2010: 62).

The apartheid conglomerates looked offshore as they sought to restore profits following the political crises and economic stagnation of the late apartheid years. The objective was to free up capital that had been trapped by exchange-control regulations and to deploy the holdings they had accrued offshore (McGregor and Zalk, 2017). However, the dismantling of apartheid era conglomerate structures – under pressure from institutional investors who demanded focus on core operations – did not reduce levels of concentration. McGregor and Zalk (2017) argue that, instead, frenetic unbundling was followed by 'an equally intense process of rebundling' in which corporates had 'consolidated single-sector control'. To illustrate the point, although Anglo's ownership of the JSE had dropped to 3.3 per cent (McGregor and Zalk, 2017), the group accounted for 54 per cent of the value of South African mining assets at the end of 2017 (Gqubule, forthcoming, 2018).

In May 1994 Anglo was South Africa's largest company with a market capitalisation of R54 billion. It controlled companies with a value of 43.3 per cent of the JSE's total market capitalisation and had minority shareholdings in numerous other companies (Chabane et al., 2003). The Anglo empire's activities stretched into almost every area of the economy with a maze of overlapping directorships, cross-shareholdings, holding companies and pyramid structures. Within the top 25 mining companies on the JSE, Anglo controlled De Beers, Minorco and JCI – the second, third and fourth largest JSE-listed mining companies. It also controlled other listed companies such as Rustenburg Platinum, Potgietersrus Platinum, Free State Consolidated Gold, Rustenburg Western Deep and Anglo Coal.

Anglo led the post-apartheid restructuring as it expanded offshore, sold assets and consolidated its South African interests in platinum and gold. In 1995, Anglo announced the unbundling of JCI into three companies: Anglo American Platinum (Amplats); JCI Limited, a mining house with interests in gold, coal, ferrochrome and base metals; and Johnnic, a diversified industrial company. Anglo retained the crown jewels in Amplats and sold shares in JCI and Johnnic to black investors. Both transactions unravelled. The once-mighty JCI eventually collapsed after endless unbundling. In 1997, Anglo consolidated all its platinum interests into Amplats. In 1998, it consolidated its gold interests into

AngloGold, which later merged with Ghanaian company Ashanti Gold to form Ashanti Goldfields in 2004.

In 1999 Anglo merged with Minorco and moved its primary listing to London. Over the next decade it sold non-core assets and became a pure mining company. In South Africa, Anglo entered the iron ore business, strengthened its position in diamonds, increased its shareholding in Amplats and exited AngloGold Ashanti in 2009. But Anglo accumulated US$13 billion of debt as it invested US$14 billion in a Brazilian iron ore project at the top of the commodity cycle and paid shareholders US$5 billion in special dividends during the good times. In 2015 the share price plummeted and the company announced a drastic restructuring to sell non-core assets and reduce debt. After 2016 Anglo's share price soared following a recovery of coal and iron ore prices and successes in reducing its levels of debt. At the end of 2017 the company was reconsidering the strategy and pace of asset disposals. After two decades of frenetic post-apartheid deal-making, Anglo remained in a strong position in the mining industry with interests in platinum, diamonds, manganese and coal. It is the only one out of the five mining houses that existed in 1994 that has survived.

THE MINING CHARTER AND STRUCTURAL CHANGE

Post-apartheid restructuring in South Africa's mining industry created some opportunity for black companies. At first, in the absence of a policy and legal framework, the private sector led the process of selling equity to black companies. The promulgation of the Mining Charter in 2004 accelerated the process.

In terms of ownership, the focus of this paper, there were two targets in the Charter: 15 per cent participation by historically disadvantaged South Africans (HDSAs) – including white women – in equity ownership or attributable units of production within five years; and 26 per cent within 10 years. The industry also agreed to provide funding of R100 billion to HDSA companies (DMR, 2002).

Not long after 2004 it became clear that the Charter was out of line with international best practices in natural resource governance. Benchmarked against other developing countries, a weakness of South

African mining policy is that it does not have a direct 'government take' through equity. If one looks at the majority of the world's top 20 mining countries, most have some state ownership. In Africa, Botswana, Zambia, Ghana, the Democratic Republic of Congo, Mali, Burkina Faso, Guinea, Gabon, Namibia and Zimbabwe have state shares of between 10 per cent and 50 per cent (Gqubule, 2015).

The Charter also fell out of line with South African government policies. It was overtaken by significant advances in the measurement of B-BBEE in the rest of the economy. The government put in place a uniquely South African system of social accounting that was contained in the B-BBEE Codes of Good Practice of 2007 and 2013 (the 'Codes'). Against this new benchmark, it became clear that the Charter betrayed every principle of empowerment that was contained in the Codes. In a word, the bar for empowerment was far lower in the mining sector than the ones that prevailed in the rest of the economy.

Under ownership, targets were not set for communities and workers. There was no indicator to measure and monitor net value. Mining companies considered the signing of a Black Economic Empowerment (BEE) transaction as the achievement of the target. However, under the Codes, the achievement of the ownership target is a 10-year process. In the rest of the economy, full compliance can only be achieved when net value has been transferred and black shareholders have been released from debt obligations. The Charter did not have a measurement system and scorecard with clear definitions to significantly reduce the possibility of different interpretations by companies, nor did it have a mechanism to independently verify the BEE contributions of companies.

The Charter had no rules on how a company could continue to earn points after a black shareholder had sold its shares. Mining companies believed that such recognition would continue in perpetuity. According to the Codes, the continuing consequences principle allows companies to keep up to 40 per cent of their ownership points under certain conditions. The black shareholder must have owned shares for at least three years. The period of recognition cannot exceed the period over which the shares were held. There must be net value created for black shareholders.

In July 2010, after a DMR mid-term review, stakeholders reached agreement on an amended charter. Under ownership it restated the 26 per cent target but did not address any of the weaknesses listed above. During 2015 a dispute erupted between the DMR and the Chamber, which landed in the courts. There were two major issues. The first was whether a company could retain its empowerment credits after a black shareholder had sold its shares – the so-called 'once-empowered always empowered' principle. The Chamber said, in its court paper, that a perpetual 26 per cent target would destroy investor confidence. However, the Codes *do* have a perpetual 25 per cent plus one share ownership target. This has not destroyed investor confidence in the rest of the economy.

The second is whether to consider the net value – after taking into account debt that was used to acquire shares – that accrues to black shareholders in calculating ownership credits. According to the Chamber, the fact that some HDSA participants had exited mining deals or that equity prices had fallen could not be held against the companies. The industry's position that ownership credits should exist in perpetuity and that net value should not be considered is not reasonable. It is against the principles of the Codes, which apply in the rest of the economy.

There were also disagreements over measurement principles. In May 2015, the DMR released its Mining Charter implementation report (DMR, 2015). The report found that the total industry simple average HDSA ownership was 32.5 per cent. But all the report's findings were meaningless because the department had made a strange decision to use employment data to weight ownership figures. The Chamber released the summarised (four page) findings from a report that had been compiled by accounting firm SizweNtsalubaGobodo and Rand Merchant Bank (Chamber of Mines, 2015). The report found that 100 per cent of the Chamber's members – who accounted for 85 per cent of the value of the mining sector – had achieved the 26 per cent ownership target. At the end of December 2014, its members had created a net value of R159 billion. This was equivalent to 26 per cent of the value of the industry on that date.

On 15 June 2017, the DMR released a final version of a third charter that had a 30 per cent target for black ownership, which had to be achieved within 12 months (DMR, 2017). Although the amended BEE Act has a trumping clause that requires charters to align with the Codes, Department of Trade and Industry (DTI) officials said that there had been no consultations with the DMR on the Charter. 'They are operating under a unilateral declaration of independence. They have not come under our fold like the other charters,' DTI Director General Lionel October said (2017). Although the 2017 charter made attempts to align with the Codes, it did not address any of the contentious issues between the DMR and the industry. On 4 April 2018 the court ruled in favour of the Chamber. The ruling could result in an end to BEE transactions in mining long before there has been meaningful transformation in the industry.

On 15 June 2018 the DMR released another version of the Charter. The new draft Charter retains the 30 per cent ownership target, but it recognises the 'once empowered, always empowered' principle for existing mining rights, which will apply to virtually the entire industry. Existing rights holders have five years to top up their black shareholding to achieve the 30 per cent target. For new mining rights, there are 8 per cent targets for communities and workers – a total of 16 per cent of the 30 per cent target – of which five per cent would be a free-carried interest. If this version of the Charter becomes final, it will equally mean an end to BEE transactions in the mining sector before the industry has been transformed.

If this happens, the blame for such an event should be shouldered by the ruling ANC, the DTI and the DMR, which provided no leadership on the issue for many years and refused to scrap the Charter and align it with the BEE Codes, which have a compromised position on the 'once empowered, always empowered' principle that was negotiated between stakeholders in the rest of the economy. However, in a press conference that followed a Mining Charter summit on 7–8 July 2018, Mining Minister Gwede Mantashe announced a new deadline of November to finalise the Charter. He said: 'We must finalise that discussion on once-empowered always-empowered. We must find a way out of that debate. In the commissions at the summit it was rejected by almost everybody' (Seccombe, 2018).

It is important to position these developments in relation to statistics on ownership in the industry. While there is significant public ownership, black ownership is well below the policy target of 26 per cent and the benefits have accrued to a minority of black companies. There is also minimal ownership by employees and communities. Three community trusts account for almost all the net value created for such beneficiaries. Overall, the Charter to date (August 2018) has failed to change the basic structure of the mining industry, despite the introduction of black shareholders. A report, *South African Mining at the Crossroads: An Update*, by the author of this chapter measures black ownership at the end of December 2017: between 1997 and 2017 there were more than 40 BEE transactions worth R130 billion within the JSE top 25 mining companies. At the end of December 2017 there was black ownership of R84.7 billion, using the flow-through principle, which strips out non-empowerment shareholders within an ownership chain.

This was equivalent to 12.5 per cent of the value of South African assets that were worth R691.2 billion by the end of 2017. This comprised R50.7 billion at the level of listed companies and R35.6 billion at the level of unlisted companies. BEE net value (after taking into account debt that was used to acquire shares) was R45.9 billion. This was equivalent to 6.6 per cent of the value of South African assets. This comprised R32.9 billion at the level of listed companies and R13 billion at the level of unlisted companies. Public ownership – by the Public Investment Corporation (PIC) and the IDC – was R111.4 billion. This was equivalent to 16.1 per cent of the value of South African assets (Gqubule, forthcoming, 2018).

A persistent feature of the mining sector is that the benefits have flowed to a minority of South Africans. At the end of 2017 African Rainbow Minerals Exploration and Investments (ARMI), which represents the family interests of Patrice Motsepe, accounted for 25.6 per cent of BEE net value. ARMI had shares in ARM that were worth R11.9 billion. BEE shareholders accounted for 25.7 per cent of BEE net value. BEE shareholders had shares in Exxaro and Kumba Iron Ore (KIO) that were worth R11.8 billion. The 15.7 per cent stake in Exxaro was worth R6.7 billion. The effective 3.2 per cent stake in KIO (held through Exxaro, which owned 20.6 per cent of KIO) was worth R5.1

billion. Therefore, two companies – ARMI and BEE special purpose vehicles (SPV) – accounted for 51.6 per cent of BEE net value (Gqubule, forthcoming, 2018).

Contrary to perceptions, the major black business interests in mining are not members of the political elite. The industry has provided minimal benefits for employees and communities. At the end of December 2017, employee share ownership plans had a net value of R2.7 billion. This was equivalent to 0.4 per cent of the value of South African assets. Community schemes and companies had a net value of R14 billion. This was equivalent to 2 per cent of the value of South African assets. But three community ownership vehicles – in Royal Bafokeng Holdings, Assore and KIO – accounted for 95 per cent of BEE net value (Gqubule, forthcoming, 2018).

Moving to the macro scale, economic growth is crucial for the success of BEE transactions. Although the structures of many transactions can be improved, the major impediments towards transferring ownership to black people are at the macro-economic level. The cost of capital has been much higher than the country's pedestrian economic growth rate since 1994, due to the rigid implementation of inflation targeting. Therefore, market timing has been the key to accumulating value. In mining, the performance of the domestic economy is not important. What matters is world commodity prices, which depend on global demand, and the value of the rand. The black companies that accumulated value acquired assets at the bottom of the cycle or soon after the start of the boom in world commodity prices in 2003.

By the end of December 2017, there was one black-owned company – Royal Bafokeng Platinum (RB Plat) – within the JSE top 25 mining companies. In addition to ARMI and BEE SPV, there were other large black shareholders within the JSE top 25 mining companies. These included the Sishen Iron Ore Company (SIOC) Community Development Trust (R4.7 billion); Royal Bafokeng Holdings (RBH) (R4.4 billion); MS 904 (R3.6 billion) and MS 350 (R2.3 billion). The SIOC Community Development Trust owns 3 per cent of the SIOC. RBH owns 52 per cent of RB Plat (R2.9 billion) and 6.3 per cent of Implats (R1.5 billion). Near-mine communities, which own 51 per cent of MS 904, have a 7.3 per cent stake in Assore (Gqubule, forthcoming,

2018). In terms of the 26 per cent ownership target of the Mining Charter, by the end of 2017, 14 of the companies within the top 25 mining companies had met the target. The shortfall for the remaining 11 companies, most of which are Anglo group companies, was between R60 billion and R70 billion (Gqubule, forthcoming, 2018).

Despite these limited gains by black shareholders, the Charter to date has failed to change the structure of the mining industry. According to EcoPartners (2011), mining debates raging in South Africa (about nationalisation and transformation) are all focused on realigning ownership. What is not yet being debated is the issue of breaking up ownership and control of ore bodies to reduce concentration in the industry. In the PGM sector, one company holds 55 per cent of reserves (EcoPartners, 2011). The 'big three' apartheid-era producers – Amplats, Implats and Lonmin – still dominate the industry. Emerging producers do not have processing, refining and sales capabilities – all of which are provided by the 'big three.' In iron ore, two companies control 100 per cent of production (Gqubule, forthcoming, 2018).

In the gold sector, three companies control 75 per cent of gold reserves (EcoPartners, 2011). AngloGold Ashanti, Goldfields and Harmony were the 'big three' when the Charter was signed. Since then, Sibanye Gold replaced Goldfields within the club of major producers after it took over most of its mines in 2013. The new 'big three' controlled 70 per cent of production in 2016 (Gqubule, forthcoming, 2018. The industry has minimal black ownership.

In coal, Anglo and BHP Billiton subsidiaries were the largest private producers during the 1990s. The two companies helped to create black-owned Exxaro in 2006. Exxaro is no longer black-owned. In 2016, the largest producers – Anglo (53.8mt), Exxaro (42.8mt), Sasol (40mt), South 32 (30mt) and Glencore (29.3mt) – accounted for almost 80 per cent of coal output.

EFFECTING STRUCTURAL CHANGE IN THE MINING INDUSTRY

Given the dismal history and ongoing structural inequality of the South African mining industry, South Africa must develop a new vision for the mining industry and take a stance on the desired future structure of the industry and its strategic sub-sectors. The state apparatus must be restructured to extract a fair share of natural resource rents and ensure that there is an equitable distribution of the proceeds to benefit the public, the host communities, employees and black investors.

In 2012 the ANC commissioned a report on the role of the state in the mining sector. It found that state ownership of mines is unaffordable. More specifically, the ANC's *State Intervention in the Minerals Sector* (2012) report said it would cost R1 trillion to nationalise the industry and R500 billion to buy 51 per cent of the mines. 'Consequently, either complete nationalisation or 51 per cent at market value would be totally unaffordable and could put our country in a situation where it could lose fiscal sovereignty' (ANC, 2012: 28). Since it was never imperative that all mines be nationalised in a single day, it is not clear why the report did not consider the affordability of the state buying them over an extended period.

More broadly, it is useful to look at international best practices to help determine a new role for the South African state in the industry. A review of international best practices in harnessing mineral resources for economic development has identified seven key issues in non-renewable natural resource governance (KIO Advisory, 2010 and Gqubule, 2014a; 2014b).

According to Chang (2007), most people agree that non-renewable (mainly mineral) natural resources should be treated differently from other sources of wealth in the economy. This is because such mineral resources are not created by the efforts of those who own the land that produces them. As a result, in most countries, private property owners, including communities, do not have the rights to mineral resources simply due to the fortuitous fact that the resources are located beneath their property. In addition, the extraction of non-renewable natural resources renders them unavailable to others, including later generations. This is why mineral resources are publicly owned in most countries. It means that countries with mineral resources tend to have larger state-owned enterprise sectors.

Few people take the same view on renewable natural resources (such as forestry and fisheries), but in most countries their exploitation is also regulated for reasons of sustainability (while in others at least a portion of such resources is state-owned). A critical issue in natural resource governance is the distribution of mineral resource rents – the difference between mineral revenues and their extraction cost – for private and public benefit.

First, a strategy for mineral resources must be located within a national vision for economic growth and development. In chapter 2, Netshitenzhe takes the view that the National Development Plan is such a vision for South Africa. Second, there must be a national vision and plan for the natural resources sector in particular. The best example is Norway's 'Ten Oil Commandments', drafted in the early 1970s. The commandments are a set of instructions given to the Norwegian government by parliament for the development of the oil sector.

Third, countries must develop the state capacity to lead the sector. State capacity is the initiator of economic growth and development. As Amsden (1989: 11–12) has pointed out 'industrialisation was late in coming to "backward" countries because they were too weak to mobilise forces to inaugurate economic development'. In more detail, the onset of economic expansion tends to be delayed by weaknesses in a state's ability to act; and if and when industrialisation does accelerate, it is due to the initiative of a strengthened state authority. Similarly, the Africa Mining Vision (AU, 2009: 18) says: 'The key element in determining whether or not a resource endowment will be a curse or blessing, is the level of governance capacity and the existence of robust institutions.' The best practice model is that of Norway, where there is clear separation between the government's policy, regulatory and commercial functions.

Fourth, the state must develop measures – a government 'take system' – to extract a fair share of natural resource rents for its citizens. In between the false dichotomies of nationalisation and privatisation there are as many ways of slicing the rent as there are countries in the world. Each model is unique. As Baunsgaard (2001) and Cottarrelli (2012) have shown, the models involve various combinations of state equity, state-owned companies, tax regimes (windfall and resource rent taxes and royalties) and contract types (production sharing agreements, joint ventures and concessions).

In extracting natural resource rents, state equity exceeded tax revenues by a wide margin during the commodities boom of the early 21st century. For example, in Chile the government collected significantly more from state-owned Codelco, with a 30 per cent share of production, than from private mining companies since 1991 (KIO Advisory, 2010 and Gqubule, 2014a). State equity can trump tax revenues because it is impossible to develop tax regimes that replicate the government tax systems in countries such as Norway and Botswana, where more than 75 per cent of natural resource rents accrue to the state. While a national oil company can appropriate all the rents in an industry, a tax regime cannot. Beyond a certain level of taxation, there will be no investors in the industry.

Fifth, there are two challenges in managing natural resource rents – the volatility of commodity prices and the so-called 'Dutch disease', whereby a resource boom can result in the appreciation of the currency and a corresponding loss of competitiveness in other sectors of the economy. Countries have used sovereign wealth and stabilisation funds to collect excess commodity revenues and to smooth expenditure to narrow the impact of boom and bust cycles. This implies that resource-rich countries should have a view on the exchange rate. Frankel (2012) argues that countries should intervene in the foreign exchange market to prevent the appreciation of a currency at the start of a commodity boom.

Sixth, countries should invest the natural resource rents in infrastructure, human capital and diversifying the economy. According to Stiglitz (2005), since extraction makes a country poorer, it is the subsequent re-investment in capital that can offset the depletion of natural resources and make a country richer.

Finally, natural resource projects tend to be capital-intensive enclaves with few linkages to the rest of the economy. The Africa Mining Vision (AU, 2009) says there is a need to diversify the economy through downstream linkages (beneficiation and manufacturing), upstream linkages (mining capital goods, consumables and services industries) and side-stream linkages in infrastructure, skills and technology development.

CONCLUSION AND RECOMMENDATIONS

Given the history of structural inequality in the South African mining industry and the argument made here for a failed post-apartheid attempt to impact on the structure of the industry, it is crucial to rethink and to strategise for a more equitable future. Drawing from the seven international best practices presented, it is proposed here that South Africa adopts a 25 per cent public ownership target *in addition* to the existing black ownership target of 26 per cent in the industry. Two vehicles should be used to achieve the 25 per cent target: a SWF and a PMC.

A feasible target over the next decade would be to achieve 51 per cent public, community, employee and black investor ownership of the country's mineral resources. This would comprise 25 per cent public ownership *in addition* to the legislated 26 per cent black ownership requirement. The envisaged public ownership is separate from and *in addition* to community and employee ownership targets. With the PIC and the IDC already owning 16.1 per cent of South African mining assets worth R111.4 billion at the end of December 2017, the achievement of the target would require an additional R61.4 billion. Within the 26 per cent BEE target, host communities and employees could be allocated 7.5 per cent each, of which 50 per cent would be a free carry. BEE investors would get an allocation of 11 per cent. With a 30 per cent target, there could be an equal distribution between the three groups. In line with the Norwegian model, there could be two investment vehicles to achieve the public ownership target. Norway has the State Direct Financial Interest – valued at $97 billion in 2014 – which mostly holds minority stakes in oil fields. Statoil – the national oil company – is an operating company that has majority stakes in most of its investments. It had a market capitalisation of $77 billion in March 2018.

The PIC and the IDC should consolidate and cede their mining assets into a new SWF that would possibly be listed on the JSE to ensure transparency and accountability. The difference with the status quo is that the SWF would be for the benefit of all South Africans and not only public servants. It would capture a rising share of rents from mining up to 25 per cent, which could be used to support national development priorities through the budget. It is noted that this proposal raises issues around the funding of pensions for government employees (see chapter 3), which are beyond the scope of this paper. The majority of civil service pension funds in developed countries, however, are unfunded or partially funded (Ponds et al., 2011). The pre-funding of pensions for South African civil servants, at 116 per cent of liabilities according to the 2016 actuarial valuation, is excessive and can be reduced substantially.

In line with the Norwegian model, a professional asset management company – possibly with former PIC and IDC employees – could be established to manage the assets. The second investment vehicle would be a professionally managed public (or state) mining company (PMC) that would develop the capacity to operate mines over the medium to long term. The SWF would identify opportunities for the PMC. The SWF would also establish an investment vehicle to manage community and employee share allocations and provide free advisory services.

In line with other mineral-rich nations of the global South, the government, on behalf of South Africans, should investigate the feasibility of Anglo becoming a South African mining champion, with 51 per cent public, community, worker and BEE investor shareholding. Anglo companies – Anglo, Anglo American Platinum and KIO – were worth R366.1 billion at the end of December 2017. This is equivalent to about 53 per cent of the value of South African mining assets at the end of 2017. With a geographic unbundling of Anglo, the PIC would have a 31 per cent stake in the new company, depending on relative valuations. The PIC and the state could finance a major portion of the community, worker and BEE participation. A century after it was established, the time has come for Anglo to come back home.

Finally, there should be a separation of the state's policy, regulatory and commercial functions. This would require the establishment of a new, professionally managed and securely financed super agency to award mineral rights and to monitor implementation by companies of empowerment and other mining policies and laws. To finance this, there should be targeted deployment of annual income from mining royalties, with 50 per cent allocated to community development funds, 40 per cent used to finance the SWF and PMC and 10 per cent earmarked for the proposed mining industry regulator. An example is Canada's Alberta Heritage Fund, which initially received 30 per cent of the province's oil royalties.

REFERENCES

African National Congress (ANC). (2012). 'State intervention in the minerals sector: Maximising the developmental impact of people's mineral assets'. Available at: www.anc.org.za/docs/reps/2012/simssummaryz.pdf. Accessed 18 December 2017.

ANC. (2018). '54th National Conference report and resolutions'. Available at: http://www.anc.org.za/54th-national-conference-reports. Accessed 30 May 2018.

African Union. (2009). 'Africa Mining Vision'. Available at: https://www.imf. org/external/np/exr/consult/2012/NR/. Accessed 18 December 2017.

Amsden, A. (1989). *Asia's Next Giant: South Korea and Late Industrialization*. Oxford University Press, London.

Anglo American. (2010). 'Thermal coal investor'. Presentation. Available at: www.angloamerican.com. Accessed 18 December 2017.

Anglo American Coal. (2014). 'South Africa Merrill Lynch investor and analyst site visit'. Presentation. Available at: www.angloamerican.com. Accessed 18 December 2017.

Baunsgaard, T. (2001). 'A primer on mineral taxation'. International Monetary Fund Working Paper. Available at: https://papers.ssrn.com/sol3/papers. cfm?abstract-id=879929. Accessed 18 December 2017.

Capp, G. (2012). 'Victim of its own success? The platinum mining industry and the apartheid mineral property system in South Africa's political transition'. *Review of African Political Economy*, Vol. 39 (131). pp. 63–84. Available at: http://www.tandfonline.com/doi/ abs/10.1080/03056244.2012.659006?journalCode=crea20. Accessed 6 January 2018.

Chabane, N., Machaka, J., Molaba, N., Roberts, S. and Taka, M. (2003). '10 Year Review: Industrial structure and competition policy'. Corporate Strategy and Industrial Development (CSID) research project. School of Economic and Business Sciences, University of the Witwatersrand. Available at: https://sarpn.org/documents/d0000875/ docs/10yerReviewIndustrialStructure&CompetitionPolicy.pdf. Accessed 19 December 2017.

Chamber of Mines. (2015). 'Chamber of Mines calculates broad based HDSA ownership of 38% and meaningful economic value transfer of > R159bn'. South African Chamber of Mines. Available at: http://www. chamberofmines.org.za/industry-news/special-features/43-chamber-research-report-on-progress-on-ownership. Accessed 2 January 2018.

Chang, H.J. (2007). 'State-owned enterprise reform'. United Nations Department of Economic and Social Affairs (UNDESA), New York. Available at: https://esa.un.org/techcoop/documents/pn_soereformnote.pdf. Accessed 2 January 2018.

Davenport, J. (2013). *Digging Deep: A History of Mining in South Africa.* Jonathan Ball, Johannesburg.

Department of Mineral Resources (DMR). (Undated). 'B1 statistical table. DMR, South Africa'. Available at: http://www.dmr.gov.za/publications/summary/149-statistics/8271-b1-stat-tables-2015.html. Accessed 18 December 2017.

DMR. (2015). 'Assessment of the Broad-Based Socio-Economic Empowerment Charter for the mining industry (Mining Charter)'. DMR, South Africa. Available at: http://www.dmr.gov.za/mining-charter.html. Accessed 19 December 2017.

DMR. (15 June 2017). 'Reviewed Broad-Based Black Economic Empowerment Charter for the South African mining industry'. *Government Gazette* 40923.

Department of Minerals and Energy. (2002). 'The Broad-Based Socio-Economic Empowerment Charter for the South African mining industry'. DMR, South Africa. Available at: http://www.dmr.gov.za/mining-charter.html. Accessed 18 December 2017.

EcoPartners. (2011). 'Synopsis of the first report of mineral resources and reserves in South Africa'. Available at: http://www.ecopartners.co.za/docs/Mineral%20Resource%20and%20Reserves%20of%20South%20Africa.pdf. Accessed 6 January 2018.

Ernst & Young. (1994–2009). 'Annual review(s) of South African mergers & acquisitions'.

Fine, B. and Rustomjee, Z. (1996). *The Political Economy of South Africa: From Mineral Energy Complex to Industrialisation.* Wits University Press, Johannesburg.

Frankel, J.A. (2012). 'The natural resource curse: A survey of diagnoses and some prescriptions'. Harvard Kennedy School, Massachusetts. Available at: https://dash.harvard.edu/bitstream/handle/1/8694932/rwp12-014_frankel.pdf. Accessed 19 December 2017.

Gqubule, D. (2014a). 'Resource nationalism in Latin America'. Unpublished paper developed for the Mapungubwe Institute for Strategic Reflection (MISTRA).

Gqubule, D. (2014b). 'Resource nationalism in Europe'. Unpublished paper developed for the Mapungubwe Institute for Strategic Reflection (MISTRA).

Gqubule, D. (2015). 'Untangling the Complex web of mineral resource ownership in Africa'. Unpublished research paper.

Gqubule, D. (forthcoming, 2018). 'Transformation of South African's mining industry: An analysis of the Mining Charter 2004–2017'. Centre for Economic Development and Transformation, Johannesburg.

Innes, D. (1977). 'The mining industry in the context of South Africa's development 1910–1940'. Collected Seminar Papers. Institute of Commonwealth Studies, University of London. Available at: http://sas-space.sas.ac.uk/4040/#undefined. Accessed 19 December 2017.

Kio Advisory Services. (2010). 'Harnessing South Africa's mineral resources for economic growth and development: Lessons and experiences from abroad'. Unpublished research paper developed for the South African Mining Development Association.

Kumba Resources. (2006). 'Prelisting statement'. Available at: www.exxaro.com/pdf/icpr/u/pdf/prelist.pdf. Accessed 2 January 2018.

McGregor, A. and Zalk, N. (28 February 2017). 'Conglomerates did not deliver on their side of the post-apartheid bargain'. *Business Day*. Available at: www.bdlive.co.za. Accessed 2 January 2018.

Mohamed, S. (2010). 'The effect of a mainstream approach to economic and corporate governance on development in South Africa'. In Edigheji, O. (ed.). *Constructing a Democratic Developmental State in South Africa*. HSRC Press, Pretoria.

October, Lionel (26 July 2017). Telephone interview given to the author.

Ponds, E., Severinson, C. and Yermo, J. (2011). 'Funding in public sector pension plans: International evidence'. OECD Working Papers on Finance, Insurance and Private Pensions No. 8. OECD Publishing, Paris. Available at: www.oecd.org/finance/private-pensions/47827915.pdf. Accessed 10 September 2018.

Seccombe, A. (8 July 2018). 'Mantashe extends public comment period on Mining Charter by a month.' *Business Live* (Online). Available at https://www.businesslive.co.za/bd/companies/mining/2018-07-08-mantashe-extends-mining-charter-commentary-by-a-month/. Accessed 12 July 2018.

Stiglitz, J. (2005). 'Making natural resources into a blessing rather than a curse'. In Tsalink, S. and Schiffrin, A. (eds). *Covering Oil: A Reporter's Guide to Energy and Development*. Open Society Institute, New York. Available at: https://www.opensocietyfoundations.org/sites/default/.../osicoveringoil_20050803.pdf. Accessed 28 July 2018.

Zuma, J. G. (2017). 'State of the Nation Address'. Parliament of South Africa, Cape Town. Available at: https://www.gov.za/speeches/president-jacob-zuma-2017-state-nation-address-9-feb-2017-0000. Accessed 5 April 2018.

FIVE

Greening South African mining through the Fourth Industrial Revolution

ROSS HARVEY

FROM THE PROVERBIAL 'right' to 'left' on the political spectrum, it is increasingly acknowledged that economic development globally has occurred through over-exploitation of natural resources. Altenburg and Rodrik (2017: 2) argue that '(h)umanity is approaching various ecological tipping points beyond which abrupt and irreversible environmental change at large geographic scales is likely to happen'. They further argue that new techno-institutional systems are needed to de-couple economic development and human wellbeing from resource depletion and carbon emissions. The ways in which energy and transport are provided will have to change entirely to disrupt currently unsustainable development pathways.

Renewable energy costs are dropping rapidly across the globe, and it is no different in South Africa. The country is optimally positioned to acquire the vast majority of its electricity needs from variable wind and solar power with occasional supplementation from open-cycle gas turbines. The mining industry could be redesigned to supply minerals and metals required for generating and transmitting renewable energy, and inputs for other products like smartphones that are a pervasive part of the Fourth Industrial Revolution.

From a productivity perspective, the industry could be transformed by the adoption of new technologies to mine differently and in a less environmentally destructive manner. This would also create upstream opportunities to produce the capital equipment required for nimbler and less environmentally invasive methods of mining. Ultimately, however, the case for adopting new technologies is not only economic; it is also social and political. Provided the transition to a low-carbon growth path is well governed, the benefits to society are likely to be widespread. For instance, the benefits of cleaner air are indivisible and reduce the social healthcare burden. Nonetheless, the benefits of technological disruption are unevenly distributed. Consequently, some groups oppose change, either through fear of job losses or because vested interests are threatened. In this respect, the current political equilibrium in South Africa may limit the adoption of new technologies and a set of supporting policy frameworks that could ignite sustainable economic growth.

The aim of this chapter is to examine the likely impacts of the Fourth Industrial Revolution on mining in South Africa and to suggest an appropriate policy framework for navigating the threats, and exploiting the opportunities, that will enhance the welfare of both the natural environment and of future generations of South Africans. The main argument is that if the South African mining industry is to recover from its current malaise, it will have to become a radical adopter of new technologies and focus on supplying minerals and metals that enhance the global technological revolution.

The chapter is organised as follows. It begins with a description of the 'Fourth Industrial Revolution' and investigates whether the term is warranted in light of the literature on the subject. It demonstrates that the potential upscaling and spread of renewable energy technologies are particularly important for the aspiration to de-couple economic wellbeing from natural resource exploitation. Second, I present a vision

145

for the future in which South Africa becomes a radical adopter of new technologies, focusing on supplying the minerals and metals required for the production of renewable energy technology. I argue that such a reorientation would serve to both address the negative legacy of mining and create sustainable linkages to other sectors of the economy. While uptake of new technologies may reduce the number of direct jobs available in the industry in the short run, it could create indirect jobs in other sectors of the economy which it fuels. Third, the implications for the existing mining industry are discussed. Commodity price movements of the future are likely to reflect the demands and opportunities entailed in the Fourth Industrial Revolution. Some commodities may become obsolete over time, which will create difficulties for a large proportion of the existing industry. Fourth, I examine political barriers to the adoption of new technologies. The governance regime for the industry along with policy connections to other industries are crucial determinants of whether the industry can recover and optimise new opportunities. This chapter concludes with a set of policy recommendations that follow from the analysis.

THE FOURTH INDUSTRIAL REVOLUTION

The Fourth Industrial Revolution is a term that was coined by Professor Klaus Schwab of the World Economic Forum (Schwab, 2016). It captures the confluence of new technologies interacting with each other and the cumulative impact this will have on the world. It is difficult to attribute a timeline to the impact of these new technologies, but it is anticipated that the changes are likely to be of a large magnitude, even if staggered. There is some debate as to whether we are witnessing a truly *fourth* industrial revolution.

The first industrial revolution spanned from about 1760 to 1840. Triggered by the construction of railways and the steam engine, it ushered in mechanical production, which made specialised, assembly line production possible. A second revolution started from the late 19th century onwards, as electricity and the assembly line made possible mass production as we know it today. The third began in the 1960s, catalysed by the development of semiconductors, mainframe and personal computing, and the internet (Schwab, 2016).

Detractors argue that what Schwab describes as a fourth industrial revolution is more accurately categorised as an extension of the third 'digital' revolution. Others accept the numerical attribution, recognising the internet as the foundational platform for a host of new technologies to interact, but critique the idea that it is *industrial*. Michael Peters (2017: 3), for instance, writes that while the new digital logic does change everything, and is the means for massive automation, it is itself not industrial. Peters argues that it is not merely the scope and scale of industrialisation that has changed, but the entire logic of the production system that now enables and rewards autonomous digital network systems.

But Schwab (2016) defends the category of a distinct, *fourth industrial revolution*, by pointing to the very velocity, breadth and depth and the cumulative systems impact of new technologies that Peters mentions, notwithstanding the fact that some particular advances are extensions of earlier technological breakthroughs. The pace is exponential, and the connectedness of the world means that each new technology is likely to beget others. Building on the digital revolution, this revolution combines multiple technologies and changes how we are wired, not just what we do and how we do it. Entire systems – from international affairs to how we run our homes – are being transformed (Schwab, 2016: 3). If production and consumption patterns will fundamentally change as a result of this transformation, it seems to warrant the term. While primarily digital in nature, its effect on all kinds of *industry* make the term heuristically useful and not overly problematic for analytical purposes.

Artificial intelligence is one of the driving forces behind these new technologies. It is increasingly used in ordinary appliances such as smartphones. Algorithms can deliver information tailored to consumer preferences. Knowledge of these preferences is ascertained through 'big data' – the ability of machines to process large volumes of data in real time to allow rapid responses.[1]

Artificial intelligence also powers self-driving or autonomous

---

1   A very simple instance of artificial intelligence is a user posting on Facebook that they desire pizza for dinner and shortly thereafter an advert for pizza and Mr Delivery appears on the user's Facebook feed.

vehicles, receiving information from the external environment and adjusting the vehicles' movements in response. Autonomous vehicles can now anticipate the likelihood of an accident more quickly and more accurately than human drivers and respond appropriately. Autonomous electric vehicles have the potential to move the world towards a low-carbon growth trajectory and reduce traffic congestion. The world's cities also have an unprecedented opportunity to limit private car usage and to move towards public transport systems that are electrified and more energy efficient, thereby reducing greenhouse gas emissions (UNEP, 2011). Another component of this revolution is the use of artificial intelligence in 'the internet of things' (IoT) – the ability to control an entire home, for instance, through the use of an internet connection and devices like Apple's Siri. Users can, for example, instruct Siri to change the refrigeration temperature and the colour of the living room lights, open the garage door and so forth, provided the relevant appliances are enabled with the required technology. Beyond the home, the IoT will impact the information available to firms' supply chain partners and how the whole supply chain operates. IoT is increasing the efficiency of business processes by providing more accurate and real time visibility into the movement of materials and products. This has a positive impact on production lines, warehousing, retail delivery and store shelving. Farming equipment manufacturer, John Deere, as well as United Parcel Service currently use IoT-enabled fleet-tracking technologies to cut costs and improve supply efficiency (Lee and Lee, 2015: 431). In the public sector, microchips in LED street lamps (that double as CCTV cameras) can transmit information about battery and bulb life that makes routine maintenance much more efficient.

The opportunities for the mining industry are similarly vast in being able to track, for instance, fleets of pick-up trucks that transport mined ore bodies to the processing plant. 3D printing provides new opportunities to manufacture spare parts in remote locations, removing productivity delays in mines, for instance, that may otherwise have to wait several days for parts to arrive.

Discussions of new technologies are particularly important in the South African context of low median income levels, poor educational outcomes, weak labour absorption prospects and low levels of formal housing. While increased access to electricity has been one of the government's major development successes, the persistently rising costs of electricity are likely to undermine economic opportunities that would otherwise exist, in addition to crowding out disposable income that would otherwise be injected into the economy. Existing inequality and recently rising poverty levels (reversing some of the gains made since 1994) create a strong policy imperative to ensure that the benefits of the technology revolution accrue in a way that provides pathways out of the middle-income trap.[2]

One of the biggest risks of the Fourth Industrial Revolution is that its benefits accrue only to those who are already wealthy and can transition into other sectors of the economy if necessary. This is particularly concerning in South Africa, given the increasing levels of violence associated with the lack of economic opportunity. It is well known that unequal societies tend to be more violent, have higher incarceration rates (proportions of the relevant populations being imprisoned) and lower levels of life expectancy than their more equal counterparts (Milanovic, 2016). Growing inequality, fomented by technological advances that displace all but the most skilled labour, poses stability risks across the globe. With the advance of new technologies, the risk is that those who failed to benefit from earlier industrial revolutions will be excluded from the mainstream economy altogether (Altenburg and Rodrik, 2017).

The risks of the Fourth Industrial Revolution, especially of increasing joblessness, are acknowledged by Schwab. He writes that the benefits of connectivity may accrue to the already privileged, and those on the periphery may become even more marginalised. As he puts it, '[t]he changes are so profound that, from the perspective of human history, there has never been a time of greater promise or potential peril' (Schwab, 2016: 2). These changes are not only affecting less-developed countries. The 2016 World Development Report – *Digital Dividends* – by the World Bank estimated that 57 per cent of jobs in the OECD alone could be automated over the next 20 years (World Bank Group, 2016).

---

2   In brief, the middle-income trap involves producing low-value products in an economy largely characterised by a small number of large firms whose taxes fund a large social welfare net.

In a recent working paper, Acemoglu and Restrepo (2017) estimate the equilibrium impact of industrial robots on local labour markets in the United States. Robots are defined as automatically controlled, reprogrammable and multipurpose machines. Their quantitative estimates show that an increase in the stock of robots – one new robot per thousand workers from 1993 to 2007 – reduces the employment to population ratio in a commuting zone with average US exposure to robots by 0.37 percentage points. In turn, average wages fall by 0.73 per cent relative to a commuting zone with no exposure to robots (Acemoglu and Restrepo, 2017: 4). This implies that each additional robot in a commuting zone reduces employment by 6.2 workers. This may also capture the labour market effects of other technological changes triggered by the adoption of robots such as digitised monitoring. However, there may also be a positive latent effect on wages and employment through productivity gains as other industries such as finance, the public sector and non-robotised manufacturing increase demand for labour (Acemoglu and Restrepo, 2017). The negative initial equilibrium impact on the labour market may therefore be mitigated by alternative employment opportunities created in the mechanisation process.

The broader context for this discussion is alluded to in the introduction of this chapter. If global economic development has been achieved through overexploitation of natural resources and human society is fast approaching various ecological tipping points, then Schwab (2016: 1) argues that new ways of using technology can help to change human behaviour and systems of production and consumption. Concurrently, with new technology comes the potential to support the regeneration and preservation of natural environments, rather than the perpetuation of hidden costs typically offloaded on the poorest communities. Reducing externalities – the difference between private returns and social costs – can also boost economic growth. As shown below, investments in green technology were, until recently, only attractive when heavily subsidised by governments, but advances in renewable energy, fuel efficiency and energy storage not only make investments profitable (boosting growth) but also contribute to mitigating climate change.

Critics argue that the major problem with the idea of low-carbon growth is that it remains embedded within a neoliberal paradigm and does not address the fundamentally flawed assumptions of contemporary economic growth models. Kate Raworth, for instance, argues for an entirely new paradigm built on the objective of meeting the needs of humanity within the means of the planet. Growth in gross domestic product (GDP) – the final value of goods and services produced in an economy in any given year – may be part of this, but Raworth insists that economists should be more agnostic about the effects of growth, as even early economists like Simon Kuznets warned that the welfare of a nation can scarcely be inferred from a measure of national income (Raworth, 2017). Raworth redraws the economy, embedding it in the earth's ecological systems and in society. She shows that its sustainability depends on the flow of materials and energy. This reflects work done in the 1970s by Georgescu-Roegen, who criticised the Solow/Stiglitz model as tantamount to believing that one could produce a cake without its raw ingredients (Georgescu-Roegen, 1975). This led to the treatment of natural capital as a free good, which incentivised economic production that eroded and polluted the planet, the very source of the resources required for production (Daly, 1997).

The inner ring of Raworth's doughnut represents a sufficiency of the resources required for a decent quality of life. The outer ring consists of the earth's environmental limits, beyond which humanity is in danger of destroying the very sources of life. The area between the two rings is the 'ecologically safe and socially just space' in which we should strive to live (Raworth, 2017: 39). This concurs with Altenburg and Rodrik's (2017: 7) notion of 'decoupling' human wellbeing and economic progress from non-renewable resource consumption and emissions. Fioramonti makes a similar case in chapter 12 of this volume and calls for a new way of thinking about value and a 'wellbeing economy' rather than a narrowly focused GDP growth economy.

Whether or not this desired shift in economic paradigm occurs, renewable energy is a particularly important component of moving the world away from the worst-case scenarios of technological unemployment and ecological tipping points detailed above. It is also crucial for South Africa, in which electricity access gains made since

1994 are at risk of being undermined by rising electricity tariffs, and the interrelated shutting down of industries previously flourishing as a function of inexpensive electricity.

Across the developing world, a significant obstacle to economic development and improvements in human wellbeing is the lack of access to electricity. Indoor air pollution, still a major cause of mortality, is predominantly a function of limited access to reliable and affordable electricity. According to Lomborg (2015: 23) one of every 13 deaths globally, or 4.3 million people annually, is due to indoor air pollution. Lomborg, however, argues against prioritising renewable energy because he sees it as an expensive way to cut a relatively small amount of carbon dioxide, giving power to fewer people with less reliability. He cites one study showing that US$10 billion can lift 20 million people out of darkness with renewables but 90 million people out of darkness with gas (Lomborg, 2015: 11). Increased access to gas-powered stoves would also significantly reduce indoor air pollution largely produced by burning charcoal and other biomass.

Notwithstanding Lomborg's concerns – shared, incidentally, by the Nobel laureates who contributed to *The Smartest Targets for the World: 2016–2030* (2015) – Altenburg and Rodrik argue that enormous efficiency leaps are now technologically feasible with the shift to renewable energies, provided there is an entirely new framework of incentivisation and pricing. In greater detail:

> *to accelerate the required technological and business model innovations, economic incentives need to be set very differently. Above all, environmental costs need to be much better reflected in prices, subsidies for fossil fuels and other unsustainable goods and practices need to be phased out and regulations must be tightened* (Altenburg and Rodrik, 2017: 7).

With regard to fossil fuel subsidies, Lomborg agrees:

> *With generous state subsidies, gasoline is sometimes sold for a few cents per gallon, mostly to the benefit of middle- and high-income groups with cars. Reducing subsidies would stop wasting*

> *resources, send the right price signals, and reduce the strain on government budgets, while also reducing CO2 emissions* (Lomborg, 2015: 10).

To achieve the levels of de-carbonisation necessary to prevent catastrophic climate change, we require deep structural change to the transport and energy sectors that have historically shaped development. Systemic changes would include a shift away from fossil fuels. As power generation moves to a low-carbon path, other sectors need to be electrified, including road traffic. Some of these changes are occurring with exponential speed. Electricity from hydro, geothermal and some biomasses can now compete with fossil fuel-based electricity. In certain favourable locations, wind and solar power can also compete with electricity from non-renewable sources. It is also expected that there will be further reductions in the costs of these new technologies. Fifteen years ago, renewable energy had a negligible role in electricity generation. According to the 2017 report on the progress of renewables, 'the world now adds more renewable power capacity annually than it adds (net) capacity from all fossil fuels combined' (REN21, 2017: 33).

Firms that invest in unsustainable energy technologies are at risk of having to write off major investments. Citing an estimate from the Economist Intelligence Unit, Altenburg and Rodrik (2017: 7) highlight that the value of the global stock of manageable assets at risk – between now and the end of the century – due to climate change, ranges from US$4.2 trillion to US$43 trillion. If the globe is to have even a 50 per cent chance of keeping peak global warming below the 2 degrees Celsius mark, then 'the global economy needs to reach net zero emissions before we have emitted cumulative carbon emissions of about 1,000 GtC' (Altenburg and Rodrik, 2017: 7). If catastrophic climate change is to be avoided, 'one third of oil reserves, half of gas reserves and more than 80 per cent of known coal reserves' need to be kept in the ground (Altenburg and Rodrik, 2017: 6). Given this, some institutional investors are beginning to withdraw from carbon assets. Market mechanisms are beginning to accelerate a structural change towards a low-carbon economy.

In light of these market shifts, Altenburg and Rodrik (2017: 9) argue that developing countries are actually in an advantageous position in so far as most of their energy and urban infrastructure has yet to be built and they can avoid costly misdirected investments in unsustainable infrastructure. Moreover, green technologies tend to carry complementary benefits. Investments in solar micro-grids in rural and remote areas, for instance, would provide clean energy at the same time as reducing indoor air pollution. Electrification in remote areas is more likely if local sources of energy can be generated and transmitted less expensively than traditional sources that rely on bulk distribution infrastructure. This vision is no longer a pipe dream. The latest 'Report on Renewables' demonstrates that renewable energy technologies are becoming increasingly cost-efficient in ever more locations, with developing and emerging economies accounting for about half of global renewable energy investments, and the market for renewables-based mini-grids is now booming (REN21, 2017).

In South Africa, as will be detailed below, scientific research shows that the entire country's electricity needs can be met with strategically positioned wind and solar plants, supplemented by the occasional injection of open-cycle gas turbines (Knorr et al., 2016). The Renewable Energy Independent Power Producer Programme (REIPPP) has procured solar and wind power at lower per kilowatt hour costs than new coal-fired power, and these costs continue to decline significantly over time (Eberhard and Naude, undated).

Altenburg and Rodrik (2017) argue that mining and power supply industries are going to be most heavily affected by structural shifts towards green energy and transport systems. This raises questions around what the future of the South African mining industry might look like, which is the focus of the remainder of the chapter.

A VISION FOR THE SOUTH AFRICAN MINING INDUSTRY

In an ideal scenario, within the next decade, the South African economy will have moved on to the low carbon trajectory envisaged by the National Development Plan (NDP):

> *By 2030 ... South Africa has reduced its dependency on carbon,*
> *natural resources and energy, while balancing this transition with*
> *its objectives of increasing employment and reducing inequality*
> *... The state has significantly strengthened its capacity to manage*
> *the ongoing internalisation of environmental costs, and to*
> *respond to the increasingly severe impacts of climate change*
> (National Planning Commission, 2012: 179).

Mining and industrial policy will have been greened to cohere together and to drive production patterns that create opportunities in other technologically driven industries. As Altenburg and Rodrik (2017: 22) put it: 'In the long term, there is no trade-off between socio-economic and environmental objectives: there is no human development on an uninhabitable planet'. But in the short term, as the transition mentioned in the NDP suggests, there are trade-offs to be managed. Pricing natural capital more appropriately[3] will place a burden on producers that have externalised these costs in the past (World Bank Group, 2017). However, the benefits of moving to greener economies through more environmentally conscious industrial policies also accrue in the short term through greater efficiency, the development of new markets and the transformation of potentially stranded assets into assets that serve ecological renewal either directly or indirectly. For example, Enel, one of Europe's largest electricity companies, is in the process of decommissioning 23 of its power stations. One will be transformed into an industrial and logistics plant, and Enel hopes that others will be turned into galleries and museums (*The Economist*, 2015). Fossil fuel companies are more resistant to change, but companies such as Norway's Statoil are making the necessary transition adjustments towards investing in renewables (Schaps, 2017). The company has won a licence to develop a wind farm off the coast of New York, and is retraining some of its oil and gas staff to work in the new wind division.

As with mining globally, the South African mining industry will be

---

3   One attempt to do this is the Wealth Accounting and Valuation of Ecosystem Services partnership led by the World Bank to ensure that natural resources are mainstreamed into development planning and national economic accounts. This process is not without its difficulties.

critical to the Fourth Industrial Revolution, but through the supply of different materials via different extraction methods. As alluded to above, meeting new demand for Fourth Industrial Revolution products would entail a move away from fossil fuel mining towards greater, exploratory investment in the minerals and metals that supply technological revolutions in transport and energy systems. A recent World Bank report sums it up well:

> *The shift to low carbon energy will produce global opportunities with respect to a number of minerals ... Africa, with its reserves in platinum, manganese, bauxite, and chromium, should also serve as a burgeoning market for these resources* (World Bank Group, 2017: xiii).

At least 16 different minerals or metals are used to produce solar panels, for instance. South Africa currently mines at least five of these: iron ore, lead, phosphate rock, silica and titanium dioxide. The metal content in wind turbines, either geared or direct drive, is also substantial, including aluminium, chromium, copper, iron, lead, manganese, neodymium, nickel, steel and zinc. South Africa is endowed with substantial chrome, manganese and iron ore deposits. In fact, as pointed out in other chapters of this volume, the country ranks first in the world for aluminosilicates (ore), platinum group metals (PGMs), manganese and chromium reserves (Chamber of Mines of South Africa, 2018). Similar metals are required for energy storage technologies, which also play a key role in the advancement of electric vehicle uptake. Electric vehicles themselves depend on a steady supply of metals such as copper and cobalt. Electric vehicles are a crucial part of a low carbon transition for both the generation and the use sides of the equation, as battery storage units (most likely dominated by lithium-ion) are required in large volumes for these vehicles to be used proficiently and at scale. If the world is on schedule to meet its climate change mitigation goals, the International Energy Agency foresees 140 million electric vehicles being in operation by 2030 (World Bank Group, 2017).

Technologies required to populate the shift towards clean energy are significantly more material-intensive than their fossil-fuel-based counterparts. The World Bank (2017: 58) predicts that metals demand will likely double for wind and solar technologies, and an even higher figure for battery storage technologies. In this volume, Joel Netshitenzhe (chapter 2) expands on this insight and highlights its relevance for the South African context.

In light of these new demand opportunities, in the ideal future scenario, tariffs and incentives would be carefully constructed to take advantage of manufacturing opportunities within new value chains. Local beneficiation would be encouraged where the linkages make sense and the resources required to realise the vision are within reach. On the supply side, the digital revolution and industrial advances would enable a range of productivity improvements in the extraction of all minerals and metals produced in South Africa.

The use of unmanned aerial vehicles (UAVs) is likely to transform geological exploration. Identifying potential ore-bearing ground will become less costly and less labour intensive. Sensors on UAVs can detect geothermal activity (Fernández-Lozano and Gutierrez-Alonso, 2016), which will help exploration firms to drill and sample only in areas where there are indications of resources. This holds promise for significant future cost savings and a minimal environmental footprint in the exploration phase of mining projects.

Robotics and automation carry significant promise in the production phase. Robots can work in hazardous environments instead of people, improving mine safety considerably. Conceivably, for instance, rock drill operators would no longer be required to extract platinum from narrow, awkwardly angled platinum stopes. Those deposits could now be accessed through relatively minor invasion, akin to laser surgery rather than open heart surgery. One vision for zero entry production areas (ZEPA) in mines is that 'all work processes should be remotely operated or automated, while special mine robots should be developed for the preventive maintenance of equipment and safe retrieval operations' (Nikolakopoulos et al., 2015: 66). Work is already being done to implement this vision where possible, and to invest in the identified areas of future research: 'The ZEPA vision is a realistic roadmap of research and development activities' (Nikolakopoulos et al., 2015: 68).

Mining involves moving and processing vast quantities of rock to extract a small percentage of valuable minerals or metal. Aside from being environmentally destructive, removing superfluous ore body from the desired metal is also expensive; it represents roughly 80 per cent of the costs of production. The industry requires economies of scale to warrant the initial, large capital investment. The greater the scale required, the larger the trucks needed and the larger the scar on the landscape. Conventional wisdom for modern mines is that plant recovery rates need to reach near 90 per cent to cover the increasingly higher costs. With declining ore grades and the need to mine deeper ore bodies, new methods, which reduce rock movement, mine more selectively and emphasise quality over quantity, are required. The Robotics 2020 Multi-Annual Roadmap for Robotics in Europe notes that the use of robotics and remote-guided vehicles is already increasing. This prospect is no longer futuristic, and the more lessons that are learned by current pilot projects, the more momentum is likely to be gained: 'There is a significant opportunity to utilise robots for extraction in order to reach more inaccessible mineral resources' ('SPARC: The Partnership for Robotics', 2015: 93). The Kankberg gold mine in Sweden is an example of what is currently possible. In collaboration between Boliden (the mining company) and Ericsson, ABB and Volvo, 'the mine has been completely automated … and plans to eventually operate with no personnel in the mine itself' (Lempriere, 2017).

One entirely new method of mining being explored is called 'in place' mining, motivated by the concept of minimal rock movement and processing at the point of extraction, rather than miles away at a large plant (*International Mining*, 2017). It promises to deliver a smaller surface footprint, reduced tailings generation, high levels of automation and low-capital-intensive mines. Mining projects could attract financing more easily and deliver returns more quickly than with the conventional model. There are three related methods that are promising for South African mines, which are increasingly marginal because of the safety risks at operating depth, along with declining quality ore bodies.

First, in line recovery (ILR) incorporates selective mining and ore upgrading to deliver only concentrated material to the processing facility. A range of upstream technologies enable ILR. South African policymakers have long spoken of the need to develop upstream and downstream linkages (National Planning Commission, 2012 and Jourdan, 2013). If South Africa is to avoid the worst job displacement effects of automation, policy-makers will need to focus on creating new opportunities in the upstream sector. This sector typically involves the production of equipment used in mining operations such as truck fleets, excavators, crushers and conveyor belts, along with chemical leaching solutions.

Emerging digital technologies in automated rock-face mapping, material characterisation and fragmentation analysis and rock preconditioning can be built into the equipment and pre-programmed for specific mines in the South African context. The machines that cut hard rock are now able to identify and exploit natural rock cleavages to make cutting more efficient. 'If rock can be cut rather than blasted, mining can become continuous, leading to process and efficiency improvements' (Vogt, 2016: 1011). Vogt also advises that despite the challenges of adopting new cutting technologies across different types of ore bodies, almost all major suppliers of mining equipment now include some level of automation, which improves machine accuracy and allows for continued operation while personnel are absent from the rock face (Vogt, 2016: 1023).

Crushing technology is also becoming nimbler, making obsolete the big crushers typically required at a processing plant. Employing upstream technology at the rock face, to selectively mine and pre-concentrate material for subsequent metal extraction, avoids the many negative environmental impacts usually associated with mining. In the case of copper mining, crushing alone is one of the largest components of a mine's energy consumption and greenhouse gas emissions. These can be drastically reduced by in-pit mobile crushing, which 'eliminates the need for trucks by having the shovel feed the run-of-mine ore directly to a continuous and dedicated belt conveyor handling system' (Norgate and Haque, 2010: 272).

Connecting a number of technologies together means that mining operations become less energy intensive, rendering the option of being solar or wind powered both more financially attractive and more operationally viable. For instance, the Kankberg mine in Sweden, mentioned earlier, uses a 5G wifi network, installed by Ericsson throughout its operations. This enabled the complete automation of the mine and in turn enabled ABB to install its SmartVentilation system. Connecting everything to the network allowed process optimisation that saved 54 per cent of the mine's energy consumption. This represents a saving of 18 MW a year on a mine that previously consumed 34 MW a year (Lempriere, 2017).

Second, in mine recovery can minimise rock movement and ore fracture (that normally occurs due to dynamite blasting and drilling). It produces a conditioned mineral block, which can be leached with chemical or biological tools to extract the valuable metal. One example of this is the Australian federal Commonwealth Scientific and Industrial Research Organisation Remote Ore Extraction System, which uses remote automated drilling and blasting (Norgate and Haque, 2010).

Third, in situ recovery (ISR) is well known to the industry and extensively used in uranium mining, for instance. In situ leaching is one form and is basically a chemical method of recovering useful minerals and metals directly from underground ore bodies (Haque and Norgate, 2014). New technologies such as glycine leaching can improve the environmental performance of ISR, as it is more selective in leaching the desired material and rejecting the rest (Oraby and Eksteen, 2014).

If the disruptive employment effects of these new technologies can be well managed through the creation of effective and competitive vertical and horizontal linkages, the industry could transform structurally and drive changes in the rest of the economy through spillover effects (from other industries following mining's lead of adopting greener technologies). It could become a driver rather than a laggard of South Africa's move towards a low-carbon growth trajectory, realising the potential of the Fourth Industrial Revolution to enable systems of production and consumption that renew rather than destroy the earth's ecological systems (Harvey, 2017).

Such a vision now needs to be located within current realities to generate a clearer picture of how exactly the mining industry in South Africa could respond to – and take advantage – of the Fourth Industrial Revolution.

CURRENT REALITIES – RISKS AND OPPORTUNITIES

Mining remains pivotal to the South African economy despite its declining direct GDP share. It is the single largest private sector employer in South Africa, supporting about 460,000 jobs directly despite losing 30,000 jobs since 2014. This represents roughly 6 per cent of total private non-agricultural employment in the country. In 2017 mining exported R307 billion worth of material – 27 per cent of the country's total exports (Chamber of Mines of South Africa, 2018). The irony, however, is that the value of these exports diminishes if the currency weakens against the US dollar. Currency depreciation makes imports more expensive, and imported products remain important inputs for all industries, including mining, in South Africa. The adoption of new technologies would not only enable cleaner mining, but may also create new products or opportunities in the upstream manufacturing space. Higher value addition in the manufacturing sector would generate potential employment and reduce the macroeconomic instability that tends to result from currency value fluctuations.

The Chamber of Mines (2015: 3) estimates that each worker employed in mining has an average of 10 dependents, i.e. an average 10:1 dependency ratio. That ratio increases in inverse relation to the skill level. In other words, lower-skilled workers tend to support more dependents than higher-skilled workers. Increased mechanisation directly threatens lower-skilled jobs in the first instance, and hence portends extensive negative second-order socio-economic effects. One obvious negative implication is that it will make more people dependent on social grants, with less disposable income circulating into the economy. One of the corollary benefits, however, is that the industry can move towards providing higher-value jobs that do not have a negative impact on people's health and general wellbeing.

Migrant labour, with its attendant social externalities, persists largely because of the demand for low-skilled labour, which is drawn from traditional labour sending areas in the former homelands. The impact on the social fabric of this system has been devastating (Wilson, 2001; Capps, 2012 and Harvey, 2014). In light of this, the idea that robotics and automation in mining operations may not eliminate jobs, but rather shift their nature, is promising (Bleischwitz, 2017).

The idea is related to the concept of occupational unbundling. The demand for rock drill operators may become obsolete (depending on the nature of the stope), but those jobs may be replaced by drone technicians, for instance. However, shifting the nature of jobs in the South African context of high inequality generates the risk that the new jobs created by technological disruption will accrue to the already privileged. The reality is that while mine safety will improve, there will be fewer low-skilled jobs available. Unskilled miners will have limited immediate employment alternatives because of the unique product space that mining occupies (Hausmann and Klinger, 2008).

According to the Chamber of Mines,[4] PGMs accounted for the highest value of exports in the precious metals and minerals category in 2017, followed by gold and diamonds. In the Chamber's base minerals category, coal dominated the exports and total sales value (in part attributable to domestic Electricity Supply Commission (Eskom)[5] supply contracts worth roughly R61.4 billion). Iron ore is next, followed by manganese and chrome (Chamber of Mines of South Africa, 2018).

Gold employment continues to decline – it now employs 112,200 people whereas in 2007 it employed about 165,000 (a 47 per cent drop). Coal employment, to the contrary, has risen on average by 33 per cent, though it is now well below its 2011 peak of 90,000 employees. PGMs employed 175,770 people in 2017 – marginally down from a 2008 peak of nearly 200,000. Chrome, manganese and iron ore are relatively small employers at the moment, though employment in these sectors is growing (Chamber of Mines of South Africa, 2018).

The dynamics of demand – and resultant prices and labour market impacts – for these minerals and metals are changing in response to climate change imperatives and the Fourth Industrial Revolution. This will have an impact on export and sales values, along with employment figures, especially as the demand for coal declines but the demand for other minerals and metals stabilises or starts to increase. The next sub-section examines these dynamics.

---

4   Rebranded in 2018 as the Minerals Council of South Africa.
5   Eskom is the state-owned power utility.

RISKS TO THE SOUTH AFRICAN COAL-
MINING SECTOR

In the long run, the most at-risk mining sector in South Africa is coal, which currently provides approximately 70 per cent of the country's primary energy and more than 90 per cent of its electricity (Eberhard et al., 2014: 1).

The Renewables 2017 Global Status Report notes that energy-related carbon dioxide emissions have been declining for the last three years consecutively. This is due to a combination of declining coal use, energy efficiency improvements and increased use of renewables (REN21, 2017: 19). Global coal production is also declining. However, fossil fuel subsidies remain a barrier to the advance of renewable energy markets, in addition to the relatively low global prices for oil and natural gas (notwithstanding an increasingly volatile Middle East). Nonetheless, subsidy reforms had been instituted in more than 50 countries by the end of 2016 (REN21, 2017: 29), indicating a long-term trend away from fossil fuels, even as consumption may increase in the short term as prices decline.

Movements towards large-scale fossil fuel divestment create the risk of stranded assets. However, the ecological and human welfare risks associated with mining more coal constitute a strong argument in favour of planned decommissioning and phasing out (Lockwood, 2012). Such policy moves are, however, likely to be resisted, especially when export revenues and a large number of jobs are at risk.

For South Africa, over-reliance on coal for electricity is a risk that the government needs to mitigate through at least two channels. First, the government needs to unbundle Eskom into at least three separate entities, as per the Independent Systems and Market Operator (ISMO) Bill that was inexplicably withdrawn from the National Assembly in 2012. Eskom is heavily indebted, largely because of corruption, poor planning and weak contract management. The reasons for this state of affairs will be explained in a later section of this chapter. The ISMO Bill called for an unbundling into generation, transmission and distribution units. Each section would be independent so as to break Eskom's current monopoly over all three. Having entrenched access to monopoly rents is part of what generated the incentive for poor governance in the first instance. It therefore constitutes a structural barrier to necessary change.

Second, the government has to increase the energy proportion sourced from renewables in the integrated resources plan (IRP) and phase out coal. The plan was updated in 2016 and distributed for public comment, the window for which closed in March 2017. The need to update the original 2010 plan was spurred by rapidly changing cost assumptions – reductions for renewables and likely increases for coal and nuclear. Neither of these changes will be easy to accomplish, for reasons explained below.

CHALLENGES TO TRANSITIONING AWAY FROM FOSSIL FUELS

Economist Daron Acemoglu and his co-authors (2016) explain the difficulty of transitioning from dirty (coal and other fossil fuels) to clean (renewable energy) technology. Dirty technologies have incumbency advantages over clean technologies as the latter must overcome high upfront research, learning and production costs. The initial gap between the two technologies tends to discourage research effort towards clean technologies (Acemoglu et al., 2016: 100). Theoretically, a combination of carbon taxes and research subsidies encourages production and innovation in clean technologies. Contrary to popular wisdom, these authors find that the optimal policy path heavily relies on research subsidies towards renewables and not primarily on carbon taxes. There are significant welfare costs associated with relying solely on carbon taxes or delaying interventions (Acemoglu et al., 2016: 101).

Delayed intervention and a lack of funding for renewables research and production is usually a function of competing budget priorities, and political will. In South Africa, two arguments against renewable energy have typically been offered, and a third recently entered the debate. First, detractors contend that coal-fired power is still less expensive than renewables. Second, they argue that new coal and nuclear power stations are required to provide stable, baseload energy, especially for energy-intensive industries like mining and smelting that require energy 24 hours a day. Third, in March 2018, unions stated that they would prevent the increase of renewables' electricity share on the grounds that it puts coal jobs directly at risk (Mahlakoana, 2018). Do these arguments bear up under scrutiny?

On the first count, Anton Eberhard and others (2014) have shown that the argument is weak. South Africa's REIPPP, chosen instead of feed-in tariffs (that were attempted but failed in 2009), has delivered remarkable results. The average solar photovoltaic (PV) tariffs decreased by 68 per cent from 2011 to 2014. During the same time, wind power tariffs dropped by 42 per cent (Eberhard et al., 2014: 1).

By 2011, a competitive bidding REIPPP process had been announced. In the first bid round, the average tariff for wind power was 114c/kWh and 276c/kWh for solar PV. By the second round, these prices had declined to 90 and 165c/kWh respectively. In the third bidding window, prices came in at 74 and 99c/kWh respectively (Eberhard et al., 2014: 14). Chris Yelland (2016) uses slightly different figures to Eberhard and Naudé (undated), but shows the fourth bid-window prices as 69 and 87c/kWh respectively. By way of contrast, coal-fired power from Medupi and Kusile, the two most recent large-scale plant constructions, is likely to cost 105c/kWh and 119c/kWh respectively.[6] The fourth round of REIPPP deals were eventually signed in April 2018, shortly after Jeff Radebe became the Minister of Energy (following Cyril Ramaphosa becoming president of South Africa).

Second, on the question of baseload availability from variable sources, the Renewable Energy 21 Report (REN21, 2017: 27) notes

---

6   The above figures represent the levelised cost of electricity (LCOE), which is the 'net present-day monetary cost per present day kWh unit of electricity delivered, which when adjusted for inflation each year over the lifetime of the plant, will recover its full costs, including the initial investment, cost of capital (including dividends and interest), fuel and other fixed and variable operating and maintenance costs' (Yelland, 2016). Yelland argues that 'one cannot in all cases meaningfully compare the LCOE from a variable source electricity (such as wind or solar PV, without storage or backup), with that of baseload generation capacity (such as nuclear or base-supply coal)' (Yelland, 2016). But it is nonetheless a useful planning tool. The LCOE is used to submit a bid price (indexed to inflation), and the independent producer has to bear the costs if they are not able to deliver at the procured price. This shifts the risk away from the state and onto private producers.

that the growth in variable renewable energy is changing the way that traditional, established power systems are planned, designed and operated; these now require greater flexibility. Traditional baseload generators such as coal and nuclear power plants are beginning to lose their economic advantage mentioned earlier. The adoption of high shares of renewable energy in countries such as Denmark and Germany have shown the potential to shift away from the traditional baseload paradigm. For the Global South, emerging economies such as Argentina, Chile, China and India are becoming attractive markets for investment in renewables (REN21, 2017: 30). China, for instance, generated a total of 258 GW from renewables in 2016, the vast majority being from wind and solar PV. The achievement of improved performance and greater cost effectiveness of technologies that allow the integration of variable energy supply into transmission grids – to ensure stable supply – will further undermine the arguments for coal and nuclear to provide baseload power.

The most important point for South Africa is that a report by the Council for Scientific and Industrial Research (CSIR) found, through simulation modelling, that it would be possible to more than satisfy the country's electricity needs from wind and solar PV energy, with occasional supplementation from open-cycle gas turbines (Knorr et al., 2016: 2). The CSIR study established the potential to generate electricity across the whole country from wind and solar PV based on data sets for five years. Energy generated by the sun through solar PV shows little seasonal fluctuation, and wind speeds and generation closely follow electricity demand over the course of a year. Wind power complements solar effectively, as it is generally at its weakest when solar provision is strongest (around noon) and strongest at night. This enables the integration necessary to stabilise variable sources for baseload power and thereby move away from coal and nuclear energy (Knorr et al., 2016: 5). The study concludes that solar and wind energy are very low-cost bulk energy providers in South Africa and the potential to move away from baseload coal and nuclear is cost effective (Knorr et al., 2016: 60).

On the third count, labour unions have created a false dichotomy between renewables and coal in terms of labour absorption. Granting the go-ahead (in April 2018) to the fourth (2016) round of successful REIPPP bidders is not going to draw jobs away from the coal sector. Jobs in the coal sector are likely to be lost in any event because coal production costs are rising and there are very few countries remaining with coal supply deficits. India is one. The stranded asset risk, combined with an impending carbon tax, mean that coal is increasingly uncompetitive, irrespective of the rise of renewables. Moreover, increasing renewable energy capacity is potentially more labour absorptive than the coal industry.

The CSIR study, and the studies by Eberhard (2014; 2017) and Yelland (2016), provide strong evidence in support of the vision articulated earlier to move South Africa away from mining and burning coal towards a low-carbon growth path. Remaining minerals and metals in South Africa, such as chrome and manganese, carry significant potential. It is almost invariably wiser to transition to a low-carbon trajectory as early as possible in spite of various obstacles, and build other industries simultaneously that can provide more sustainable jobs in the long run.

Platinum group metals

As covered at length by Ritchken in chapter 3 of this volume, South Africa is home to over 80 per cent of the world's known PGM reserves, with over 63,000 tons (World Bank Group, 2017: 49). Platinum is an as yet irreplaceable catalyst for hydrogen fuel cell technology, capable of powering automobiles (Makhuvela et al., 2013: 5). Fuel cells have now demonstrated their versatility as a reliable and clean power generation source in a number of sectors. Cells produce energy electrochemically, essentially providing a low-to-zero emission option. In addition to being small and versatile, cells are scalable, with utility scale units providing multi-megawatt power. Fuel cells utilise hydrogen, which can be extracted from virtually any hydrogen-containing source, including fossil fuel and renewable sources. An interesting development is power-to-gas as a method of storing excess power generated by wind or solar technologies. Hydrogen is generated by passing electric current into water to split it into oxygen and hydrogen gas. The latter can then be stored underground for later use or injected into a natural gas pipeline to be transported downstream.

Whether platinum demand increases will be a function of which energy storage technology dominates the transport revolution. 'If hydrogen fuel cell vehicles predominate, then demand for platinum group metals could increase rapidly' (World Bank Group, 2017: 20). Or, as the World Economic Forum put it in 2015: 'As the fuel cell industry continues to grow, potential new markets and partnership opportunities will arise for allied energy industries that share the industry's vision of a clean energy future' (Markowitz, 2015).

Gold

Gold is unlikely to play a significant role in new supply chains associated with the Fourth Industrial Revolution, but it will greatly benefit from technological improvements to mining and metallurgical processes. These will generate cost savings that are likely to attract greater investments into South Africa's gold sector. Gold remains an attractive store of value in the context of global economic uncertainty. This demand, combined with technologically driven cost savings, will likely sustain South Africa's gold-mining industry in the medium to long term. In the short term, gold-mining profitability is likely to continue to decline.

Chromium and manganese

Chromium is a key component of both geared and direct-drive wind turbines. According to the World Bank's best-case scenario for the world meeting its climate change mitigation targets, the median demand for chromium (for supplying wind technologies) is likely to increase by nearly 250 per cent between now and 2050. In a more realistic scenario, it would increase by slightly more than 150 per cent (World Bank Group, 2017: 12). This is significant for South Africa, as the country currently produces half of the world's chromium – 15,000 metric tonnes in 2015. Kazakhstan has slightly more reserves than South Africa, but its production is still less than a fifth of South Africa's (World Bank Group, 2017: 34). Chromium is also used in carbon capture and storage installations, along with combined cycle gas turbines, which portends even greater future demand. The CSIR (2016) report recommends gas turbines as the optimal form of electricity to supplement solar PV and wind when required. Moreover, because of chromium's ability to resist high temperatures and corrosive environments, it will also be used in plant construction for cleaner coal-fired electricity generation.

The picture is almost exactly the same for manganese in terms of a likely increase in demand. It is also a key component of geared and direct-drive wind turbines. Unlike chromium, it is also used in lithium-ion batteries, which will likely play a significant role in the future energy storage technology market.[7] Manganese requirements for carbon capture and storage installations are 3.761 kg/MW – about 10 times higher than that of chromium. In high-intensity battery-powered electric vehicles, manganese requirements are about 91.5 kg per vehicle (World Bank Group, 2017: 70). The US Geological Survey estimates that South Africa alone contains 75 per cent of manganese resources worldwide with an estimated 200,000 metric tonnes (World Bank Group, 2017: 45). It produced 6,200 metric tonnes in 2015, a third of the global total. Again, with no satisfactory substitute in its major technological applications, the opportunities for South Africa are significant. Taking advantage of downstream opportunities for developing electric car and lithium-ion battery components, along with wind turbine components, would be highly strategic.

Copper

Copper reserves in South Africa are relatively small, but the metal is also likely to see substantial demand increases, given its use in geared and direct-drive wind turbines, as well as in copper indium gallium selenide (CIGS) 'thin-film' solar technology (World Bank Group, 2017: 9). This technology is receiving global attention as these thin-

---

7   See the important discussion on pages 20–21 of the World Bank report (2017), which argues that lithium-ion batteries will dominate the market if the vehicle fleet of the future is fully electric, whereas if hydrogen fuel cell vehicles predominate, demand for PGMs would increase rapidly. The report does not, however, seem to consider the possibility that this may be an unnecessary dichotomy – there is a possibility (likely in my view) of fuel cell and fully electric vehicles co-existing.

film cells have achieved efficiencies 22.8 per cent higher than crystalline silicon wafer-based solar cells. They are also less costly to manufacture (Ramanujam and Singh, 2017: 1306). While South Africa does not have large reserves, the opportunities for regional co-operation with Zambia are significant. Supply agreements could, for instance, be arranged with Zambian mines to supply a consistent amount of copper ore to South Africa for use in manufacturing various components in the value chains of products such as CIGS solar cells, wind turbines, batteries and electric vehicles. Together with a steady supply of chromium and manganese, similar opportunities may be available.

FULFILLING THE VISION: WHAT WILL IT TAKE?

Future prospects for South African mining are generally positive but subject to a number of caveats. On the supply side, new technologies offer significant cost savings in mining and metallurgical processes. As elaborated in detail in the chapters of Netshitenzhe and Ritchken (chapters 2 and 3 respectively), there are numerous upstream and downstream opportunities. Downstream, South Africa could produce components of the fuel cell value chain, for instance, or parts to fit solar PV panels or wind turbines. Upstream, manufacturers of mining equipment could develop technologies that both increase production efficiency and spawn other industries. On the demand side, South Africa's obvious comparative advantage lies in PGMs, chromium and manganese. Coal can be phased out in the medium to long term, and chromium and manganese used to retrofit current power plants to ensure that they burn coal more cleanly. Beyond that, value chains associated with wind turbines, electric vehicle and energy storage technologies present myriad opportunities for a number of manufacturing and service sectors.

Greening the industry yields higher benefits to society as a whole than to private interests, and private interests may therefore attempt to block reform. This poses a challenge for the public sector. As Altenburg and Rodrik (2017: 22) underline, 'structural change always requires a proactive public sector.' This involves supporting research and development, subsidising entrepreneurial cost discovery, facilitating information sharing and learning and co-ordinating investment.

In South Africa, then, it remains to identify the potential public sector obstacles to the adoption of green industrial policies that will best serve the mining industry, human welfare and the environment in the long run.

BARRIERS TO THE ADOPTION OF A GREEN INDUSTRIAL POLICY IN SOUTH AFRICA

South Africa faces a significant challenge in driving structural transformation that minimises job losses, at the same time as decoupling economic progress from environmental degradation. These transformations always create winners and losers. Understanding interest groups and their power resources is critical to enabling transformation towards the low-carbon growth path envisaged in the NDP and as outlined in this chapter.

According to Acemoglu and Robinson (2006: 115), new technologies are disruptive and may erode the political advantages enjoyed by incumbent elites. These elites are therefore likely to block the adoption of technologies they perceive as possibly undermining future economic rents derived from political incumbency. In South Africa, elites may include politically connected insiders, along with embedded political and business interests. This does not preclude competition between interests, however, especially in an unstable or rapidly changing political economy environment.

New technologies may enrich competing groups which pose a potential threat to the power held by incumbents. Elites are unlikely to block development when there is a high degree of political competition, or when they are highly entrenched. If competition is limited but power is threatened, they are more likely to block new technologies. When the political stakes are high and external threats are absent, blocking is also more likely. 'These considerations make politically powerful groups fear losing power and oppose economic and political change, even when such change will benefit society as a whole' (Acemoglu and Robinson, 2006: 115). Union attempts to block the REIPPP, for instance, are a good example of this. Similarly, Eskom under its previous board, during the tenure of former president Jacob Zuma, was similarly obstructive of the adoption of new technologies.

This theoretical framework of 'political losers' as potential barriers to economic development is useful for understanding South Africa's challenges to managing the risks and embracing the opportunities associated with the Fourth Industrial Revolution. The framework emphasises what Acemoglu and Robinson (2006) call 'a political replacement effect' in which political elites block beneficial economic and institutional change that threatens to destabilise the existing system and make it more likely that the elites will lose political power and future rents. The emphasis in this theory is on political power, as this is the channel through which incumbent elites acquire and protect their rents. It is therefore not primarily about elites protecting their economic rents, but the power that allows those rents to accrue in the first place.

Eskom is one of the key institutions that, until recently, persistently blocked a domestic energy revolution. It played a central role in the anatomy of 'state capture' (Swilling et al., 2017). This is one particular manifestation of the political replacement effect. Using their positions of authority in the government and Eskom to protect their political influence and political allies, actors like Lynne Brown (the former Minister of Public Enterprises) and Brian Molefe (the former CEO of Eskom) allegedly used the state utility as a means of channelling rents to politically connected insiders. These key agents have been replaced since President Cyril Ramaphosa acquired power in February 2018, generating widespread expectations that Eskom would be transformed, especially when the majority of the board was replaced.

Dovetailing with this well-documented rent-seeking was a move in August 2016 to block the fourth round of the REIPPP bidding window (Eberhard and Godinho, 2017). Eskom cited an oversupply of electricity from its own generation plants and what it considered to be the high tariffs agreed upon with IPPs. The argument appears flimsy in light of the evidence. In the fourth bidding window (BW4), the average tariff for awarded wind projects was 62c/kWh, and an average tariff for solar PV at 79c/kWh. Eberhard and Naudé (undated: 27) reported that the Department of Energy (whose minister was also replaced by Radebe) was delayed in bringing BW4 to a close. Eberhard described Eskom's actions as highly irresponsible and malicious (Van Biljon, 2017).

The use of Eskom as a vehicle for rent-seeking and to block the entrance of new players in an REIPPP, which has seen a total investment from the private sector of R192.6 billion and installed capacity of 6,382 MW, provides a picture of Acemoglu and Robinson's (2006) theoretical predictions. A significant portion of this rent-seeking network has been displaced since President Ramaphosa came to power, though there remains some risk that other rent-seeking behaviour will continue (Friedman, 2018). This change has increased the probability of attaining the vision articulated in the NDP despite numerous challenges including inequality, poverty, an increasing debt to GDP ratio and an arguably unaffordable public sector wage bill.

CONCLUSION

As it exists in 2018, the mining industry faces social and environmental contradictions; its existence holds the promise of jobs and latent wealth creation, but those jobs are often unsafe and impose health costs on workers. The natural environment is also invariably polluted, which compounds the negative social effects that tend to affect mining communities. This chapter shows that the Fourth Industrial Revolution entails significant opportunities for the transformation of not only South Africa's mining industry, but also the broader economy that the industry supports. The confluence of a number of technologies suggests that the world is on the cusp of energy and transport revolutions. But a mining strategy needs to be decoupled from the demand for coal-fired electricity. An unbundling of Eskom, combined with a determined strategy to phase out coal as a primary electricity source, should aid this endeavour.

Given that renewable energy costs are dropping rapidly across the globe as well as in South Africa, according to the best scientific evidence available, the country is optimally positioned to acquire its electricity needs from variable wind and solar power with occasional supplementation from open-cycle gas turbines. This again will help to de-couple energy generation from mining, as coal, for instance, will no longer have to be mined to feed coal-fired power stations. There will never be a complete de-coupling, however, as wind turbines and solar panels require a large amount of minerals and metals, but those are mostly cleaner to mine than coal and are not burned

The mining industry could be restructured in light of this de-coupling and new technological realities to supply the minerals and metals required for generating and transmitting renewable energy, and for other products like electric vehicles, solar and wind power technologies and energy storage products that are integral components of the Fourth Industrial Revolution. It could also provide materials like platinum to feed a fuel cell technology downstream industry, or at least manufacture components that feed into the global fuel cell value chain. On the production and metallurgical processing side, the mining industry could be transformed by the adoption of new technologies. This would also create upstream opportunities in the capital equipment supply chains required for nimbler and less environmentally invasive methods of mining. There is a potential upstream and downstream feedback loop that could be exploited where downstream products such as fuel cells are manufactured locally and provide upstream inputs to mining vehicle fleets.

Finally, the 2018 political equilibrium in South Africa crowded out some portion of rent-seeking incumbents, who were a barrier to the adoption of new technologies. The new set of political players in mining, energy and public enterprises may be more amenable to the adoption of new technologies. They may also be capable of, and committed to, better governance. This bodes well, not only for the project of transitioning to a low-carbon growth path but also for that of rendering justice to the majority of South Africans who have yet to experience the material fruit of South Africa's rich mineral endowment.

REFERENCES

Acemoglu, D., Akcigit, U., Hanley, D. and Kerr, W. (2016). 'Transition to clean technology'. *Journal of Political Economy*, Vol. 124(1), pp. 52–104.
Acemoglu, D. and Restrepo, P. (2017). 'Robots and jobs: Evidence from US labor markets'. National Bureau of Economics Research Working Paper

No. 23285. National Bureau of Economics Research, Massachusetts.

Acemoglu, D. and Robinson, J. A. (2006). 'Economic backwardness in political perspective'. *American Political Science Review*, Vol. 100(1), pp. 115–131.

Altenburg, T. and Rodrik, D. (2017). 'Green industrial policy: Accelerating structural change towards wealthy green economies'. In Altenburg, T. and Ashman, C. (eds). *Green Industrial Policy: Concept, Policies, Country Experiences* pp. 1–19. German Development Institute and Partnership for Action on Green Economy, Geneva.

Bleischwitz, R. (13 November 2017). 'The greening of the miners'. Project Syndicate, London. Available at: https://www.project-syndicate.org/commentary/mining-resource-nexus-low-carbon-economy-by-raimund-bleischwitz-2017-11. Accessed 10 September 2018.

Capps, G. (2012). 'A bourgeois reform with social justice? The contradictions of the minerals development bill and black economic empowerment in the South African platinum mining industry' *Review of African Political Economy*, Vol. 39(132), pp. 315–333.

Chamber of Mines of South Africa. (2015). 'South African mining industry industrial relations environment'. Chamber of Mines of South Africa, Johannesburg.

Chamber of Mines of South Africa. (2018). 'Facts and figures 2017'. Chamber of Mines of South Africa, Johannesburg. Available at: http://www.mineralscouncil.org.za/industry-news/publications/facts-and-figures. Accessed 10 September 2018.

Daly, H.E. (1997). 'Georgescu-Roegen versus Solow/Stiglitz'. *Ecological Economics*, Vol. 22(3), pp. 261–266.

Eberhard, A. and Godinho, C. (12 September 2017). 'How corrupt power captured Eskom and helped pull the plug on growth'. *Business Day*. Available at: https://www.businesslive.co.za/bd/opinion/2017-09-12-how-corrupt-power-captured-eskom-and-helped-pull-the-plug-on-growth/. Accessed 12 September 2017.

Eberhard, A., Kolker, J. and Leigland, J. (2014). 'South Africa's renewable energy IPP Procurement Programme: Success factors and lessons'. Public-Private Infrastructure Advisory Facility, Washington DC.

Eberhard, A. and Naudé, R. (2017). 'The South African renewable energy IPP Procurement Programme'. Cape Town. Available at: https://www.gsb.uct.ac.za/files/EberhardNaude_REIPPPPReview_2017_1_1.pdf. Accessed 1 August 2018.

The Economist. (2 November 2015). 'Anyone want a power station?' Available at: https://www.economist.com/news/business/21678218-italys-largest-power-company-faces-up-stranded-assets-problem-anyone-want-power. Accessed 20 November 2017.

Fernández-Lozano, J. and Gutierrez-Alonso, G. (2016). 'The use of UAVs (unmanned air vehicles) in geology'. Conference Paper. Available at:

https://www.researchgate.net/publication/311495417_The_Use_of_
UAVs_Unmanned_Air_Vehicles_in_geology. Accessed 1 August 2018.

Friedman, S. (15 May 2018). 'How a deal with provincial strongmen is haunting South Africa's ruling party'. *The Conversation*. Available at: https://theconversation.com/how-a-deal-with-provincial-strongmen-is-haunting-south-africas-ruling-party-96666. Accessed 23 May 2018.

Georgescu-Roegen, N. (1975). 'Energy and economic myths'. *Southern Economic Journal*, Vol. 41(3), pp. 347–381.

Haque, N. and Norgate, T. (2014). 'The greenhouse gas footprint of in-situ leaching of uranium, gold and copper in Australia'. *Journal of Cleaner Production*, Vol. 84(1), pp. 382–390.

Harvey, R. (2014). 'Minefields of Marikana: Prospects for forging a new social compact'. Occasional Paper 183. South African Institute of International Affairs, Johannesburg.

Harvey, R. (2017). 'Envisioning a more equitable and sustainable future'. *South African Journal of International Affairs*, Vol. 23(4), pp. 541–544.

Hausmann, R. and Klinger, B. (2008). 'South Africa's export predicament'. *Economics of Transition*, Vol. 16(4), pp. 609–637.

International Mining. (1 June 2017). 'In place mining – A transformational shift in metal extraction'. *International Mining*. Available at: https://im-mining.com/2017/06/01/place-mining-transformational-shift-metal-extraction/. Accessed 29 June 2017.

Jourdan, P. (2013). 'The optimisation of the developmental impact of South Africa's mineral assets for building a democratic developmental state'. *Mineral Economics*, Vol. 26, pp. 107–126.

Knorr, K., Zimmermann, B., Bofinger, S., Gerlach, A.-K. and Bischof-Niemz, T. (2016). 'Wind and solar PV resource aggregation study for South Africa'. Centre for Scientific and Industrial Research, Pretoria.

Lee, I. and Lee, K. (2015). 'The internet of things (IoT): Applications, investments, and challenges for enterprises'. *Business Horizons*, Vol. 58(4), pp. 431–440.

Lempriere, M. (24 September 2017). 'The minerless mine: Ericsson's Kankberg project is a glimpse into the future of automation'. *Mining Technology*. Available at: https://www.mining-technology.com/features/featurethe-minerless-mine-ericssons-kankberg-project-is-a-glimpse-into-the-future-of-automation-5925612/. Accessed 16 October 2017.

Lockwood, A. (2012). *The Silent Epidemic: Coal and the Hidden Threat to Health*. MIT Press, Cambridge.

Lomborg, B. (2015). *The Nobel Laureate's Guide to the Smartest Targets for the World 2016–2030*. Copenhagen Consensus Center, Copenhagen.

Mahlakoana, T. (14 March 2018). 'Unions line up to block deal with Independent Power Producers'. *Businessday Live*. Available at: https://www.businesslive.co.za/bd/national/labour/2018-03-14-unions-line-up-to-block-deal-with-independent-power-producers/. Accessed 15 March 2018.

Makhuvela, A., Msimang, V., Perrot, R., Ferraz, F., Ferreira, V., Stone, A., Merven, B. and Senate, M. (2013). *South Africa and the Global Hydrogen Economy: The Strategic Role of Platinum Group Metals*. Mapungubwe Institute for Strategic Reflection (MISTRA), Johannesburg.

Markowitz, M. (12 June 2015). 'How fuel cells are transforming energy markets'. World Economic Forum. Available at: https://www.weforum. org/agenda/2015/06/how-fuel-cells-are-transforming-energy-markets/. Accessed 29 June 2017.

Milanovic, B. (2016). *Global Inequality: A New Approach for the Age of Globalization*. Harvard University Press, Cambridge, MA.

National Planning Commission. (2012). 'National Development Plan: Vision for 2030'. National Planning Commission, Pretoria.

Nikolakopoulos, G., Gustafsson, T., Martinsson, P.E. and Andersson, U. (2015). 'A vision of zero entry production areas in mines'. *IFAC-PapersOnLine*, Vol. 28, pp. 66–68.

Norgate, T. and Haque, N. (2010). 'Energy and greenhouse gas impacts of mining and mineral processing operations'. *Journal of Cleaner Production*, Vol. 18, pp. 266–274.

Oraby, E.A. and Eksteen, J.J. (2014). 'The selective leaching of copper from a gold-copper concentrate in glycine solutions'. *Hydrometallurgy*, Vol. 150, pp. 14–19.

Peters, M.A. (2017). 'Technological unemployment: Educating for the Fourth Industrial Revolution'. *Educational Philosophy and Theory*, Vol. 49(1), pp. 1–6.

Ramanujam, J. and Singh, U.P. (2017). 'Copper indium gallium selenide based solar cells – A review'. *Energy & Environmental Science*, Vol. 10, pp. 1306–1319.

Raworth, K. (2017). *Doughnut Economics: Seven Ways to Think Like a 21st Century Economist*. Chelsea Green Publishing, White River Junction.

REN21. (2017). 'Renewables 2017 Global Status Report'. REN21 Secretariat, Paris.

Schaps, K. (1 August 2017). 'European oil majors seek to harness U.S. offshore wind'. Reuters. Available at: https://www.reuters.com/article/ us-oil-majors-offshore-wind/european-oil-majors-seek-to-harness-u-s-offshore-wind-idUSKBN1AH3R2. Accessed 20 November 2017.

Schwab, K. (2016). *The Fourth Industrial Revolution*. World Economic Forum, Geneva.

SPARC: The Partnership for Robotics. (2015). 'Robotics 2020 multi-annual roadmap'.

Swilling, M., Bhorat, H., Buthelezi, M., Chipkin, I., Duma, S., Peter, C., Qobo, M. and Friedenstein, H. (2017). 'Betrayal of the promise: How

South Africa is being stolen'. Available at: https://pari.org.za/wp-content/uploads/2017/05/Betrayal-of-the-Promise-25052017.pdf. Accessed 23 May 2018.

UNEP. (2011). 'Towards a green economy: Pathways to sustainable development and poverty eradication: A synthesis for policy makers'. United Nations, Geneva.

Van Biljon, D. (29 June 2017). 'Energy: ESKOM get your house in order'. *Financial Mail*. Available at: https://www.businesslive.co.za/fm/features/2017-06-29-energy-eskom-get-your-house-in-order/. Accessed 10 September 2018.

Vogt, D. (2016). 'A review of rock cutting for underground mining: Past, present and future'. *The Journal of the Southern African Institute of Mining and Metallurgy*, Vol. 116(11), pp. 1011–1026.

Wilson, F. (2001). 'Minerals and migrants: How the mining industry has shaped South Africa'. *Daedalus*, Vol. 130(1), pp. 99–121.

World Bank Group. (2016). 'World development report 2016: Digital dividends'. The World Bank Group, Washington, DC.

World Bank Group. (2017). 'The growing role of minerals and metals for a low carbon Future'. The World Bank Group, Washington, DC.

Yelland, C. (22 July 2016). 'Understanding the cost of electricity from Medupi, Kusile and IPPs'. *Daily Maverick*. Available at: https://www.dailymaverick.co.za/article/2016-07-22-understanding-the-cost-of-electricity-from-medupi-kusile-and-ipps/#.WeW4rEx7EUE. Accessed 17 October 2017.

SIX

Formalising artisanal and small-scale mining

Problems, contradictions and possibilities

HIBIST KASSA

ARTISANAL AND SMALL-SCALE MINING (ASM) has increasingly become globally recognised as an important livelihood strategy. Particularly within the contexts of increased impoverishment of rural populations, the search for alternative livelihood strategies and volatile commodity prices, ASM is in some instances displacing traditional sectors such as agriculture (Awumbila and Tsikata, 2010; Hilson and Gatsinzi, 2014).

In the 1980s and 1990s, a wave of formalisation in the Global South sought to more effectively integrate ASM into national economies. This occurred within the frame of structural adjustment, expanded large-scale mining (LSM) due to favourable conditions such as tax breaks and attempts to eliminate so-called criminal ASM activities. Either specific legislation on ASM was passed, as in Ghana in 1989, or general mining laws were reformed as in the Democratic Republic of Congo (DRC) (Bugnosen, 2005: 6).

Intensified interest in the informal economy, as a route to poverty alleviation, led to a wave of legislative measures, aimed at regulating ASM, in the 1980s and 1990s as a first step to formalisation.[1] These interventions, aimed at formalising ASM, were presumed to mark the end of criminalisation. However, the case in Ghana shows how these measures have instead created new challenges, with ASM remaining largely criminalised, and the deployment of the military to implement a moratorium on the sector. It is this understanding that has informed the call for decriminalisation in South Africa.

Regulation has tended to sharpen and entrench mechanisms of exclusion, mainly by impeding access to more profitable concessions, which in turn leads to persistent informality and criminalisation. Policy interventions also tend to undermine the transformation of ASM operations into safer, more sophisticated or specialised ventures.

In South Africa, abandoned and active mines have become sites of contestation over natural resources. Mine tailings in Ekurhuleni Municipality, for instance, have been at the centre of reprocessing, with artisanal miners operating in the shadow of a mining corporation encroaching on this work.[2] The 2018 allocation of licences to artisanal miners, known as zamazama,[3] in Kimberley, is an example of how to integrate these operations. More generally, ASM in precious and semi-precious minerals tends to be undertaken clandestinely or associated with illegality. The existing legal framework in South Africa has not explicitly defined ASM as illegal. Instead, it is at best 'non-legal'.

It is proposed here that it is critical to reflect on international experiences of regulation while taking cognisance of the specificities of the South African case. Ghana presents a compelling case of relevance

---

1   Formalisation of ASM 'involves property rights resolution and enforcement, land-use planning, fiscal regulation, and, more broadly, the implementation of environmental and social norms' (Salo et al., 2016: 1058).

2   In Vlakfontein, zamazama started reprocessing in 1994 and in this period several companies entered the mine tailing.

3   *Zamazama* means 'to take a chance' in *isiZulu* and is now commonly used to refer to artisanal miners operating in abandoned and active mining concessions.

to South Africa. This chapter therefore begins by investigating the challenges of defining ASM. It then traces the historical processes of dispossession, dislocation of indigenous mining and the emergence of large-scale mining in South Africa. The final section explores policy gaps and the politics of policy interventions.

DEFINING ARTISANAL AND SMALL-SCALE MINING

The problem of defining ASM has plagued researchers, policy-makers and civil society for over three decades (Hentschel et al., 2002). The use of concepts such as 'pre-capitalist production', 'peasant mining', 'indigenous mining' and 'African diggers' points to the different ways in which this form of mining is understood (Lanning and Mueller, 1979; Graulau, 2006). In part, these terms also reflect how mining, responding to the impulses of local economies, had its own character and orientation prior to the colonial encounter.

It has also been suggested that categories can be developed with reference to the scale of ASM and the degree (if any) of mechanisation: the number of workers, the techniques deployed and the level of capitalisation (Noestaller, 1987). Hentschel et al. (2002) suggest that these do not need to be present in all cases but can be combined to take account of the country context. However, Noestaller (1987) explains that even the number of people working at a site cannot be used to draw conclusions about the actual scale of an operation. The capital intensity and the nature of deposits can shape a mine operation and distort an understanding of its scale.

Mutemeri and Petersen (2002), referring to the South African Small Business Act (1996) argue that the Act's categories of micro, very small, small and medium businesses are not helpful when it comes to mining. Also, ASM is not easily categorised as being either formal or informal, legal or illegal. Even in contexts where regulation has been attempted, as in Ghana, ASM is still primarily criminalised.

The South African Chamber of Mines (renamed in 2018 the Minerals Council of South Africa) has sought to make a distinction between 'criminal networks' operating in ASM and those operations that are based in communities, understood as informal mining (SAHRC,

2015). In South Africa, as elsewhere on the continent, ASM does not only include precious and semi-precious minerals and metals but also non-precious minerals such as clay and kaolin. The latter, however, do not generate as much interest in controlling ASM operations, due to the low level of returns (Graham, 2017).

Debrah et al. (2014), similar to Ledwaba and Mutemeri (2017: 3), in a comparative study of South Africa and Ghana, add that there is 'a lack of context and an understanding of the continuum' of ASM, including the role of junior miners. They therefore distinguish between artisanal mining, on one hand, and small-scale mining, on the other. The former 'refers to unorganised mining activity that does not make use of sophisticated machinery, whereas 'small-scale' is used in the context of organised miners that may not necessarily use sophisticated machinery but have a higher revenue turnover' (Debrah et al., 2014: 914). They therefore propose looking at each of the two forms individually.

The presumption by Debrah et al. (2013) that artisanal miners are 'unorganised' needs to be questioned. The lack of sophisticated machinery or technology in an operation does not preclude some form of organisation. This may not be a formal structure, such as a committee, co-operative or an association, but there are forms of co-operation that guide the way in which work is organised in informal mining operations.

In the case of artisanal miners in gold production, an agreement on where and how to process the ore or reprocess the waste from mine dumps is required. A basic principle of mutual respect and recognition of the right to work an area is needed to ensure a group can operate in an area, even if each individual operates autonomously. The interaction with middle men, where the precious minerals are sold, also needs to be negotiated and agreed upon by the teams collaborating in an operation, as found in Khutsong and Vlakfontein. These forms of co-operation can lay the basis for structures to emerge.

The question of safety and risk underground also depends on sharing information based on experience built up by artisanal miners. All this takes a degree of collaboration and agreement. These practices and principles, which enable some form of self-regulation of work and inter-personal relations in ASM operations, however fragmented they may appear, hold potential for autonomous self-organisation by artisanal miners. Depending on the context, this can be in some form of interaction with communities affected by mining or other civil society groups. Nyoni (2017: 147) makes a similar case in relation to artisanal miners in Roodepoort, where he shows how the zamazama self-regulate access to mine shafts as part of a broader 'morally defined ... code of conduct'.

According to the South African Human Rights Commission (SAHRC), the mining regime in South Africa appears to cater for small-scale miners but not artisanal miners (SAHRC, 2015). The Commission underlines the need to understand the differences amongst artisanal miners. Within operations, those sponsors and supervisors in charge of more profitable operations – usually men – are more likely to access capital, technology, information on deposits and capacity building services. This has a bearing on who can obtain licences. The SAHRC concludes that a lack of definition has in fact undermined interventions and more precise classifications are needed.

Geenen (2013: 2) attributes this challenge of crafting appropriate policy for ASM in the DRC mainly to legal and policy limitations resulting from neoliberal reforms that led to large tracts of land being offered by the state to LSM operations at the expense of ASM producers. In these conditions, the latter are unable to compete for concessions of value. Also, in conditions of increased competition, LSM operations are unwilling to let go of even marginally profitable concessions.

Dichotomising the formal and informal sectors undermines an understanding of the linkages between ASM and LSM. The contributions of ASM to the formal economy can be significant. In Ghana, diamond production is carried out solely through ASM.[4] ASM also contributed 31 per cent to national gold production in Ghana in 2016 – an increase from 2.2 per cent in 1989 (Government of Ghana, 2017).

---

4   Ghana Chamber of Mines reports a decline of production of diamonds from 185,376 carats in 2015 to 141,005 carats in 2016 mainly due to a shortfall in production from small-scale miners. No explanation is provided for the shortfall. Available at https://ghanachamberofmines.org/wp-content/uploads/2016/11/Performance-of-the-Mining-Industry-in-2016.pdf

All these complexities have left research on ASM in a rather ambiguous state, without a clear definition of the sector. However, given the complexities, perhaps arriving at a firm definition is not actually helpful. It may instead be more appropriate to arrive at context-specific descriptions, situated within the history of mining and the production and reproduction processes shaping operations. This would entail unpacking the configuration of relations between the state, transnational mining companies and ASM operations. Also required is an understanding of the structure of operations, the level of technology, the type of commodity and the nature of ore deposits. Other categories of inquiry should include the type of mineral, methods of production, nature of linkages with the local economy and the social spaces within which the operations take place. More broadly, such an approach would bring into view the social, political and economic factors relevant to formulating policy interventions and anticipating likely outcomes.

A key determinant of the variations in ASM is the commodity being mined. Precious and semi-precious minerals such as gold, coltan, diamonds and other gemstones tend to attract greater attention from regulators (Graham, 2017). This can entail regulations not only on concessions but also inputs. In the case of gold, this requires regulations on possession of mercury, a key and toxic element used in amalgamation (Munakamwe, 2017). Similar attention is not directed to regulating non-precious minerals like kaolinite clay, or salt mining.

The methods deployed can cover rudimentary and low-cost tools and techniques or more mechanised and capital-intensive operations. This gradual shift towards more capital intensive and mechanised operations has been ongoing as the ASM sector develops, as in the case of Ghana. Therefore, the scale of the operations can be related to the degree of capitalisation and mechanisation, which can either lead to an expansion in the amount of labour required or more specialised skills. Similarly, in South Africa, more sophisticated processing methods are gradually being integrated in gold ASM operations for tasks like crushing rock ore. This evolving sector therefore requires regulations that seek to provide the extension services required for operations especially given the high risks borne by ASM miners, both male and female.

It is also important to understand what ASM, in all its variations, means in contexts where LSM has historically dominated the market. The state tends to be heavily skewed towards supporting, enabling and facilitating LSM operations. Given this, the focus of state regulation in South Africa tends to centre on where, how and what is mined. The focus on precious minerals, in particular, dominates the emerging discourse on ASM and contestations around regulation and criminalisation. It is to the South African case that this chapter now turns.

DISPOSSESSION, DISLOCATION OF INDIGENOUS MINING AND THE RISE OF LARGE-SCALE MINING IN SOUTH AFRICA

South Africa is a case of extreme dispossession and attempts at undermining African petty commodity production (Lanning and Mueller, 1979). In the pre-colonial period, mining was supplementary to subsistence farming and tended to be a store of wealth, while also indicating status and providing spiritual and cultural value. Iron, copper and salt were the key commodities traded at the time. The quality of iron produced was of such a standard that the heavy concentration of carbon led some to categorise it as steel. Well into the 19th century iron wrought in Africa was considered to be of a higher quality than that produced in Europe, with the latter considered by Africans as being 'rotten' (Lanning and Mueller, 1979: 27–28). This gives a strong indication of the forms of indigenous production which were displaced with the entry of mining firms, aided by financiers of the metropoles and the colonial state (Lanning and Mueller, 1979).

Charles Feinstein (2005: 51, 55–59) outlines how dispossession itself was insufficient to free up labour for exploitation in the mines. This dispossession had to also uproot people from the soil. So long as Africans could secure their livelihoods through agriculture, there was a need for more aggressive measures to ensure labour was available. This included the use of slave labour, taxation and restrictions on movement.

African resistance remained resilient even after the establishment of the Cape Colony and the emergence of the Transvaal as a separate Afrikaner colony. African petty commodity producers sought to protect their land and cattle from appropriation. African agency to resist the dispossession and purchase of land spurred legislative interventions to restrain Africans. Beinart and Delius (2015) provide a useful exploration of the 1913 Natives' Land Act and the contestation that preceded it.

Lanning and Mueller (1979) trace how the discovery of diamonds was a crucial turning point. The discovery catalysed a convergence of interests in eliminating African indigenous production. As Lanning and Mueller put it, 'fifty years after the discovery of diamonds at Kimberley ... through a combination of force and trickery, the mining companies had broken down the earlier resistance of African rulers to the foreign exploitation of African minerals' (Lanning and Mueller, 1979: 52). In gold mines, when the surface workings were exhausted, simple methods of mining were no longer adequate to penetrate depths below 40 metres. By 1892, the introduction of the potassium cyanide-based extraction – developed by MacArthur and Forrest in Scotland – improved extraction techniques. This required 2,009 litres of water to process a tonne of ore, for which dams were constructed. The shift to mining deeper underground was also dependent on advances in technology, finance, a stable work force and expertise. These circumstances made it necessary for agglomerations to occur leading to the emergence of the group system (Bonner and Nieftagodien, 2012: 6–8).

Financiers with links to the metropoles emerged, and a process of merging claims began. Nonetheless, there was a concerted attempt to ensure small claims as units of production were protected. No more than three claims per digger were allowed. Taxes were raised to support the new colony of Griqualand West, while there was also an attempt to regulate claims to secure the protection of small diggers from proprietors. For instance, in Kimberley, more than half of the workforce were African diggers. The diamond mines of Kimberley alone had 3,600 claims (Lanning and Mueller, 1979: 33–34).

Further undermining artisanal mining, the British colonial government considered the small diggers to be an obstacle for raising taxes and creating a stable environment for future investments. Mismanagement of the colonial administration of Griqualand West, in addition to other factors, fuelled a financial crisis which worsened and put at risk protections for small diggers. These diggers organised under the Kimberley Defence League and became an armed threat to the Griqualand West colony (Lanning and Mueller, 1979: 33–39).

These desperate conditions were manipulated by the large claim owners, financiers and company representatives, who swiftly escalated the situation and ensured the Griqualand West colony capitulated to British control by 1871. The transformation of the diamond industry was consolidated by barring Africans from becoming skilled workers, holding claims or even from 'washing debris' (Lanning and Mueller, 1979: 38–39). The same authors note that a pass system was also introduced and was later entrenched with the construction of the mine compounds. Hence the structure was built for the South African mining industry as we know it.

Wolpe (1972) contends that the creation of reserves laid the basis for entrenching a system of cheap wage labour to meet needs of white mining capital, in which women were key to reproduction. This is contested by Lipton (2007: 43), who points to how the reserves served the interests of plantation farmers while mining companies shifted more to migrant labour from beyond the borders of South Africa. This dependence on migrant labour continues to be reflected in the contemporary operations of LSM in South Africa, but also ASM.

In addition to private interests, there was a concerted effort by the state from 1935 and into the 1970s to restrain black small-scale enterprises (informal or backyard production) in the reserves with discriminatory measures. Favouring large-scale enterprises, the state sought to counter the development of a black industrial class. Each reserve had its own development corporation which set up industrial parks for black-owned small-scale enterprises in the reserves, but these still faced bureaucratic obstacles and did not receive similar backing to large enterprises. The combination of the undermining of vocational training institutions and production through lack of incentives and other restrictions, meant that African industries were not able to compete with European trades in urban areas (Da Silva, 1987: 39–44, 47–48, 171). These shifts provide some explanation for the ways in which segregation, later entrenched under apartheid by the National

Party, undermined African petty commodity production. They also show how ASM was very quickly crushed by large-scale financiers and mining firms, with the support of the state, as part of a broader push to dispossess Africans and create a cheap wage labour system. It also emerges that the nature of the geological deposits of gold and diamonds required capital-intensive operations and underground mine shafts. Even at their nascent stage, LSM operations took a form that ensured that ASM operations, where they occurred, remained largely marginal.

This history of exclusion at the centre of dispossession and accumulation in South Africa has positioned ASM in the contemporary period as representing, even if in symbolic terms, an indigenous alternative to LSM or the accumulation of capital in the hands of the dispossessed. It is against this backdrop that the importance of integrating and enabling the growth of ASM is affirmed. Ledwaba (2017: 34) and Mutemeri and Petersen (2002: 288–289) highlight how policy shifts after 1994 marked an attempt to support historically disadvantaged South Africans in the mining sector which included ASM. The obstacles to ASM operations were identified. These centred on access to markets, finance, deposits and mineral and land rights.

The challenges are technical in the first instance, but only at a superficial level. ASM raises political and economic questions about the nature of the state's relation to LSM and how this relationship has shaped the legal framework which continues to largely exclude or criminalise ASM operations. The nature of integration in the domestic market has also facilitated profit-making based on precarious labour in mining in an extreme way. Gold is sold directly to buyers who are based in the neighbouring areas, who in turn resell in Johannesburg. For instance, a gram of gold sold in Langaville for R420 can be resold in Johannesburg for R460 or R500. Munakamwe (2017: 171–172) traces the trade networks which feed into the formal market. At the top end of the chain, gold is resold for R1,100 and R1,200. This disparity points to the need for an open market to enable zamazama miners to gain fairer prices.

The challenges ASM miners face in overcoming the obstacles to legalising their operations are consistent throughout the Global South. This is the case in countries like Ghana where regulations, seeking explicitly to integrate ASM into the formal economy, have instead led to intensified criminalisation. To grapple with this, it is necessary to consider the underlying logics and politics of policy interventions, which the next section explores.

POLICY GAPS IN ARTISANAL AND SMALL-SCALE MINING: LESSONS FROM GHANA

Initially, ASM was left out of poverty alleviation programmes and strategies since it was regarded as being too marginal and risky by international development agencies and governments, irrespective of the role it played in the mining sector and rural economies of the Global South (Davidson, 1993: 315; McQuilken and Hilson, 2016). In the 1980s it had been assumed that artisanal mining was a product of individual choice, adventurism and enterprise units (Noetstaller, 1987). The hardship imposed under structural adjustment in the 1980s is often cited as unleashing greater economic instability in rural economies and households, creating even greater need for alternative livelihood strategies (Hilson, 2001: 2–3; Bugnosen, 2005: 15). Artisanal mining came to fit into this and expanded rapidly in the 2000s mining commodity boom. In the 1980s, the World Bank initially adopted a position of treating artisanal mining in the same policy framework as LSM. Altogether, these measures have been found to be largely inappropriate and too rigid to meet the needs and the interests of artisanal miners (Davidson, 1993: 315–317).

ASM is now understood as being largely driven by unemployment and poverty (Davidson, 1993; Hilson and Gatsinzi, 2014; Geenen, 2013; McQuilken and Hilson, 2016). In countries like Ghana, smallholder farmers are shifting to ASM because agriculture has been suffering a decline since the 1980s. In addition, throughout the South, retrenchments from LSM operations and from the public sector and wider conditions of deindustrialisation have contributed to a livelihood crisis (Awumbila and Tsikata, 2010; Hilson and Gatsinzi, 2014). Institutions like the World Bank and the International Labour Organisation thus came to recognise ASM as an alternative livelihood strategy, though as mentioned, policy gaps remain in both national and international arenas (McQuilken and Hilson, 2016).

Barchiesi (2010: 72–75) explains that in most African economies, the informal sector grew rapidly as a 'response to adjustment'. Informal sector activity in South Africa today reflects the expansion of precarious work and jobs, which resulted from government measures to create a 'business-friendly environment' after 1994. Casualisation, which undermines the conditions of permanent workers, has positioned the informal sector as an alternative to permanent jobs. Formal sector workers have been pushed to explore possibilities outside of wage work, especially through the setting up of small enterprises. Some hold second jobs in individual enterprises as self-employed workers. The conditions of precarity, and their inability to provide for household needs, have also encouraged women to seek work beyond unpaid care work.

Throughout South Africa, the reprocessing of mine tailings is an instance of work in the informal sector. Largely clandestine ASM operations have emerged. In some cases, artisanal miners in mine tailings find themselves in competition with LSM operations keen to reprocess mine waste. In chapter 12 of this volume, Fioramonti rightly points to the reprocessing of mine tailings by artisanal miners as an important frontier for rethinking mining work.

A forum hosted by the Ekurhuleni Municipality in January 2003 reflects this thinking (Cezernowalow, 2003). In attendance were representatives from the Department of Mineral Resources, the Department of Water Affairs and Forestry, mining companies in the municipality, owners of private dumps and a committee of small-scale miners. The forum explored the potential for expanding and formalising artisanal and small-scale mining as part of a broader programme of linking national programmes to local needs. This was a space created for government and the mining industry to assess social plans, environmental impacts and the decline of LSM, in addition to support for clandestine artisanal and small-scale mining (Cezernowalow, 2003).

In the words of Karuna Mohan, Executive Director of Local Economic Development of the Ekurhuleni municipality, the aim was to transform criminalised activity into a livelihood option. In an optimistic reading of a then proposed mining law, the legislation was taken to provide for the possibility for mineral rights to be inclusive of 'historically disadvantaged people and women' (Cezernowalow, 2003). There was also optimism about widening the tax base of the municipality. A co-ordinating company was proposed to support the development of small-scale mining and also manage broader resource management. Some of the activities proposed included ensuring access to finance, training, and monitoring; rehabilitation of areas destroyed by large-scale mining; supporting upstream and downstream linkages and providing technical support in the form of geological information (Cezernowalow, 2003).

Instead of this initiative being taken forward, DRDGOLD South Africa and Mintails Limited reopened the Ergo surface treatment plant in Brakpan, Ekurhuleni in 2007. DRDGOLD is a South African enterprise reprocessing mine tailings and underground mines in South Africa and Zimbabwe. Formerly known as Durban Roodepoort Deep Mining Company, it was established in 1895, hence it is clearly a large-scale mining operation with a significant history of accumulation. Given this, DRDGOLD was in a position to enter into a joint venture with Mintails Limited and purchase the Ergo plant from the previous owner, AngloGold Ashanti, thereby eliminating the possibilities for artisanal miners in the area. Apart from DRDGOLD, other mining companies are encroaching similarly. For instance, in Vlakfontein (Kwa-Thema) mine tailings, zamazama miners report being displaced from preferred sites of work by Benoni Gold.

This displacement in Ekurhuleni ultimately brings out what Bugnosen (2005: 13) explains as persistent conflicting relations between LSM and ASM. These are prevalent when artisanal or small mine operations are set up in areas where large-scale miners hold prospecting or exploration rights, but do not make use of them. In Ghana, conflicts with LSM firms are not uncommon, with the state intervening with damaging consequences, not only for ASM operations, but also the local economy (Hilson et al., 2007).

In the early 2000s, the National Security Council of Ghana co-ordinated military intervention against ASM operations, in response to pressure from the Chamber of Mines. Soldiers were mandated to stop unlicensed ASM, operating on the concessions of LSM companies. This led to mass arrests, destruction of ASM equipment and eventually a resettlement programme (Hilson et al., 2007).

The initial sweep of ASM operations began in Prestea (Western Region), moving to Obuasi (Ashanti Region) and finally to Ntronang (Eastern Region). The resettlement was to be implemented in Prestea, where ASM miners were operating on the concession of Golden Star Resources Limited. The artisanal miners were promised that they would be relocated and reorganised in co-operatives, with licences to mine specific sites. About 6,000 people were to be relocated to three areas in Japa (Western Region), Adjumadium (Western Region) and Winneba (Central Region). The project failed due to many factors including an inability to register Prestea based miners, a lack of detailed geological survey and the existence of ASM operations on the new sites (Hilson et al., 2007).

This failed relocation project was preceded by another attempt in 2003 at a negotiated sharing arrangement in Prestea between Golden Star Limited and ASM operators in Prestea, Himan and Bondai. The ASM operators were to be the sole operators within a site known as Number Four Bungalow. It was agreed that only 100 people and 100 pits would be worked on the site. Lastly, negotiations were to be key in any interventions, especially displacement. In practice, the agreement was not respected by Golden Star Resources Limited (Hilson et al., 2007).

These examples point to regulation and negotiation as strategies to displace or disrupt ASM operations in the interests of LSM operated by foreign investors. These strategies were also used in Obuasi, another mining town in Ghana, where the entire concession is under AngloGold Ashanti (AGA) and legal ASM is thus precluded.

In the early 2000s, a mass arrest of 117 people triggered the formation of an ASM committee in Obuasi. This happened at a time when ASM surface mining had been undermined in the process of a reclamation exercise undertaken by AngloGold Ashanti as required by the Environmental Protection Agency. The artisanal and small-scale miners therefore needed alternative areas in which to operate (Abubakar, interview, 24 November 2017).

Obuasi Municipality is mostly within the concession of AGA, and most ASM operations in Obuasi are therefore within the concession.

In such instances, the law gives the authority to determine where ASM can take place to the minister. Typically, however, areas which are mineral rich are not allocated as concessions for ASM operators (McQuilken and Hilson 2016: 27). This becomes an obstacle for artisanal and small miners attempting to operate within the legal framework. According to an ASM licence holder, Alhaji Ben Abdulai (Interview, 23 November 2017):

> *There is a difference between licence and concession so if you have a licence that alone does not mean you can enter someone's concession and start mining. We also have some people with a concession, but they don't have a licence to mine, therefore if you have a licence you need to also get your own concession before you can operate ...*
>
> *There was a time I went for a parcel of land at Asante Akyem to mine, even though concession belonged to a mining company called Golf Cost; they had left it untouched for 25 years. The moment they heard of our presence on the land they also came out to work. So I have had to struggle with them, but I had the backing of the community because I had employed 850 people as my workforce while the so-called multinational company had only nine workers with only three locals so between me and such a company – who is helping build our nation?*

In April 2017, Ghana's Minister of Lands and Natural Resources, John-Peter Amewu, announced a six-month moratorium. This was in response to pressure from the Chamber of Mines, supported by Accra-based media and a nationwide campaign against environmental damage resulting from alluvial ASM. The moratorium, which was extended indefinitely and was still in effect at the time of writing (August 2018), also affects ASM of rock ore deposits. In effect, the moratorium seeks to limit continued development of local capital in gold mining, if not eliminate it completely. This seems to be affirmed by the indefinite extension of the moratorium until 'illegal' ASM is eradicated (Ghana News Agency, 2018).

If not facing comprehensive barriers, as in these attempts of the

Ghanaian state acting for what it puts forward as environmental protection, ASM faces state policies that constrain growth considerably. In an overview of ASM-related legislation in the Global South, Bugnosen (2005) argues that existing legislation tends to limit the upgrading and transformation of small-scale mining. For instance, limits are placed on the depth of operations, sanctions are applied to the use of explosives and advanced technology is prohibited. Similarly, where artisanal and small-scale miners are given access to land, it is to plots that are not lucrative (Bugnosen, 2005).

In South Africa, Ledbawa (2017: 34) and Mutemeri and Petersen (2002: 288–289) highlight how policy shifts after 1994 marked an attempt to support the entry of historically disadvantaged South Africans. ASM was identified as one of the sectors to receive support. As in other countries, the obstacles centre on access to markets, finance, deposits and mineral and land rights. The Minerals and Petroleum Resources Development Act of 2002 initially enabled licences to be issued for operations not exceeding one and a half hectares. By 2011, about 1,030 small-scale miners were registered in South Africa (Mine and Health Safety Council, cited in Ledwaba, 2017: 35).

Section 2.8 of the 2017 Mining Charter makes provision for micro-scale players in mining. It categorises mining firms as operating on a micro-scale if the company has a turnover of less than R1 million. These actors are exempt from targets around mining community development, employment equity, ownership and procurement. The draft 2018 Mining Charter (released at the time of writing) emphasises the need to redress and counter historical injustice. All of this demonstrates attempts to accommodate small-scale players in mining (Ledwaba, 2017; Debrah et al., 2013; Mutemeri and Petersen, 2002). This is more likely opening space for emerging black capitalists attempting to position themselves in the mining sector.

However, in practice, this does not create a supportive policy space for smaller-scale producers – ranging from petty commodity producers (primarily based on self-exploitation or family labour) to other small operations employing wage labour. Moreover, ASM operations in South Africa are operated by both South African workers and documented and undocumented migrants. The narrative of these migrants is one of exclusion from better-paying work, especially in the formal sector. In interviews conducted in Vlakfontein (Kwa-Thema) and Carletonville, male miners had previously worked in the construction sector or were retrenched from the large-scale mines. Similarly, it is unclear whether the South African legislation which makes allowance for small players in mining is inclusive of non-South African artisanal miners.

Access, however, is not the only area where legislation or other forms of state intervention can be useful. ASM occurs in conditions of great risk and danger. ASM gold miners, for instance, track seams of deposits which run along underground mine shafts below 4 kilometres. These shafts tend to be more unstable and prone to rock falls. Poisonous gas and flooding also pose grave dangers. Similarly, mining of tailings poses serious health risks with exposure to toxic chemicals such as cyanide, mercury, lead and uranium (Benchmarks Foundation, 2017). Even if granted access as artisanal and small-scale miners, these workers require health and safety protection if such mining is to constitute a viable alternative livelihood.

Davidson (1993: 315) explores the issue of transformation and upgrading of artisanal mining operations to improve their potential for profit and to minimise drawbacks. The potential can be explored through value addition and supply of raw materials for local industries. The reconfiguration of artisanal mining can take different forms depending on the interest of the actors. Munakamwe (2015: 10–11; 2017: 173) illustrates how zamazama in underground gold mining are supplying local refineries, such as the Rand Refinery, in South Africa. The context of criminalisation in South Africa makes it challenging to estimate how much of the gold is supplied to refineries by the zamazama, unlike in the case of Ghana (cited earlier).

For artisanal miners in South Africa, decriminalisation is a key intervention. This should not only be focused on permits to access and sell minerals but also include access to finance and support for marketing and appropriate technology. In addition, access to mining permits must be extended to undocumented artisanal miners in the country which is not the case in Kimberley where the process of decriminalisation has begun. According to Davidson (1993), some ASM groups may focus on setting up organisational structures (co-operatives or associations) to facilitate upgrading and yet others may focus more broadly on building an indigenous viable mining sector. Davidson adds that experience of local participants in small-scale mining can provide a good basis on which to craft policy.

The picture becomes even more complex when reflecting on gender and the implications of policy interventions which have an impact on the cultural, political and social barriers faced by women. With regard to mining broadly, traditional systems of land ownership and tributary and customary rights largely favouring males are the framework for land appropriation, control of royalties and compensation (Amanor-Wilks, 2009: 11).

These in turn shape relations in ASM, even where the state is seeking to support women. In Tanzania, for example, the state has granted mining permits to women, but men and chiefs refuse to recognise their right to operate mines. Not only are women undermined, in extreme cases they are perceived as a threat and face violence. All of this occurs within a context in which women are marginalised in auxiliary services or processing, and thereby excluded from greater shares of the surplus extracted. Even worse, some women in ASM are not even paid a daily wage by men in control of production, and instead receive gifts (such as clothes) or nothing at all (Tallichet et al., 2005: 191–192).

This draws attention to the dynamics of inequality in ASM operations which deepen in conditions of extreme precarity. In spite of this, women working in ASM see it as an opportunity for an alternative to low incomes in subsistence agriculture, as in the case in Northern Ghana (Awumbila and Tsikata, 2010). Valiani and Ndebele in chapter 9 of this volume make a similar point regarding mining in Kimberley where women who engage in diamond production view their work positively – not only in relation to high unemployment and undervalued female work, but even in comparison with women employed underground in large-scale mines.

Along these lines, Lahiri-Dutt (2011) argues for an approach that goes beyond highlighting women as key actors in mining production, protest or reproduction, and rather asks 'what mining means to communities'. She sums it up well:

Informal mining is yet to be deeply understood and theorised ... a new feminist epistemology of mining would focus on women and men's lives in informal extractive practices, where they neither own the land and the minerals, nor are they 'exploited' in the conventional sense as a 'working class' (Lahiri-Dutt, 2011: 11).

In Africa, where customary laws have persisted – despite being subject to change or contestation – and converged to produce an array of gendered outcomes, a holistic assessment is needed to inform policy-making. An essential precondition to crafting policy on ASM is the decriminalisation of ASM and an opening up of public discourse that begins by taking cognisance of the inequalities embedded in ASM operations.

CONCLUSION

Since 1994, South Africa has undergone a wave of liberalisation that has had a direct impact on manufacturing, but also on mining and the labour regime more broadly. The casualisation of work and the consequent expansion of the informal sector have resulted in an increasingly precarious existence for working people, with women continuing to carry the burden of work.

Policy intervention in ASM continues to be highly contested. Internationally, attempts at regulation have raised further problems and exposed tensions and contradictions between ASM and LSM. The continued criminalisation of ASM in contexts where formalisation has been attempted highlights this point. Examining the effects of policy interventions on women also exposes further contradictory outcomes. In order to promulgate effective policies, it is necessary to take on board the perspectives of ASM miners, while building an understanding of the nature of the deposits, the labour process of mining operations, linkages within the local economy and power relations within ASM operations. These can inform recommendations for policy that would work in the interest of ASM miners and so avoid deepening conditions of precarity.

ASM offers the possibility of developing specialised skills and of integrating previously excluded groups, including women and undocumented migrants. However, interventions that seek to formalise or integrate the sector into the minerals economy do not necessarily lead to progressive outcomes. Instead, there is a risk that they will deepen criminalisation and entrench existing exclusions.

What is required is a holistic understanding of ASM operations. The importance of ASM, as a space for accumulation and establishing linkages within local economies for historically excluded groups, cannot be ignored. Therefore, support for existing operations, especially those run by groups such as women and undocumented migrants, is critical. This can take the form of technical support, access to markets, finance, work or/and residence permits, mineral and land rights, alongside prioritising health and safety. In particular, ensuring the health and safety of ASM miners calls for access to affordable technology that can facilitate this. It is also critical to enable the zamazama to trade directly in an open market to gain a fairer price while securing the inclusion of undocumented migrants. However, in the long run the nature of the support offered, as well as how it is controlled and by whom, can either undermine the interests of ASM miners or affirm them by promoting social ownership of technologies and mineral rights. The absence of such an approach to interventions will amount to a continuation of exclusions and criminalisation.

REFERENCES

Abubakar, M. (24 November 2017). Interview with Hibist Kassa and Justice Mensah, Obuasi.

Alhaji Ben Abdulai. (23 November 2017). Interview with Hibist Kassa and Justice Mensah, Obuasi.

Amanor-Wilks, D.E. (2009). 'Land, labour and gendered livelihoods in "peasant" and "settler" economy'. *Feminist Africa*, Issue 12, pp. 31–50.

Ashman, S. (2013). 'Systems of accumulation and the evolving South African minerals-energy complex'. In Fine, B., Saraswati, J. and D, Tavasci (eds). *.Beyond the Developmental State: Industrial Policy into the 21st Century*. Pluto Press, London. Available at: https://doi.org/10.1093/afraf/adm038. Accessed 1 July 2018.

Awumbila, M. and Tsikata, D. (2010). 'Economic liberalisation, changing resource tenures and gendered livelihoods: A study of small-scale gold mining and mangrove exploitation in Ghana'. In Tsikata, D. and Golah,

P. (eds). *Land Tenure, Gender and Globalisation: Research and Analysis from Africa.* Asia and Latin America International Development Research Centre, Ottawa.

Barchiesi, F. (2010). 'Informality and casualization to South Africa's industrial unionism and manufacturing workers in the East Rand/Ekurhuleni region in the 1990s'. *African Studies Quarterly*, Vol. 11(2 and 3), pp. 67–85.

Beinart, W. and Delius, P. (2015). 'The Natives Land Act of 1913: A template but not a turning point'. In Cousins, B. and Walker, C. (eds). *Land Divided, Land Restored: Land Reform in South Africa for the 21st Century.* Jacana Media, Johannesburg.

Benchmarks Foundation. (27 August 2017). 'Policy Gap 12: Soweto report: "Waiting to inhale": A survey of household health in four mine-affected communities'. Available at: http://www.bench-marks.org.za/press/a_survey_of_household_health_in_four_mine_affected_communities.pdf. Accessed 4 October 2017.

Bonner, P. and Nieftagodien, N. (2012). *Ekurhuleni: The Making of an Urban Region.* Wits University Press, Johannesburg.

Bugnosen, E.M. (2005). 'Small-scale mining legislation: A general review and an attempt to apply lessons learned'. In Hilson, G.M. (ed.). *The Socio-Economic Impacts of Artisanal and Small-Scale Mining in Developing Countries.* Balkema Publishers, Lisse.

Cezernowalow, M. (29 January 2003). 'Ekurhuleni Metro sets up mining Forum'. *Mining Weekly.* Available at: http://www.miningweekly.com/print-version/ekurhuleni-metro-sets-up-mining-forum-2003-01-29. Accessed 2 July 2018.

Da Silva, M. (1987). 'Small-scale industry in black South Africa'. MA dissertation submitted to the Faculty of Arts. University of Witwatersrand, Johannesburg.

Davidson, J. (1993). 'The transformation and successful development of small-scale mining enterprises in developing countries'. *Natural Resources Forum.* pp. 315–326.

Debrah, A.A., Watson, I. and Quansah, D.P.O. (2014). 'Comparison between artisanal and small-scale mining in Ghana and South Africa: Lessons learnt and the way forward'. *Journal of the Southern African Institute of Mining and Metallurgy*, Vol. 114, pp. 913–921.

DRDGOLD. (2017). Corporate profile. Available at: http://www.drdgold.com/about-us/corporate-profile. Accessed 25 June 2018.

Ekurhuleni Metropolitan Municipality. (2010). 'Ekurhuleni growth and development strategy 2025'. Germiston. Available at: https://www.ekurhuleni.gov.za/456-development-guide-2010-lowres/file. Accessed 10 September 2018.

Feinstein, C. (2005). *An Economic History of South Africa: Conquest, Discrimination and Development.* Cambridge University Press, Cambridge.

Geenen, S. (2013). 'Dispossession, displacement and resistance: Artisanal miners in a gold concession in South-Kivu, Democratic Republic of Congo'. Institute of Development Policy and Management, University of Antwerp.

Geenen, S. (2017). 'Explaining fragmented and fluid mobilization in gold mining concessions in Eastern Democratic Republic of the Congo'. *The Extractive Industries and Society*, Vol. 4(4), pp. 758–765.

Ghana Chamber of Mines. (Undated). 'Performance of the mining industry in 2016'. Location of publication unstated. Available at: https://ghanachamberofmines.org/wp-content/uploads/2016/11/Performance-of-the-Mining-Industry-in-2016.pdf. Accessed 10 September 2018.

Ghana News Agency. (8 March 2018). 'Ban on illegal mining extended again'. *Graphic Online*. Available at: http://www.ghananewsagency.org/print/129601. Accessed 10 March 2018.

Government of Ghana. (2017). 'Project appraisal and implementation document for the multilateral mining integrated project'. JMK Consulting, Accra.

Graham, Y. (2017). 'Concluding remarks at a workshop on neoliberal restructuring, and primary commodity dependence, radical political economy, economic strategy and industrialisation in Africa'. Accra.

Graulau, J. (2006). 'Global processes, local resistances: Gendered labour in peripheral tropical frontier: women, mining and capital accumulation in post development Amazonia'. In Lahiri-Dutt, K. and Macintyre, M. (eds). *Women Miners in Developing Countries: Pit Women and Others*. Ashgate Publishing, London.

Hentschel, T., Hruschka, F. and Priester, M. (2002). 'Global report on artisanal and small-scale mining', January, No. 70. International Institute for Environment and Development and World Business Council for Sustainable Development, England.

Hilson, G. (2001). 'Introduction'. In *The Socio-Economic Impacts of Artisanal and Small-Scale Mining in Developing Countries*. Balkema Publishers, Tokyo.

Hilson, G. and Gatsinzi, A. (2014). 'A rocky road ahead? Critical reflections on the futures of small-scale mining in sub-Saharan Africa'. Faculty of Business, Economics and Law, Guildford.

Hilson, G., Yakovleva, N. and Banchirigah S. M. (2007). 'To move or not to move: Reflections on the resettlement of artisanal miners in the Western Region of Ghana'. *African Affairs*, Vol. 106(424), pp. 413–436.

Lahiri-Dutt, K. (2004). 'Informality in mineral resource management in Asia: Raising questions relating to community economies and sustainable development'. *Natural Resources Forum*, Vol. 28(2), pp. 123–132.

Lahiri-Dutt, K. (2011). Digging women: Towards a new agenda for feminist critiques of mining. *Gender, Place and Culture*, pp. 1–20.

Lahiri-Dutt, K. (2013). 'Gender in mining: Feminising an ancient human endeavour'. Delivered at the Centre of Social Change, University of Johannesburg.

Lanning, G. and Mueller, M. (1979). *Africa Undermined: A History of the Mining Companies and Underdevelopment in Africa*. Penguin Books, New York.

Ledwaba, P.F. (2014). 'Tiger's eye in the Northern Cape Province – Potential for employment creation and poverty creation'. *The Journal of The Southern African Institute of Mining and Metallurgy*, Vol. 114, pp. 881–885.

Ledwaba, P.F. (2017). 'The status of artisanal and small-scale mining in South Africa: Tracking progress'. *The Journal of The Southern African Institute of Mining and Metallurgy*, Vol. 117, pp. 33–40.

Ledwaba, P. and Mutemeri, N. (2017). 'Preliminary study on artisanal and small-scale mining in South Africa'. Open Society Foundation for South Africa, University of Witwatersrand, Centre for Sustainability in Mining and Industry, Johannesburg.

Lipton M. (2007). *Liberals, Marxists and Nationalists; Competing Interpretations of South African History*. Palgrave Macmillan, New York.

McQuilken, J. and Hilson, G. (2016). 'Artisanal and small-scale gold mining in Ghana: Evidence to inform an "action dialogue". Background research to inform an action dialogue'. IIED, London.

Munakamwe, J. (2015). 'The interface between the legal and illegal mining processes: Unpacking the value chain of illegally mined gold'. 10th Global Labour University, Washington DC. Available at: https://www. global-labour-university.org/fileadmin/GLU_conference_2015/papers/ Munakamwe.pdf. Accessed 10 September 2018.

Munakamwe, J. (2017). 'Livelihood strategies, mobilisation and resistance in Johannesburg, South Africa'. In Nhemachena, A. and Warikandwa, T. V. (eds). *Mining Africa: Law, Environment, Society and Politics in Historical and Multidisciplinary Perspective*. Langaa RPCIG, Cameroon.

Mutemeri N. and Petersen F.W. (2002). 'Small-scale mining in South Africa: Past, present and future'. *Natural Resources Forum*, Vol. 26, pp. 286–292.

Noetstaller, R. (1987). 'Small-scale Mining: A review of the issues'. World Bank Technical Paper No. 75. The World Bank, Washington, DC.

Nyoni, P. (2017). 'Unsung heroes? An anthropological approach into the experiences of zamazamas in Johannesburg, South Africa'. In Nhemachena, A. and Warikandawa, T.V. (eds). *Mining Africa: Law, Environment, Society and Politics in Historical Perspective*. Langaa Research and Publishing Group, Bamenda.

Oosthuizen, M.A., John, J. and Somerset, V. (2010). 'Mercury exposure in a low-income community in South Africa'. *South African Medical Journal*, Vol. 100(6), pp. 366–371.

Salo, M., Hiedanpää, J., Karlsson, T., Ávilac, L., Kotilainend, J., Jounelaa,

P. and García, R. R. (2016). 'Local perspectives on the formalization of artisanal and small-scale mining in the Madre de Dios gold fields, Peru'. *The Extractive Industries and Society.* Vol. 4, pp. 1058–1066.

South African Human Rights Commission (SAHRC). (2015). 'Report of the SAHRC investigative hearing: Issues and challenges related to unregulated artisanal underground and surface mining activities in South Africa'. SAHRC, Johannesburg.

Tallichet, S.E., Redlin M.M. and Harris R.P. (2005). 'What's a woman to do? Globalized gender inequality'. In Hilson, G.M. (eds). *The Socio-Economic Impacts of Artisanal and Small-Scale Mining in Developing Countries.* Balkema Publishers, Tokyo.

Verweijen, J. (2017). 'Luddites in the Congo?' *City 21,* Vol. 21(3–4), pp. 466–82.

Wolpe, H. (1972). 'Capitalism and cheap labour-power in South Africa: From segregation to apartheid'. *Economy and Society,* Vol. 1(4), pp. 425–456.

Yakovleva, N. (2007). 'A case study of Birim North District of Ghana: Resources Policy Perspectives on Female Participation in Artisanal and Small-scale Mining'. *Resources Policy,* Vol. 32, pp. 29–41.

Section Two

The Future Contextualised in the Industry's Continuing Past

SEVEN

Private property?

Village struggles over meanings of land and mining revenues on South Africa's platinum belt[1]

SONWABILE MNWANA

THERE IS MOUNTING OPTIMISM that South Africa can bring back the glory days of economic success through its promising platinum mining industry. South Africa holds more than 80 per cent of the world's global platinum reserves, thanks to the Bushveld Igneous Complex, an enormous igneous rock formation which holds significant quantities of platinum group metals (PGMs) including platinum, palladium and

1 The Open Society Foundation for South Africa funds this research through the Mineral Wealth and Politics of Distribution research project (Project number: OSF-SA 03629). Previously this research was funded by the Ford Foundation under the Mining and Rural Transformation project at SWOP, Wits University.

a range of other minerals. It is on this vast stretch of platinum-rich land, loosely called the 'platinum belt', that South Africa's huge mining operations are located.

Even the persistent decline in global demand for PGM commodities, and the recent spates of labour unrest in the platinum sector, do not seem to dampen hope for a glorious, platinum-driven economic future. The leading global platinum producers have been investing in platinum-using fuel cell technologies – a move which is expected to significantly boost global demand and, in return, stabilise South Africa's ailing economy (Liedtke, 2018)[2]. In chapter 3 of this volume, Edwin Ritchken points to the PGM sector as the main driver of South Africa's economy in the 21st century, just as the gold industry was in the 20th century, even though the two industries have slightly different impacts.

However, this optimism overlooks the complexities of local social milieux and the intricate dynamics of rural landholding. This phenomenon is more pronounced in South Africa's platinum belt where African families hold customary rights to land. This system of access to land is often vaguely described (by mining capital, state and other powerful actors) as 'communal'. Customary rights, however, are often undermined through mining capital's collusion with local chiefs and the state, affording mining capital easy entry to communal land.

This chapter's contribution primarily shows how this collusion has not only led to an additional form of dispossession in rural communities on the platinum belt but has also tended to privilege mining capital when struggles over rural land ensue. Essentially, the empirical argument in this chapter demonstrates the fluidity of local meanings of 'land', particularly in the face of dispossession at different historical moments. It argues that the ongoing, mining-led dispossession and land-related conflict in rural South Africa epitomise one such moment of dispossession.

Land is central to the livelihoods and sustenance of more than 60 per cent of African families in sub-Saharan Africa (Boone, 2017: 2). In South Africa, the livelihoods of the approximately 18 million black

2   See http://www.miningweekly.com/article/amplats-invests-in-high-yield-energy-technologies-2018-04-18.

residents in the former 'homelands' remain in many ways connected to land. Yet the property rights – property in land – of the majority of African families in the countryside remain legally poorly defined and, at best, ambiguous. The African National Congress (ANC)-led government has failed to legally empower the rural poor by ensuring their strong access and control over land and natural resources (including minerals) since it first came into power in 1994. Instead, post-apartheid government legislation and policies have established traditional authorities using almost exactly the same political and geographical boundaries used by apartheid governments to set up so-called 'tribal' authorities. Thus, the 'democratic' government has ironically redefined Africans in the countryside as members of 'traditional communities' – effectively 'tribes' under local chiefs. The surge in chiefly power has taken an interesting twist if one closely examines contemporary land struggles on South Africa's platinum belt.

The platinum belt has been marked by a rapid expansion of mining over the last two decades. Platinum has increasingly displaced gold as an employer and as a part of the crucial mining sector. But the platinum belt is largely located in the old 'homelands', on 'communal' land controlled by 'traditional' authorities (chiefs). These 'tribal' authorities lease mining rights and land to large private corporations in return for payments. Large mines operate amidst impoverished villages in overcrowded areas, where generations of dispossessed and impoverished African families have eked out a precarious existence through farming and other means. It is with a significant degree of caution that I make this point about dependence on land: in the case of South Africa's former homeland areas, the land was never sufficient to sustain a thriving, fairly autonomous African peasantry. It had to be supplemented with migrant labour income – however minuscule (Magubane, 1978).

The social shifts and struggles that have been produced by the rapid expansion of mining in former homeland areas have not been adequately examined. This chapter takes a step towards addressing the gaps. It examines some of the emerging local struggles over control of land and mining revenues. I argue that these struggles attest to the reality that ordinary villagers receive very limited benefits from the mining deals that their chiefs sign with mining capital. The struggles also epitomise contestations over the meaning that actors attach to land and mineral wealth in rural South Africa. I demonstrate the agency of the rural poor who use competing histories of land and politics, and contest the very meaning of land in order to resist the control and power of chiefs and dispossession by mining capital. In claiming land rights and associated social identities, the rural poor challenge the power wielded by chiefs over both land and mineral revenues. This chapter draws on archival materials collected from the study villages (personal archives of respondents), the national archives (Pretoria) and the North West provincial archives. It also draws on detailed ethnographic material collected in the study villages.

LAND CONFLICT

It is well established that rural conflict in Africa is mostly connected to disputes over land and landed resources (Peters, 2004). Despite the persistent narrative of an abundance of farming land in rural Africa, conflict over land is at a new peak. A scarcity of land and sharp increases in its value tend to lead to significant conflict, mainly due to large-scale, commercial, land-based investments (land grabs) for food production and exports of biofuels, mineral resources and timber (Peters, 2004). The effects of climate change and rapid population growth cannot be ruled out in contemporary land struggles in Africa. Boone observes that conflict on a significant scale in different regions of Africa – among others Sierra Leone, Liberia, the Democratic Republic of Congo, Rwanda and Kenya – is strongly rooted in severe inequalities and the dispossession of smallholder farmers in the countryside (Boone, 2017: 5).

The perceived ambiguity and vulnerability of African landholding systems also present a conducive environment for elite capture of communal resources, land grabbing and conflict. The colonial state, particularly in southern Africa, totally rejected pre-colonial African landholding systems and excluded Africans from holding land as private property. Thus, since the colonial period, the rights to rural land held by African families are perceived to be 'communal' in character and sanctioned by indigenous ('traditional') institutions defined under customary law. As Berry puts it:

> *The results [of colonial conceptions of African property rights] were racialised systems of property rights, under which non-Africans owned land as private property, while Africans held theirs collectively, as members of customary communities, or 'tribes'* (Berry, 2002: 642).

But the very meaning or character of 'held' is never conclusively defined. For instance, in South Africa, most of the land that is 'held' by Africans in the former homeland areas is, in essence, owned by the state. This land was controlled by the South African Native Trust (SANT)[3] – a state 'trustee' body that was constituted under the Department of Native Affairs during colonial and apartheid periods. The SANT, together with the regulations and processes enforced by the Native Administration Act No. 38 of 1927, Natives' Land Act No. 27 of 1913, Native Trust and Land Act No. 19 of 1936 and the Bantu Authorities Act No. 68 of 1951, not only significantly enhanced the perception of chiefly custodianship, but effectively enabled state control ('indirect rule') over the vast African population in the countryside (Delius and Chaskalson, 1997; Mamdani, 1996). The former SANT land – the 'Trust Land' – now falls under the current state's Ministry/Department of Rural Development and Land Affairs. Therefore, the state remains the owner and local chiefs remain the assumed custodians of this land on behalf of their 'communities' ('tribes') – the customary rights holders. Customary ownership of land tends to prove challenging for the rural poor mainly because the definition and 'content' of custom itself remains elusive. The dominant 'communal' categorisation of land tenure rights in South Africa is problematic. There is strong evidence that pre-colonial land tenure in Africa was 'both communal and individual' in nature (Cousins, 2007: 284). Rights of access were derived from membership of a group, or allegiance or affiliation to a particular political authority – usually an authority of a traditional ruler

---

3   Established by the Native Trust and Land Act No. 19 of 1936.

(chief) (Shipton and Goheen, 1992; Cousins, 2007). To a significant extent, the colonial experience distorted the content and character of pre-colonial indigenous institutions, especially customary law. Mamdani (1996) comprehensively tackles this argument. For Mamdani (1996), dispossession of Africans and the weakening of their property rights were central to the colonial state's indirect rule over the African population in the rural periphery, and to the control mechanisms used. The colonial, distorted versions of customary law rendered the people's rights to land and landed resources ambiguous, at best. Some have argued that customary property rights are negotiable and flexible – as such, they are able to guarantee legally secure land tenure to the poor and vulnerable. Scholars also observe that even within the socially embedded (customary) landholding systems forms of exclusion and inequality prevail (Cousins, 2007). The fact that such systems of land tenure are negotiable and flexible offers opportunities for customary held land and landed property to be controlled by powerful local actors, at the exclusion of the majority of residents who are often poor and vulnerable. Thus, Peters (2004) suggests that researchers working on land-related conflict in Africa need to place more focus 'on who benefits and who loses from instances of "negotiability" in access to land'. For Peters, such a shift of theoretical focus enriches the understanding of 'situations and processes ... that produce inequality at a local level – processes that limit the flexibility of custom for some social groups and categories' (Peters, 2004: 270). Custom could be central to an understanding of how meanings over land change – and custom is not static. During moments of rapid social change custom also undergoes significant transformations and contestations. The early colonial officials could not fully comprehend this in their pursuit of 'legitimate' African property and political regimes that would enable their control over land and Africans. Berry (2002) observes:

> *Far from the timeless web of accepted practice that colonial officials imagined (or hoped for), 'custom' proved in practice to be a shifting kaleidoscope of stories and interests that eluded codification. Officials' efforts to get the customs right – by*

*inventing them, if necessary – were often as destabilizing as they were oppressive (*Berry, 2002: 642).

But custom was increasingly codified in the 1930s and beyond. Much of what went into the codification project was based on the work of anthropologists. However, at that stage the fundamental challenge with the codifying of African customary law was that it was way too late in the day for this process. Since the early 19th century, colonial administrators had been experimenting with their own versions of custom. The most dominant version was the one that favoured chiefly control over communal land. Following the deliberate failure by the colonial state to recognise pre-colonial African systems of tenure, the accounts that were available when custom was being recorded were those of colonial officials and powerful local chiefs (Delius and Chaskalson, 1997). Consequently, as Delius and Chaskalson point out, the voices of African peasants were largely excluded when codified customary law was constructed in southern Africa. This gave significant leverage to chiefly control of land and landed resources.

I argue that these historical processes shape contemporary struggles over land and mining revenues on the platinum belt. At different historical moments, rural residents have been resisting the power of chiefs over communal property. Essentially, struggles of the present not only epitomise shifting meanings over land, especially in areas where rural land is targeted for the extraction of mineral resources. During moments of rapid social change meanings attached to land change, as do institutions that regulate access, control and distribution of land rights. Therefore, land struggles are also struggles over meaning.

ON MEANINGS

Land carries different meanings and local contexts play a critical role in the social construction of meanings around property relations. The legal interpretations of power over rural land often differ from the meaning attached to control of the land by its African owners, occupiers and users. This contribution draws on various studies that conceptualise rural land and struggles over natural resources as contestation over meaning (Peters, 1984; Shipton and Goheen, 1992; Moore, 1993). It argues that mining-led dispossession reveals not only the vulnerability of land rights on the platinum belt but also processes of contestation over the meaning of land. From this perspective struggles over resources are 'engaged via a struggle over meaning' (Peters, 1984: 35). As Peters (1984: 35) observes the fact that 'a category of person or act becomes defined in one way rather than in another is clearly a victory of one meaning over another'. Of course, meanings of land are often fluid and uneven, as such, and could encourage domination of the powerless by local, powerful actors (Moore, 1993).

Shipton and Goheen in their discussion on different meanings attached to land ask some critical questions:

What does land mean, and to whom? How are land and its resources defined and categorised in local cultures, and what kinds of translation problems arise? What kinds of social affiliations affect land use and control? (Shipton and Goheen, 1992: 309)

These questions are relevant to understanding the shifting meanings of land in rural South Africa, particularly in the context of mining expansion and rapid shifts in the value and meaning of land. Large tracts of agricultural land have been fenced off for mining operations. Local small-scale farmers in the impoverished villages nearby are left divided and destitute (Mnwana and Capps, 2015). Specifically, the following questions guide this analysis: What have been the historically dominant meanings attached to land in the villages where mining occurs on the platinum belt? How have these meanings shifted (if at all) with the advent of platinum mining and, most importantly, to whose benefit or detriment? Which local institutions and groups ('social affiliations') have historically defined the character of land rights and how (if at all) have these groups changed in the modern, rapid expansion of platinum mining? Such questions take into account both historical and contemporary analyses of the meanings attached to land. But understanding land struggles through shifting meanings 'over time' requires a strong historical thrust. Thus, Moore elaborates:

Historical patterns of access to, control of, and exclusion from resources emanate from and, in turn, mould competing meanings and cultural understandings of rights, property relations, and entitlements (Moore, 1993: 383).

[...] To understand both [the] material actions and their symbolic meanings, we need to move toward a historical perspective which also highlights the role of social memory (Moore, 1993: 397).

A detailed focus on the remembrance of past systems of landholding can bring us closer to an understanding of the current land struggles in the countryside. Recent research on rural struggles on the platinum belt has shown that people make claims over land – platinum-rich farms – as groups (Mnwana and Capps, 2015; Mnwana, 2016). In such intense struggles against powerful chiefs and mining companies, individuals rely on other group members to back, support and secure their claims. As will surface in the discussion below, these land-claiming groups often use the law and the courts to battle for land.

Elsewhere, I interrogate the antinomy of the upsurge of the power of chiefs in post-apartheid South Africa, particularly their control over land and mining revenues on the platinum belt. In this contribution I look closely at how the rapid expansion of mining over communal land has not only led to significant land dispossession of small-scale farmers in villages but is also producing intense struggles at village level. Such struggles mainly take the form of competing group land claims. I draw on the detailed case study of the Lesetlheng village land claim. It is an extraordinarily organised land claim, and provides the ideal raw material for a detailed analysis of the shifting meanings of land as village farmers attempt to reclaim their land in the face of mining-led dispossession.

METHODS TO EXPLORE THE IMPACT OF PLATINUM MINING ON RURAL COMMUNITIES

I have spent more than a decade studying the multiple impacts of platinum mining on rural communities in the North West and

Limpopo provinces of South Africa. This study went through various phases. At times, I conducted the fieldwork alone, and at other times, I led a research team comprising researchers and research assistants. The research methods used included in-depth interviews, observations, analysis of documents, archival research and detailed oral histories. Participants were purposively selected and included village farmers, activists from the land-claiming groups, youth, mine workers, local headmen, government officials and knowledgeable village elders (women and men who have lived and farmed in the area).

The empirical analysis in this contribution examines closely the dynamics and shifts in social relations of production at a local level. I detail two land claims driven by groups in two villages that fall under the Bakgatla-Ba-Kgafela traditional authority area (henceforth Bakgatla area) in the North West province. Both claims are rooted in the history of group land purchases by the dispossessed Africans in what was then the colonial Transvaal province. Many African groups purchased land through missionary intermediaries in the districts of Marico, Pretoria and Rustenburg (Delius and Chaskalson, 1997). Group land buyers in the Transvaal from the late 1800s up to the second half of the 20th century had to find the nearest 'recognised' chief to mediate the transaction on their behalf. They also had to categorise themselves as 'tribes' for the sake of fulfilling the requirements of buying land (Claassens and Gilfillan, 2008: 298–299; Capps, 2010).

Claassens and Gilfillan also concur that:

> [t]he requirement that groups of purchasers should constitute themselves as 'tribes' was indicative of government assumptions about the nature of rural African society. Such a requirement ignored the reality of the ways in which groups of people constituted themselves in favour of stereotypes about timeless tribal identity. In fact, people sometimes joined the land-buying groups precisely to escape the 'tribal' context (Claassens and Gilfillan, 2008: 298).

The 19th-century struggles of African farmers are well documented, particularly the denial of their entry into the colonial land markets and their subsequent 'indirect' purchasing of land via white (mainly missionary) intermediaries (see Delius and Chaskalson 1997; Claassens and Gilfillan, 2008; Capps, 2010; Mnwana and Capps, 2015). African farming communities in the villages that ended up constituting the modern Bakgatla chiefdom in the Rustenburg region were very much involved in group land purchases. Contemporary struggles over land significantly reflect this history.

THE BAKGATLA AREA AND A HISTORY OF GROUP LAND PURCHASES

Elsewhere I discuss how Bakgatla chiefs enjoyed significant power and control over purchased land, ostensibly as assumed custodians of tribal property (Mnwana, forthcoming). The administrative process involved in the purchasing of land by Africans granted chiefs enormous leverage to define and impose customary rights on their 'subjects' who, quite often, were the actual buyers of land which became registered as tribal property. Group land buying was a dominant feature in the Bakgatla area in the late 19th century (Mnwana, forthcoming). The arrival of a Dutch Reformed Church missionary, Rev. Henri Louis Gonin, to start a mission station in Pilanesberg in 1862 provided an opportunity for Africans in the Bakgatla area to start purchasing a few farms or portions as private group properties. These were registered in the name of Rev. Gonin, since the Transvaal government prohibited Africans from holding land titles in their names. By the year 1900 Gonin had already mediated African group purchases of four farms or portions: Saulspoort 269 (a portion), Kruidfontein 649, Modderkuil 565 (a portion) and Holfontein 593. The earliest recorded group land purchase was for the farm Saulspoort 269, the present Moruleng village, in 1868. This farm was sold by Paul Kruger. A group of Bakgatla under Kgosi Kgamanyane contributed towards this purchase. According to Mbenga (1996: 210) this Bakgatla group contributed £360 for a portion of this farm while Rev. Gonin contributed £540 for the remaining portion.[4] The title deeds to the missionary-mediated purchases were later transferred to a state trustee (Commissioner of Native Affairs) during Retrocession Transvaal.[5] During the hearings of the Native Land Commission in Pilanesberg, Saulspoort[6] in 1906, Rev. Gonin confirmed his role as an intermediary in some of the land purchases:

I did not come here to speculate ... The natives have grazing rights over the portion of the farm [Saulspoort] held by me, but they must not encroach on the gardens and the lands ... The government has never given any land in this part of the country to the Bakhatla [sic] Tribe. I have made no profit in purchasing these farms for the natives (Gonin, 1906).

It is crucial to note that different senior members of the ruling Pilane dynasty organised for the purchase of the farms that Rev. Gonin held in trust. It is however not very clear whether these purchases were always conducted purely through the aegis of the entire Bakgatla tribe. Tensions among the political elite often led to 'cessations' in 1850 and 1860, which meant that groups of the Bakgatla were scattered all over the Pilanesberg on different white-owned farms. On his arrival, Rev. Gonin first encountered smaller groups of Bakgatla under the leadership of different sons of the Kgosi Pilane, the founder of the current ruling dynasty. Although Kgosi Kgamanyane Pilane was

---

4   The total price for this farm was £900. Gonin later (in 1898) sold much of his portion of this farm to the Bakgatla tribe under Kgosi Lenchwe (Kgamanyane's son and successor). Mbenga (1996: 207–214) notes that the Bakgatla nearly forfeited the £360 they contributed towards this purchase when Kruger reneged on his earlier decision to sell the property to the Africans (the Bakgatla who were residing on it) due to the pressure from the *Volksraad* (Afrikaner Transvaal parliament). It was Rev. Gonin's intervention which secured this purchase. The farm was registered in Rev. Gonin's name in August 1869 (Mbenga, 1996: 210).

5   A period roughly between 1881 and the late 1890s, marked by colonial treaties (the Pretoria Convention of 1881 and the London Convention of 1884) between the British and the Boer/Afrikaner settler communities that gave effect to gradual (albeit contested) Boer self-rule in the former Transvaal republic.

6   Now Moruleng village.

regarded as the 'big chief' at the time (Gonin, cited in Mbenga and Morton, 1997: 150) with many followers, his dissident brothers exercised independence and authority over their respective followers. After Kgosi Kgamanyane's departure to Botswana, some of his brothers and their followers purchased farms through Rev. Gonin. The histories of the two disputed farms – discussed below – detail this phenomenon. In light of the dispossession, division and fragmentation of the Bakgatla polity in the late 19th century, it is difficult to confirm that the missionary-mediated purchases were conducted purely on a tribal basis. Moreover, although standard historiography on the Bakgatla chieftaincy[7] point towards the tribe as the only unit through which land purchases were made, contemporary struggles challenge this position – it is hard to sustain this view if one follows the detailed oral histories of land purchases (see below).

Accounts such as the one above also allude to tribal purchase of land. However, this study shows that contemporary idioms and narratives of group land contestation in the Bakgatla area are rooted in a history of diverse, private group land buying. Quite often, discrete African syndicates purchased land on a private basis. In the former Pilanesberg region the Bakgatla chiefs mediated the purchase of many farms during the first half of the 20th century. The purchases were registered under a 'state trustee' on behalf of the Bakgatla chiefs and the tribe. This feature also explains the massive geographical expansion of the Bakgatla territory during the same period. Subsequently, in 1953, when the apartheid state established the Bakgatla-Ba-Kgafela Tribal Authority in terms of the Bantu Authorities Act No. 68 of 1951, there were 27 African group-purchased farms that resorted under the Bakgatla territory.

---

7   See Mbenga (1996: 207–214), Mbenga and Morton (1997: 150) and Morton (1998: 87).

LESETLHENG VILLAGE IN THE PLATINUM-RICH
BAKGATLA AREA

The village of Lesetlheng lies in the north-eastern foothills of the Pilanesberg Mountains. It is one of the 32 villages that constitute the now 'platinum-rich' Bakgatla-Ba-Kgafela traditional authority area (henceforth the Bakgatla area) under Kgosi (chief) Nyalala Pilane. Lesetlheng occupies the farm Kruidfontein 649, north of Moruleng – formerly called Saulspoort (the main/administrative village in the Bakgatla area). As one of the oldest villages in the area, Lesetlheng is a fairly large village and many of its Setswana-speaking residents claim a common historical origin and identity. They trace their ancestry to a group that broke away from the leadership of Kgosi Kgamanyane Pilane before the latter moved his group of Bakgatla to Saulspoort in 1861. This group followed Tshomankane Pilane, Kgamanyane's brother from Kgosi Pilane's third house, to a place called Boipitiko – now the present-day Sun City Resort. Around 1877, while at Boipitiko, Tshomankane organised for the purchase of the farm Kruidfontein 649. Tshomankane died in Boipitiko and was buried there. In 1885 Tshomankane's son, Ditlhake Pilane, took over the leadership of his father's group of Bakgatla. In 1888 Kgosi Ditlhake led his followers to Kruidfontein where they settled and established the current village of Lesetlheng (Morton, 1998: 87).

Although Tshomankane's group became an autonomous political unit after seceding from Kgamanyane's Bakgatla polity, it is said that this group later rejoined the Bakgatla tribe (Schapera, 1938: 306). Standard literature sources do not specify the actual date of this reunion (e.g., Mbenga, 1996; Mbenga and Morton, 1997; Morton, 1998). This uncertainty can be attributed to the colonial administrative measures that increasingly facilitated the absorption of smaller African groups into a few 'recognised' chiefdoms. Tshomankane never enjoyed the colonial state's recognition as a chief of Bakgatla. But his son, Ditlhake Pilane, did rule as a chief of Bakgatla in South Africa for a short period from the late 1800s to 1903. Apparently, Ditlhake's rule took place after his father's breakaway group of Bakgatla had been reintegrated into the 'big' tribe in Moruleng.

Kgosi Lenchwe Pilane I,[8] through the help of the British colonial state, managed to remove Ditlhake from the position of chieftaincy in 1903. Lenchwe I installed Ramono Pilane, his own brother from Botswana. Ramono's appointment as Lenchwe's 'representative ... at Saulsport *[sic]*,'[9] was approved by the Native Commissioner in Rustenburg in February 1903. The Commissioner then wrote to the secretary of the Native Affairs Department reporting that his approval was based on his observation that Ditlhake was 'a weak man and does not command the respect of the people' whereas Ramono commanded 'very great respect.'[10] More instructive was the Commissioner's subsequent comment that, he was 'much pleased' with Ramono's 'expressions of being prepared to work with the (g)overnment'.[11] Such remarks hint at the colonial state's administrative mode of indirect rule which could not function if local chiefs showed strong signs of accountability and closeness to their people. Colonial rule replaced such chiefs with local despots who could be used as instruments of administrative control over African people (Mamdani, 1996).

Lesetlheng has grown significantly over the past two decades. The north-western side of this village is flanked by open pit operations of the Pilanesberg Platinum Mines (PPM) and the Sedibelo Project – a Bakgatla-owned mining project. Lesetlheng has also become a magnet for many migrant mine workers seeking rental accommodation. The PPM began digging its open pit around 2008. The Itireleng Bakgatla Mineral Resources (Pty) Limited (IBMR) Sedibelo Project began development in 2007. The impact of these operations is already visible. Huge tracts of communal farming land have been fenced off for mining activities. Villagers blame the chief and the mines for not consulting them before these projects were initiated. As a result, there is an ongoing land dispute which also involves the nearby mineral-rich farm Wilgespruit 631 (2JQ).

The farm Wilgespruit is currently registered as a tribal property.

---

8   He was the Bakgatla paramount chief (*kgosikgolo*) in Botswana at the time.
9   Pretoria National Archives: NTS, Vol. 372, Ref. 36/13/F1237.
10 Ibid.
11 Ibid.

The Minister of Rural Development and Land Reform holds the title deed in trust for the Bakgatla-Ba-Kgafela traditional community. This farm contains rich and easily accessible PGM reserves. IBMR – a 100 per cent Bakgatla-owned holding company – acquired mining rights to this farm. Kgosi Pilane and his traditional council has since sold some of these rights to PPM. In 2012, IBMR sold Sedibelo West to PPM for US$75 million for the expansion of its open-cast mine (the Tuschenkomst pit) (Mnwana and Capps, 2015). This expansion enters deep into the farm Wilgespruit. This transaction has not only displaced Lesetlheng farmers but also intensified the land dispute between the claimants and the chief (see below).

WILGESPRUIT – A CLAIM TO PRIVATE LAND ACQUISITION IN THE HEART OF PGM TERRITORY

There is an ongoing dispute about the ownership of Wilgespruit. A group of residents in Lesetlheng claim that their forefathers bought it. According to the claimants, buyers were constituted as a private syndicate from 13 clans: Kgosing, Pheto, Mogorosi, Ramolefe, Tlagadi, Huma, Matshego, Serema, Rampedi, Morema, Ramolome, Ramolemane and Botalaota. All of them resided in Lesetlheng between 1916 and 1919 when the farm was bought. When we conducted this study, there was an ongoing application for the adjustment of the title deed. The Lesetlheng claimants made this application in 2012 through the Department of Rural Development and Land Reform. A commissioner was appointed to investigate the application in terms of the Land Titles Adjustment Act No. 111 of 1993. But, as we shall see later in this discussion, the claim to Wilgespruit has intensified land tensions at village level.

The land claimants in Lesetlheng identified themselves as Bakgatla-Ba-Kgafela. According to them their forefathers who bought Wilgespruit were affiliated to the Bakgatla tribe under Kgosana Ditlhake's leadership. After settling in Kruidfontein in the late 1880s, the people of Lesetlheng leased some of the neighbouring European-owned farms for grazing and ploughing, including the farm Magazynskraal 3JQ. Kruidfontein was not suitable for ploughing and it was not big enough to use for livestock grazing. Ditlhake was directly involved in mobilising for the purchase of this farm. He collected the money, cattle and other items from the buyers to raise funds for the purchase price:

> He [Ditlhake] told them [the buyers] that if they could buy and own land it would save them from having to pay rental after harvest (Interview One, Lesetlheng, 31 October 2013).

The process of purchasing the farm began in 1916. Oral traditions of the elders in Lesetlheng reveal that buying Wilgespruit was quite challenging. There were several collections of cattle, crops, money and other items until the farm was finally registered in 1919. It was bought from N. Gluckman and E Judes. Gluckman and Judes bought the farm for £1,560 in 1914.[12] It was hardly two years before they sold it for a staggering £2,600.[13] According to Lesetlheng informants, the process of purchasing this farm was economically demanding, confusing and emotionally draining (see Mnwana, 2018).

The deed of sale for Wilgespruit is dated 16 March 1918. This was a critical period of intensification of the colonial segregation policy under the new formed state – the 'Union' of South Africa – which granted rights only to white citizens. As such, the buyers were now compelled to register the land through the aegis of the officially recognised Bakgatla chief. The Union government in South Africa regarded (although did not officially recognise) Kgosi Lenchwe Pilane I of Bakgatla in Botswana, as the senior leader – a 'paramount chief', *kgosikgolo* – of Bakgatla in South Africa and Botswana. As a result, a number of farms purchased by African groups in Pilanesberg were registered as 'tribal' purchases in the names of the Botswana-based chiefs of Bakgatla, as assumed trustees of 'tribal' property. Kgosi Lenchwe's son, Isang Pilane, signed the deed of sale for Wilgespruit on behalf of his father, Kgosi Lenchwe I, who was suffering from a long illness during that period which led to his death in 1924. He delegated

---

12 Pretoria National Archives, Register, RAK 3020.
13 Pretoria National Archives, NTS, 254, 1137/16/F596.

his son, Isang, to act on his behalf. Ditlhake Pilane co-signed the deed of sale together with Dialoa K. Pilane who was the acting chief of the Bakgatla in Pilanesberg. The transfer for this purchase was registered in 1919 under the 'Minister in charge of Native Affairs ... in trust for the Bakgatla tribe under Chief Linchwe K. Pilane'.[14] This transfer was delayed owing to the evidently inflated purchase price and the hardships that the buyers experienced in raising funds. Seemingly, state officials never took notice of the buyers' financial difficulties. If they did, they showed little sympathy. For instance, the sub-Native Commissioner in Rustenburg wrote in a letter to the Secretary of Native Affairs dated 8 February 1919 that the delay in this transfer was simply caused 'by the natives' tardiness in collecting money to pay the costs'.[15]

THE 'LAW OF WILGESPRUIT' AND INCLUSIVE DISTRIBUTION OF LAND RIGHTS

Few written records exist that attest to oral traditions on group land buying in the Bakgatla area. But the Matshego clan in Lesethleng has carefully kept a record in what they call the 'Old Preserved Book'. Members of the Matshego clan have passed this book on from generation to generation. Though not detailed, the book contains some records about members of the Matshego clan who contributed towards the purchase of Wilgespruit and other farms. When collections were made for purchasing Wilgespruit a member of the clan recorded these contributions. The information is written in Setswana in both ink and pencil. Although there are no precise dates for most recordings, it is a source of important information. On one of the pages, there is an interesting record titled: '*Molao oa Polasa sa* Wilgespruit' – 'The Law of the farm Wilgespruit'. There is a brief explanation about its purchase. Of key interest is a declaration made in May 1919 by a certain Raiyane Pilane, who was apparently an acting *kgosana* in Lesetlheng. Pilane (undated) declares:

---

14 Ibid.
15 Pretoria National Archives, NTS, 254, 1137/16/F596.

'*Ke le segela Polesa. Ka li kgoro o sa rekang ge baabo ba mokoba. Molato ga se oa kgosi.*' – *I'm demarcating the farm. The clans that did not contribute should not come to the kgosi if they are chased out of the farm [by the buyers]. It will not be chief's problem.*[16]

This 'declaration' was made less than three months after Wilgespruit was transferred to the Minister of Native Affairs in trust for the Bakgatla chief and his Bakgatla tribe. This declaration confirms most of the informants' claims that Wilgespruit was divided among the clans (*dikgoro*) that purchased it. Clearly, a *kgosana* (headman) was involved in early decisions about the demarcation of the farm, but later the members of the purchasing clans seem to have exercised greater independence. Power over distribution of land rights remained within the members of the buying clans. Each clan of buyers was allocated a huge ploughing plot called a *panka* (plural *dipanka*). Each *panka* was further divided into smaller portions of cultivated plots called *diakere* (singular *akere*). Each *akere* belonged to one of the families of the buying clan (*kgoro*). Some informants claimed that the size of the plots given to different clans was determined by the specific contributions that they made when the farm was purchased. It was each clan's prerogative to grant portions of ploughing plots within their *dipanka* to non-buyers. These included relatives, friends, neighbours and immigrants who became attached to certain clans in Lesetlheng. The following response captures the intricate process of distribution of land rights:

The farm portions continued to be used by each clan until now. New members joined the clans through marriage and adoption as 'refugees' from other villages or tribes. These families were allocated pieces of land within the portion of the adoptive clan to cultivate and feed the children (Interview Two, Pretoria, 10 November 2013).

---

16 Quotation from a personal archive – a detailed record of meetings and family contributions to farm purchases, kept by one of the members of the Matshego clan in Lesethleng and called the 'Old Preserved Book'.

Wilgespruit was one of the most productive farms in the Bakgatla area (Mnwana, 2018). Oral histories established that farming families used to harvest countless bags of sorghum, maize, beans and many other crops. Every winter after harvesting, the farmers at Wilgespruit opened the farm for communal grazing. An elderly man explained:

> *We were ploughing and staying there. After ploughing when we just reap the crops, the cattle would stay there. All the villagers were ploughing. It was a very big agricultural farm* (Interview Three, Lesetlheng, 6 September 2013).

However, when this study was conducted (from 2013 to 2017) most of the agricultural land at Wilgespruit had been fenced off for mining operations. Families who still had cattle were using the remaining land for grazing. It was becoming increasingly difficult for Lesetlheng famers to visit their cattle posts since entrances were often guarded by heavily armed private security officers.

There was some physical evidence that this farm was once a piece of productive agricultural land. The small mud and corrugated iron structures where people used to live during ploughing season were still there. It was also common to see old rusty ploughs and other farming implements lying around. Some of the former ploughing plots had small (hand-dug) dams that various clans dug to water their crops. Some of these dams were still there although most of them had dried up.

Mining has also introduced rapid socio-economic shifts in the rural villages that fall under the Bakgatla area. New forms of tensions and divisions have begun to surface among Lesetlheng residents. The following discussion traces these lines of communal fracture at village level.

'IT IS PRIVATE LAND!'

> *Now we wonder … What we are talking about is that, we saw the company [PPM] installing fence and prospecting on our land without any consultation or a community resolution. Then we*

*asked the chief what is going on there on our land because it is
a private land without our knowledge. Why do the people just
come and start mining without any consultation?* (Interview
Three, Lesetlheng, 6 September 2013).

The rapid escalation of mining activities on Wilgespruit and other
adjacent farms has significantly affected the agrarian lifestyle of
Lesetlheng's residents. Village farmers claimed that Chief Pilane did
not consult with them before allowing PPM to fence off huge portions
of Wilgespruit – their farming land. The land claimants in Lesetlheng
argue that their names, as the descendants of the original buyers, should
be on the title deed for Wilgespruit, not the name of the chief and the
tribe. They view the registration of the title deed for Wilgespruit in the
name of the tribe (and the chief) as a form of dispossession. This group
demands to engage directly with PPM and other mining companies that
operate on the farm. According to them, their ancestors purchased this
farm as a private group, not the Bakgatla tribe as a whole. In turn, they
contend they should receive direct revenues from the mine. Clearly
the Lesetlheng claimants refute the chief's custodianship over mining
revenues. As shown in the aforementioned interview quote, they argue
that Wilgespruit is private property.

The emphasis on the private character of rights to this farm has
produced new forms of local struggle. The land rights on this farm,
although historically allocated to specific clans, have subsequently been
distributed, inherited and defended on a customary basis. Families of
non-buyers, who were allocated ploughing fields at Wilgespruit, have
held those rights unconditionally and their children inherited such
rights without any challenges.

Things took an unexpected shift in 2012 when the Land Title
Adjustment Application Commission began its investigation. The
commissioner requested the claimants' family trees. This required the
claimants to validate their claims by demonstrating who the original
buyer was in each family and how they are related to members of the
original land-buying syndicate. Sensitive issues started to surface.

Some of the elders had always known that not every family that
was ploughing on Wilgespruit was descended from the original

buyers. The requirement for land claimants to produce family trees raised two contentious issues. First, the descendants of non-buyers were not going to submit family trees since their ancestors did not contribute to the farm's purchase, and as such, they were excluded. This exclusion discouraged some of the village members who did not belong to one of the 13 clans of buyers. The land claim exposed divisions between the 'buyers' and 'non-buyers' in Lesetlheng. It also revealed some splits within the land claimants. Disputes around the processes through which each family could legitimise its claim started to emerge. New group identities also surfaced. Some began to claim that the direct descendants of the families that bought Wilgespruit – called *dibeso*, or *sebeso* (singular for *dibeso*) – are the main beneficiaries in the *kgoro* (a clan) (Mnwana, 2018). Clans are more inclusive and, as such, would render the historically insignificant meaning of land – as a private property – less fulfilling for those in the group who want to engage directly with mining capital as the *de jure* owners of this land (Wilgespruit). Such a finding raises the question: why is land now claimed exclusively as a private property (of a group) when that meaning had been avoided for almost a century?

Clearly, land has formed the basis of the diverse livelihood strategies that Africans in rural South Africa have historically been dependent on, along with other resources (selling their labour, migrant remittances, etc.). However, in the case of Lesetlheng, rural residents seem to have resisted commodification of land in the form of privatisation of property rights until recently. The free distribution of land rights to families and individuals outside the clan of buyers epitomises this phenomenon. This happened despite the land in question being historically acquired through a private group purchase (whether by a 'tribal' [Bakgatla] group or a distinct syndicate). A classical agrarian political economy analysis of social relations of production would primarily define land as 'property' or as a commodity – a basic means of production. Such a meaning is clearly summed up in Bernstein's four key questions on capitalist social relations of production in relation to landed property, namely who owns what, who does what, who gets what and what do they do with it (Bernstein, 2010: 22)?

Such a meaning envisions land as a productive asset, with a hierarchy of rights attached to it and aimed towards the accumulation of surplus by a few powerful actors. The agrarian lifestyle of Lesetlheng villages produced a significant degree of inclusiveness and social cohesion and allowed peasants to resist commodification of the land. This should also be seen against the backdrop of unparalleled land dispossession and the systematic destruction of a once-thriving African peasantry, by the colonial and apartheid states. As evidenced in the case of Lesetlheng, land has meant more to the people than a productive asset from which to derive a livelihood. It has been central in communal cohesion and a means of resisting landlessness and dispossession. Therefore, the material role of land and its capitalist social and economic hierarchal meanings that ascribe status and power to a few individuals – e.g., 'owners' – were successfully ignored or suppressed for several decades. Under the current mining dispensation, mining capital, the state and local chiefs tend to elevate this materialistic meaning of land – land as private commodity. Hence, material inequalities and communal divisions at village level have escalated.

The meaning of land as a family, group and communal resource may not have been fully egalitarian (see Mnwana and Capps, 2015; Mnwana, 2018), but it kept the social bonds at a micro level (family, clan and village) intact for almost a century. It also kept landlessness at bay. This meant that dispossessed African peasants could share whatever small areas of land they managed to access. It is through this meaning that agrarian livelihoods survived in villages like Lesetlheng, in the North West province. Nevertheless, the advent of platinum mining challenges the inclusive meanings of land in an unparalleled way.

As the gold industry was instrumental in turning African peasantry into cheap migrant labour, the platinum industry seems to be driving the last remnants of Africans out of land and agrarian livelihoods. Thus, the question remains: is the rural-based mining industry heading towards destroying the remaining vestiges of African peasantry through yet another layer of dispossession?

CONCLUSION: WHEN LAND BECOMES PRIVATE PROPERTY!

In previous contributions (see Mnwana and Capps, 2015; Mnwana, 2016) I have hinted at how some of the land claims on the platinum belt have led to a series of court battles, conflict and gender exclusion at village level. Land disputes also connect to prolonged resistance to mining capital, chiefs and the state. In recent years, villagers have held a series of protests against the local mines and the chief in the Bakgatla area, demanding compensation for loss of land and jobs. We detail these protests elsewhere (see Mnwana and Capps, 2015).

This chapter highlights how current, exclusive property claims over land and shifting meanings over the mineral-rich farm Wilgespruit could be pointing to not only new forms of fracture but also to the shifting meanings of land in the context of mining-led dispossession.

The early, fairly egalitarian mode of distribution of land rights between the 'buying clans' and families of 'non-buyers' on Wilgespruit is nothing new in African systems of landholding and resource distribution. For instance, Peters (1994), in her work on elite capture (through privatisation) of common resources (water/boreholes and pastoral land) in the Botswana-based Bakgatla community, details how the elite 'syndicates' that controlled the borehole development scheme were somewhat inclusive in their distribution approach. She explains:

> [n]one of these syndicates included only kinsman ... Even where most syndicate members came from a single ward, they were not all kinsmen ... some members came from other wards through connections of kinship, marriage, friendship or patronage (Peters, 1994: 72).

Such a representation connects well with the 'private' control and distribution of rights on Wilgespruit among the buying clans. The rules or social processes that have historically governed access and distribution of land rights were evidently more inclusive and largely drawn from shared customary norms. The 'land buyers' of Wilgespruit farm were, arguably, more accommodating of the families of non-buyers than the Botswana borehole syndicates. The generous distribution of 'privately' acquired land rights, which operated on a customary basis for almost a century in Lesetlheng, points towards one less-examined aspect of African agency in the context of colonial dispossession: African farmers wanted to reclaim their land. If purchasing it back through missionaries or chiefs (as intermediaries) was the way to acquire land, then they were willing to do whatever it took to enter colonial land markets. Hence, group purchases were not merely about Africans wanting to own private property as groups.

The case of Wilgespruit's inclusive and generous distribution of land reveals at least two critical issues about African agency in the face of dispossession. The first is that African farmers bought land to defend and reclaim their indigenous landholding systems, not merely to 'own' private property. As explained earlier, the colonial state failed to recognise African landholding systems and institutions. The farmers in Lesetlheng revived, defended and kept these customary processes of landholding for more than a century after colonial conquest. For instance, the struggle over the title of the farm is a very recent phenomenon. Never before have the farmers in Lesetlheng claimed the title of the farm, although the history of private group purchase was always well known. Then again, this accommodating character did not lead to undifferentiated property relations between the members of 'purchasing clans' and families of 'non-buyers'. There were some tensions and historical inequalities. These could not be covered at length in this chapter. However, they are detailed elsewhere (see Mnwana, 2018).

The second issue is analytical: mining expansion represents another form of dispossession. However, one may argue that since the Bakgatla chieftaincy gains control over mining revenues through exclusive claims to land and tribal identity, the land claimants see no other alternative but to shift their meaning of land towards exclusive group claims and histories. Again, one can argue, this is a response to yet another form of African land dispossession. Nonetheless, mining is yet to feature prominently in scholarship on African land dispossession.

*As this manuscript was going to print (in October 2018), the Constitutional Court of South Africa passed a significant judgement in favour of the Lesetlheng land-claiming group in the Bakgatla area. The implications of this judgement could not be analysed and discussed in time for publication. I will do this in a future paper, currently in progress.

REFERENCES

Bernstein, H. (2010). *Class Dynamics of Agrarian Change*. Fernwood Publishing, Halifax.

Berry, S. (2002). 'Debating the land question in Africa'. *Comparative Studies in Society and History*, Vol. 44(2), pp. 638–668.

Boone, C. (2017). 'Legal empowerment of the poor through property rights reform: Tensions and trade-offs of land registration and titling in sub-Saharan Africa'. WIDER Working Paper No. 2017/37.

Capps, G. (2010). 'Tribal-landed property: The political economy of the Bafokeng chieftaincy, South Africa, 1837-1994'. Unpublished PhD thesis. London School of Economics.

Claassens, A. and Gilfillan, D. (2008). 'The Kalkfontein land purchases: Eighty years on and still struggling for ownership'. In Claassens, A. and Cousins, B. (eds). *Land, Power and Custom: Controversies Generated by South Africa's Communal Land Rights Act*. UCT Press, Cape Town.

Cousins, B. (2007). 'More than socially embedded: The distinctive character of "communal tenure" regimes in South Africa and its implications for land policy'. *Journal of Agrarian Change*, Vol. 7(3), pp. 281–315.

Delius, P. and Chaskalson, M. (1997). 'A historical investigation of underlying rights of land registered as state owned'. Report commissioned by the Tenure Reform Core Group. Department of Land Affairs, Pretoria.

Gonin, H.L. (20 December 1906). 'Proceedings of the Native Location Commission'. Pretoria National Archives, TA C27.13. Saulspoort.

Liedtke, S. (2018). 'Another hydrogen boost for platinum fuel cells'. Available at: http://www.miningweekly.com/article/amplats-invests-in-high-yield-energy-technologies-2018-04-18. Accessed on 18 April 2018.

Magubane, B. (1978). 'The "native reserves" (bantustans) and the role of the migrant labor system in the political economy of South Africa'. In Idris-Soven, E. and Vaughan M.K. (eds). *The World as a Company Town: Multinational Corporations and Social Change*. Mouton Publishers, The Hague.

Mamdani, M. (1996). *Citizen and Subject: Contemporary Africa and the Legacy of Late Colonialism*. Princeton University Press, Princeton.

Mbenga, B.K. (1996). 'The Bakgatla-Baga-Kgafela in the Pilanesberg District of the Western Transvaal from 1899 to 1931'. Unpublished PhD thesis. University of South Africa, Pretoria.

Mbenga, B. and Morton, F. (1997). 'The missionary as land broker: Henri Gonin, Saulspoort 269 and the Bakgatla of Rustenburg District, 1862–1922'. *South African Historical Journal,* Vol. 36(1), pp. 145–167.

Mnwana, S. (2018). 'Mining, rural struggles and inequality on the platinum belt, South Africa'. In Khadiagala, M., Mosoetsa, S., Pillay, D. and Southall, R. (eds). *New South African Review 6: The Crisis of Inequality.* Wits University Press, Johannesburg.

Mnwana, S. (forthcoming). 'Chiefs, land and distributive struggles on the platinum belt: A case of Bakgatla-ba-Kgafela in the North West Province, South Africa'. In MISTRA (ed). *Traditional Leadership in Contemporary South Africa: Towards a Pioneering Publication.* Mapungubwe Institute for Strategic Reflection (MISTRA), Johannesburg.

Mnwana, S. and Capps, G. (2015). ' "No chief ever bought a piece of land!" Struggles over Property, Community and Mining in the Bakgatla-ba-Kgafela Traditional Authority Area, North West Province, South Africa'. Working Paper: 3. Society Work and Development Institute, University of the Witwatersrand, Johannesburg.

Moore, D.S. (1993). 'Contesting terrain in Zimbabwe's eastern highlands: Political ecology, ethnography, and peasant resource struggles'. *Economic Geography,* Vol. 69(4), pp. 380–401.

Morton, F. (1998). 'Cattle holders, Evangelists, and socioeconomic transformation among the baKgatla of Rustenburg District, 1863–1898'. *South African Historical Journal,* Vol. 38(1), pp. 79–98.

Peters, P.E. (1984). 'Struggles over water, struggles over meaning: Cattle, water and the state in Botswana'. *Africa,* Vol. 54(3), pp. 29–49.

Peters, P. E. (2004). 'Inequality and social conflict over land in Africa'. *Journal of Agrarian Change,* Vol. 4(3), pp. 269–314.

Pilane, R. (Undated). 'Old preserved book'. Personal archive of the Matshego clan in Lesethleng.

Schapera, I. (1938). *A Handbook of Tswana Law and Custom: Tribal Legislation among the Tswana of the Bechuanaland Protectorate.* London.

Shipton, P. and Goheen, M. (1992). 'Introduction. Understanding African land-holding: Power, wealth, and meaning'. *Africa: Journal of the International African Institute,* Vol. 62(3), pp. 307–325.

EIGHT

Cleaning up after mines are long gone

Understanding the complex dimensions for inclusive development

SHINGIRIRAI S. MUTANGA

MINING HAS PLAYED A VITAL ROLE in the economy of South Africa for over 100 years. In 2015 the mining industry contributed R286 billion towards South African gross domestic product (GDP) representing 7.1 per cent of overall GDP. Despite the descending trend from its peak some decades ago (declining from the 21 per cent contribution to GDP in 1970), the mining sector still makes a valuable contribution to national economic aggregate output (KPMG, 2017). The sector produces direct and indirect employment and fiscal linkages through royalties, taxes and foreign exchange earnings (Mujuru and Mutanga, 2016). However, the net positive impacts have been beset by socio-economic and environmental pressures that are associated with the excavation of mineral resources.[1]

Both underground and surface mining internationally has ushered in a plethora of challenges to the environment and natural ecosystems. The major challenges are around disused or resource-depleted mines which are causing tremendous damage to the environment through the acidification of water bodies, degraded soil quality, biodiversity loss and the obliteration of natural landscapes (Xiao Wu et al., 2011). These factors all have ripple effects on human wellbeing. This is compounded by the challenges of identifying, and holding to account, those responsible for disused or derelict mines.

South Africa has a track record of having been the world's leading producer of gold, with the world's largest reserves of manganese and platinum group metals (PGMs). In addition, South Africa has considerable deposits of diamonds, chromite ore, vanadium, iron ore and coal. Invariably there are multiple disused mines, estimated to be a total of 5,906 (Auditor-General, 2009: 4).

Mining has a wide range of impacts; however, this chapter confines its focus to the complexities associated with environmental restoration of disused mines and the social ramifications of this restoration in post-mining areas. A number of questions help couch the discussion. To begin with, there is the apportionment of liabilities, which has remained a cause for concern. Can past owners still be found and held liable for liabilities they escaped previously? Not all derelict mines are ownerless, but what about those that are ownerless? For ownerless mines, the impasse is whether government has to shoulder liability, or whether this should be shared by members of the mining fraternity, based on the benefits to them derived from mining in South Africa and their responsibility to care about the current and past side effects of such benefits.

Secondly, there is a need to contribute to achieving Sustainable Development Goals (SDGs) on clean water, intertwined with the need to provide good health and wellbeing for all, in relation to the restoration of mining areas, particularly in South Africa. The question therefore is

---

1   In the past the mining industry would look at maximising profit without necessarily reserving capital investment towards the environment during mineral exploration. The extent of pressure and costs associated is context specific.

can we develop systematic and integrated approaches that may aid the extractive industry through engagement, partnership, and dialogue between the industry, government, civil society and the communities?

Thirdly, how can we upscale innovations to meet the targets of the SDGs identified as central to this discussion? The next section thus provides the nexus between the SDGs and the clean-up after mines.

NEXUS BETWEEN CLEAN-UP AFTER MINES AND SDGS

SDGs represent a paradigm shift in global sustainability thinking. They recognise the complex interactions between the social, economic and the environmental dimensions of sustainability. Though there are no dedicated goals or targets for extractives, the Columbia Center for Investment has tracked the impact of mining or the extractive industry on each of the 17 SDGs.

The focus of this chapter is on the impact mining has on the physical and bio-physical pillars of the environment. Mining activities typically have impacts on land, water and air, which are key pillars required to attain optimum ecosystem function. In addition, mining ramifications are a serious threat to human health. SDGs 6 (clean water and sanitation), 15 (life on land) and 3 (good health and wellbeing) have thus been selected to introduce the overarching chapter's main focus as it grapples with the challenge of restoring mining areas after the cessation of mining.

Essentially SDG 6 is about clean water and sanitation, and is closely linked to SDG 15, which focuses on life on land. SDG 15 seeks to protect, restore and promote sustainable use of terrestrial ecosystems, ensure sustainably managed forests, combat desertification, halt and reverse land degradation and halt biodiversity loss. The rationale for selecting this SDG is the alarming statistics which indicate that more than 40 per cent of the global population is affected by water scarcity (UNDP, 2016a). Over 1.7 billion people are currently living in river basins whose water use exceeds the basin's recharge capacity (UNDP, 2016b).

Within this global challenge, South Africa has been declared a water-scarce country. Juxtaposing the water scarcity challenge with mining points to the tremendous threat posed by abandoned mines, which are pumping out acid mine drainage (AMD) to the limited fresh water bodies available. This damage is exacerbated by current operational mines which are discharging highly polluted and toxic effluent (waste) into streams. Cleaning up after the mines thus becomes critical – in light of the need to achieve SDG 6 on clean water and sanitation. SDG 15 acknowledges that there is unprecedented land degradation and loss of arable land at 30 to 35 times the historical rates (UNDP, 2016a). To quantify the severity of the challenge, a case in point is that of India where land degradation shows a high stripping ratio for the coal industry. A study by Sahu and Dash (2014) revealed that Coal India Ltd removed approximately 500 million cubic metres (mcum) of overburden (OB) to produce 260 mt of coal in the year 2003–2004 at an average stripping ratio of 1.92 mcum of OB per tonne of coal production.

On one hand mine development requires access to land and water; on the other, it results in significant damage to the very same natural resources that are critical for sustainable terrestrial ecosystems and human wellbeing. The interconnectedness of global issues drove the system design of the SDGs, resulting in strong themes of universality, integration and transformation. Similarly, the need for integrated land and water resources management remains a cornerstone for achieving the SDGs.

THE MULTIFACETED CHALLENGES ON POST-MINING SITES

Large-scale mining can lead to the collapse of tailing facilities and a reduction in productive land for agriculture as well as urban living. According to Gu, Tan and their collaborating authors (Gu et al., 2012), the negative features of mineral exploitation include opencast excavations increasing the number of mine dumps and ground subsidence which can lead to landslides, spontaneous combustion, explosions and cracks. These constitute mining disasters which raise potential risks for urban development and public safety. Gao et al. (2012) underscore this by pinpointing mine-related geological disasters as an important branch of geology.

Land destruction

Subsidence is one of the major geological disasters that have led to land destruction. Subsidence can be defined as a time-dependent process induced by human or natural activities and characterised by the lowering of the ground surface following the removal of gas, liquid or solid matter (Jennings et al., 2008). In China, 92 per cent of coal is derived from underground mining which has left behind numerous underground wall panels and thick coal seams that have led to tremendous subsidence (Hu, 1996). Deformation of the rock mass may lead to surface subsidence which may include the local, lateral and upward displacements of rock above unmined areas (near mine boundaries or barrier pillars) caused by the downward movement of overburden into mine cavities. Strains induced by mining and transmitted through intervening strata to the surface may be compressive or tensile and may have both horizontal and vertical components. In simpler terms, mining can have an effect on the geological properties which may result in the types of subsidence outlined.

In South Africa the subsidence phenomenon has been widespread with an estimated record of 2,500 sinkholes and subsidence events (Oosthuizen and Richardson, 2011). A significant majority (98 per cent) have occurred in Gauteng. Gauteng province has three prominent areas of sinkhole and subsidence development: the Far West Rand, the area south of Pretoria and the East Rand (Council for Geoscience, 2011). Beyond South Africa's borders, other cases in point are diamond mining in Angola and the Democratic Republic of Congo, where mining has led to the destruction of terrestrial ecosystems (UNEP, 2011).

Apart from reduced arable land, sinkholes cause costly damage. Figure 1 illustrates the 55-metre diameter sinkhole that swallowed a West Driefontein mine crusher machine. The estimated cost of the damage due to sinkholes was in excess of R1.3 billion (Oosthuizen and Richardson, 2011: 1)

Figure 1: Illustrations of the effect of sinkholes
*Source: Oosthuizen and Richardson, 2011. The photograph on the left
is the West Driefontein sinkhole, illustrating how the earth's surface and
infrastructure can easily be destroyed. The aerial photograph on the right of
the Wonderfonteinspruit Valley, Venterspost shows multiple sinkholes on a
farm, suggesting reduced availability of farmland.*

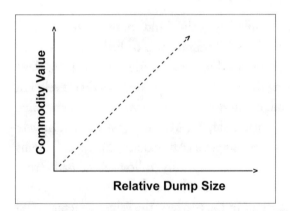

Figure 2: Relationship between commodity value and relative dump size
*Source: EcoPartners, 2016. This graph illustrates the direct relationship
between commodity value and mine dump size. The size of the dump increases
proportionally with greater commodity value.*

Apart from causing subsidence, mineral resource exploitation also
causes damage through solid waste or mine dumps. More recently
labelled 'residue stockpiles' or 'residue deposits', these mine dumps are
a physical feature of all commodities excavated. The question of who
owns mine dumps is contentious and remains unresolved. The South
African government is the leading 'proprietor' of approximately 6,000

abandoned and ownerless mines and their associated dumps across all mineral commodities (Gao et al., 2012: 8). Mines have closed once economically and technically exploitable ores have been exhausted leaving eyesores and polluted land. The South African case provides concrete evidence of how large-scale mining created the residue stock piles and deposits which are detrimental to the environment. However, artisanal small-scale mining also contributes to major environmental and health challenges, especially negative impacts on the physical environment (river siltation and lands not reclaimed) and on miners' health, as a result of exposure to mercury and cyanide (for gold miners) (UNECA, 2011). The report also cites widespread air, soil and water pollution in Zambia's copper belt with extensive 'digging, pumping and disposal of large volumes of waste water, and smelting operations that emit sulphur dioxide'. Land destruction from gold mining in the Wassa West District in Ghana is also highlighted (UNECA, 2011: 8; see also chapter 6).

The link between commodity value and dump sizes is well established (EcoPartners, 2016). Commodities of little value tend to leave minor dumps while high value commodities, mined from deep shafts, tend to yield larger dumps. Figure 2 illustrates a direct relation between dump size and commodity value. Invariably, high value commodities such as copper, gold, PGMs and coal have relatively higher dump sizes over other minerals (Abubakary, 2014). This stems from the process whereby mines typically follow a set path from prospecting, to development, to extraction and finally closure as the finite resources are exhausted. In the process, the mines generate solid waste (one type of which is known as tailings) stored at or near the mine site itself. Planning for mine closure should thus take into account the nature, type and characteristics of the mineral exploration, keeping in mind the potential waste generated in the process.

The challenge of AMD in post-mining areas

AMD arising from gold mining has been the most commonly documented challenge in South Africa due to the volumes involved (there has been massive decanting or pouring of acidic waste). While the source of AMD formation is largely abandoned mines and their associated waste dumps, the problem of acid water spreads far beyond the immediate surroundings of the mines and mine dumps. A case in point is the West Rand where AMD has exposed residents to health hazards, radon exhalation, radiation, dust and other tailings-related hazards from the old slimes dams (Adler and Rascher, 2007).

The environmental health impacts of AMD in South Africa have not been systematically surveyed (SAHRC, 2016). However, there is evidence to suggest that long-term exposure to AMD has resulted in increased rates of cancer, decreased cognitive function, skin lesions, health problems in pregnant women, neural problems and possible mental retardation (Claassen, 2006).

One of the social impacts of AMD has been the displacement of 10,000 households in Khutsong, in the West Rand, at great human and financial cost (Mujuru and Mutanga, 2016: 72). Illustrating the connections between water and land contamination, and between rural and urban impacts, farms rendered unproductive by salts emanating from AMD cannot be sold for urban expansion purposes. This is because the land in question cannot support urban building due to remaining underground mine tunnels. (This is happening in Fleurhof in Gauteng, posing a serious risk to many people.)

Acidification of water resources

Acid water flows either on the surface or underground and eventually flows into natural waterways such as streams and rivers (see Figure 3). There is an estimated 62 megalitres per day post-closure decant from coal mines in the Highveld Coalfield in Mpumalanga and around 50 megalitres per day of AMD discharging into the Olifants River catchment. This reduces the quality of water for irrigation and municipal consumption and damages freshwater ecosystems and biodiversity, as illustrated in Figure 3. The so-called Western Basin[2] area of Mpumalanga illustrates the worst of the effects of AMD pollution.

---

2   The Western Basin covers the area that has been worst affected by AMD. The mining shafts started decanting as far back as August 2002, thereby polluting the Tweelopiespruit that drains into the Krugersdorp Game Reserve.

Figure 3: AMD damage to freshwater ecosystems caused by coal mining in the Mpumalanga Western Basin
Source: Author

The discovery of the decanting (pouring of acidic waste) can be traced back to August 2002, when the polluted Tweelopiespruit began draining into the Krugersdorp Game Reserve (CSIR, 2009). Essentially, the residues of both existing and non-operational mines have been nonpoint sources of AMD[3] into the mine shafts. Old mine shafts, namely the Black Reef Incline, Number 17 and Number 18 Winzes, have experienced tremendous AMD decants (DWA, 2013). On average, 20 million litres per day (Ml/d) of AMD decant have been recorded, with measured peak volumes reaching as high as 60 Ml/d during the wet season. Of these recorded volumes, approximately 12 Ml/d is partially treated, while the rest flows freely into the Tweelopiespruit (see Earthlife Africa, 2010). An estimated 27 Ml/d of AMD needs to be treated in this basin to retain the water below the environmental critical level (ECL). The ECL is the level that the water in the shafts and voids should not exceed in order to protect the environment, and in particular, groundwater (Mujuru and Mutanga, 2016). Interventions have been put in place to address the challenges. An assessment of progress made to date is needed.

The enormity of the problem is a source of despair for a number of stakeholders. The legislative directive to mines within the area to pump

---

3   Nonpoint source pollution refers to land runoff, precipitation, atmospheric deposition, drainage, seepage or hydrological modification (rainfall and snowmelt) where tracing pollution back to a single source is difficult.

and treat AMD has yielded little result. Advocate-researchers such as David van Wyk (2012) argue that mine owners are reluctant to manage and mitigate environmental impacts of disused mines because it is not profitable to do so. As much as social and labour plans and related legislation require mining companies to plan for mine closure, disused or abandoned mines remain a big challenge. In some cases, operating companies have been shown to have limited capacity and ability as the main drivers, leaving AMD as an intractable environmental challenge (Mujuru and Mutanga, 2016).

This has forced the South African government to mobilise resources and create a fund for AMD clean-up. An estimated R653 million rands have been allocated for this, through the Department of Water Affairs, Department of Mineral Resources and the Department of Trade and Industry (Dyosi and Banda, 2016).

In sum, AMD has been described as a ticking time bomb. Mujuru and Mutanga (2016) underscore that South Africa is a water-scarce country. The degradation of water quality from AMD exacerbates the situation by reducing the availability of clean water, posing a threat to ecosystem health, as well as to economic development in the country. Pollution of rivers and groundwater resources makes water unsuitable for several uses unless there are suitable water treatment processes in place. Given the severity of the threat posed by the unused mines, the question arises of how South Africa arrived at this point.

THE GENESIS OF POST-MINING LANDSCAPE DECAY IN SOUTH AFRICA

The enormity of post-mining degeneration in South Africa can be explained by two issues that have impacts in tandem. Firstly, the extent of the post-mining challenge in South Africa can be attributed to a lack of awareness about these problems and a lack of willingness to tackle them, even once they have been identified. A review of the history of environmental legislation that governs the mining industry highlights important aspects of this relationship (see Figure 4).

Mujuru and Mutanga (2016) examine the state's early attempts to regulate the mining industry's impact on the environment, concluding that despite historic pollution from mines and the existence of laws establishing liability for this pollution, mine owners were not held liable for the rehabilitation and remediation of mine waste. A case in point is the Transvaal Mining Laws of 1903, which focused on the so-called 'safe-making of mining operations'. This legislation sought to promote health and safety at the mines. Between the 1930s and 1950s, there was a focus on 'minimum distance to structures', as enshrined in regulations relating to mines, works and machinery. This regulation prescribed a minimum distance between the mining sites and physical structures, be they commercial or residential. The basic planning for environmental recovery started in the 1950s, with the introduction of the Mines and Works Act. The Act was promulgated in 1956. Essentially, it provided for rehabilitation planning, as well as soil and vegetation recovery planning, given the massive destruction of vegetation which had been occurring during the set-up or establishment of mines. The promulgation of the Minerals Act of 1991 was the first attempt at featuring rehabilitation as an express policy objective of mineral legislation (Feris and Kotzé, 2014). The Act made provision for operating mines to provide funds for environmental and social rehabilitation upon mine closure.

The post-apartheid government, in some respects, paid greater attention to environmental challenges such as AMD, both tacitly and explicitly. The AMD challenge is largely articulated in Section 24 of South Africa's Constitution (Constitution of the Republic of South Africa, 1996). The section is about the environment and the health and wellbeing of people in the country. Emphasis is put on environmental protection for the benefit of present and future generations (Lwabukuna, 2016). With the urgent need to find resources to meet the growing demands for ecological restoration and environmental improvement – particularly on post-mining landscapes – the following Acts and regulations were promulgated and amended:

- Minerals Act No. 50 of 1991,
- National Environmental Management Act (NEMA) No. 107 of 1998
- Minerals and Petroleum Resources Development Act (MPRDA) No. 28 of 2002, and
- Water Resources Management Act No. 11 of 2009.

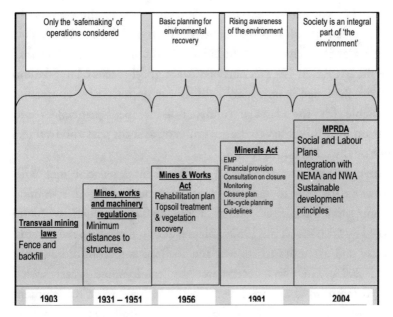

| Only the 'safemaking' of operations considered | Basic planning for environmental recovery | Rising awareness of the environment | Society is an integral part of 'the environment' |

Figure 4: A sample of mining-related legislation in South Africa, 1903–2004
Source: Mujuru and Mutanga, 2016

The Acts and regulations provided a framework for estimating the cost of the closure operations to be carried out by the mining companies (see Swart, 2003). A commitment to revisit the estimation annually was made by the government through the NEMA (Act No. 107 of 1998), as well as the National Water Act (NWA, Act No. 36 of 1998), to ensure adequate provision of funds.

Despite this legislation, ceased mining operations remain a challenge since most of them are regarded as ownerless, hence the legal 'polluter pays' principle has not had the desired effect, even though Clause 28 of NEMA specifies a duty of care of the environment. In addition, Section 28(1) of the National Environmental Management Amendment Act No. 14 of 2009, strengthens the remediation obligation for historical liability in South Africa.

However, tracking down past owners of mines and holding them liable for AMD formation is a physically and legally insurmountable task for private companies and organisations. This leaves the government, with its access to public resources, as the only body with the means to address the challenge (Feris and Kotzé, 2014). This corroborates the view of Lwabukuna (2016) who acknowledges that while the MPRDA, NEMA and NWA are good pieces of legislation, the main challenge is enforcing liability for pollution by companies responsible for the damage (under polluter pay principles) and apportioning this liability to the government, as most past mine owners cannot be traced in South Africa.

Attempts to reform the legal environment have not met with success, partly due to lack of interest in addressing post-mining environmental challenges. Some of the challenges such as AMD emerged as far back as the 1950s (Blaine, 2013). However, meaningful response and attempts to address the scourge began unfolding only around 2002. This was necessitated by acidic mine water which emerged from an abandoned shaft in the Mogale City/Randfontein area of the Western Basin. The flow had been predicted by scientists in 1996 (Mujuru and Mutanga, 2016). Predicted challenges of floods and other hazards related to AMD raised environmental awareness in many sectors. This included the tourism industry, where there were in particular concerns about the tourist attraction of Gold Reef City.

Given the enormity of the challenges posed by resource-depleted mines, including with regard to achieving SDGs 6 and 15, the next section explores complexity science as a solution-oriented tool to elucidate the dimensions of post-mining landscape restoration.

THE ROLE OF COMPLEXITY SCIENCE IN POST-MINING LANDSCAPE RESTORATION

Complexity theory is destined to be the dominant scientific trend of the 21st century as it cuts across several disciplines (Manson, 2001). Sustainable mining has inherent complex adaptive systems theory as its focal principle. In other words, sustainability is a property of a system. The required transformation towards sustainability pushes us to go beyond our traditional segmentation of knowledge into separate disciplines. This entails having a deeper understanding of the systems that require transformation and the interrelationships between these systems. Scholars such as Clark and Dickson (2003) and Clark et al. (2010) observe the need to study the dynamic exchanges between society and nature through an interdisciplinary lens in order to find innovative solutions to complex challenges.

Essentially, mining landscape systems are characterised by interacting feedback loops, long delays and causes. The observed symptoms may also come from a completely different part of a system (Franklin, 2005; Maluleke and Pretorius, 2016) which could be exogenous factors around the mine. There is a large body of literature which offers robust ways of understanding how mineral resource extraction interacts with ecological processes. This literature includes integrated resources management (Cairns and Crawford, 1991); ecosystems approach (CBD, 2004) and Millennium Ecosystems Assessment (MEA, 2005). However, there is still only limited understanding of ecosystem health, its interacting factors and associated value chains for environmental sustainability.

Given the gravity of the complex environmental challenges associated with post-mining areas, approaches are required that broaden our understanding of the system innovations necessary for sustainability (Amponsah-Dacosta and Mhlongo, 2016).

One way of doing this is to use systems thinking, which is a structured way of analysing complex interrelationships that are problematic or simply of interest to humankind (Kaggwa et al., 2017). At the heart of systems thinking is the recognition that factors behind problematic situations are interdependent, and that causal effect between these various factors is often two-way, and that the impact of action is neither instantaneous nor linear (Mutanga et al., 2016). The factors contributing to the problems of reconstructing post-mining landscapes are often associated with a one-way articulation of the challenge. Often interventions are predicated on knowledge that does not incorporate an understanding of two-way causal effects, or of the notion that processes are neither instantaneous nor linear over time.

Systems thinking is thus a formal, abstract and structured cognitive endeavour to think about systems in general (Cabrera et al., 2008). It makes explicit the causal-effect assumptions that come into play between related variables in a system, enabling independent assessment and improvement of the mental models behind particular thinking. Similarly, this can operate within the realm of proposals for post-mine clean up initiatives. Systems thinking focuses on the results of the interplay of feedback loops – these could be either reinforcing or balancing loops (Mingers and White, 2010). This, in turn, incorporates interlinkages between different agencies within the proposed frameworks and interventions for post-mine rehabilitation. The methodology is effective in increasing understanding of the observed phenomenon and in establishing the possible consequences of the different options available at a decision point (Sterman, 2000).

Systems thinking is also a precursor to the development of systems dynamics models which are probabilistic in the soft systems sense (Sterman, 2002). Forrester explains:

> *we cannot assume a perfect model in which every relationship is known exactly. Therefore, we are committed to models in which every decision function has at least in principle, a noise or uncertainty component'* (Forrester, 1961: 124)

Multiple stability regimes are a stable ensemble of a complete system and can be exposed by running systems dynamics models in different scenarios, particularly as they preserve non-linear feedbacks.

Complexity theory thus provides the foundation for the integration of different theoretical approaches to a sustainable transition. What follows is a discussion of the qualitative dimension of systems dynamics application in mineral exploration, with a quest to drive sustainability particularly in post-mining areas.

MENTAL MODELLING ON MINERAL EXPLORATION

In order to plan for and achieve the required sustainable transformations, we need to work with questions which cannot be answered through traditional segmentation of knowledge into various disciplines. In order to achieve sustainable solutions, players need an understanding of the systems that need transformation and the interrelationships within these systems. In this case, in order to understand the complexity of restoring post-mining areas, it is necessary to analyse the causal effects in mineral exploration that lead to resource-exhausted mine areas.

The causal loop concept adopted from Roberts et al. (1983) uses arrows to depict causality with polarity nodes (+) and (-). The (+) indicates that the variable at the head of the arrow changes in the same direction as the variable at the tail. The (-) denotes that a change at one end leads to a change in the opposite direction at the other end. Figure 5 illustrates the causal relationships and feedback loops in the life cycle of mineral extraction.

Essentially, mineral extraction may determine the available mineral resources at a point in time. The availability of mineral resources can have a positive impact on production levels, and continuous extraction reduces the minerals available, which may lead to depletion, creating a reinforcing loop as illustrated in Figure 5. Depletion of commodities leads to additional resource-exhausted mines, which creates yet again a reinforcing loop between mining production and the number of abandoned or resource-exhausted mines. Advanced technologies may lead to further mineral exploration and extraction on the same mine sites. The entire causal effect between mineral exploration, available resources and production creates a reinforcing loop.

This therefore requires intervention to advance ecological sustainability. This may appear simplistic, but there are several factors that have a bearing on the relationships between the different variables. The feasibility of extracting available minerals may be influenced by the availability of the capital required to undertake exploration, and often this is characterised by delays such as technological efficiency, policy or regulatory environment and unforeseen incidents such as disasters during operation. This is one of the greatest tragedies of many developing countries, apart from the fact that the commercial viability of most mining plants has a net effect on mineral extraction. There is a plethora of factors involved, among which are fiscal policies, political stability and human capital, enabling policy environment and mineral quality. Equally, exogenous factors have an effect on production.

Most third world countries have failed to capitalise on the stock of minerals they own to create economic growth; rather, these economies have suffered as a result of mineral extraction. According to Alaoma and Voulvoulis (2018), the mining industry left landmarks, i.e. mine dumps, in communities and regions of mineral wealth. As such, the 'organs' of nature are wrested and stripped away with a disproportionate impact on humanity and ecosystems in these regions. The feedback loops illustrated in Figure 6 provide, on the one hand, the upstream, downstream and side-stream mining value chain linkages, while on the other hand the negative externalities are common in most post-mining regions. This corroborates Forrester's (1969) principle of relative attractiveness, which has been adopted to unpack the socio-economic, effects on development in the mining area. There are also unintended negative externalities, such as illicit financial flows, human health impacts (Voulvoulis et al., 2013), socio-cultural impacts (Kitula, 2006), ecosystem and biodiversity destruction as well as human health effects (Azapagic, 2004).

The negative impacts of mining come in many forms, as elaborated in Figure 5, including loss of biodiversity, deforestation, acidification of water resources and soil erosion. Ecological challenges may prompt the state to introduce new policies or to amend existing policies. Increased regulatory instruments may reduce the degradation of the environment or enforce regulations for post-mine management and closure, creating a balancing loop between extraction, policy and environment. Policy would impact on players with the responsibility to protect and restore the environment in depleted/abandoned mine areas in South Africa.

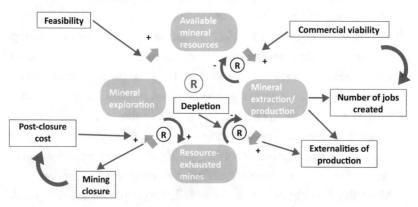

Figure 5: Causal relationships: Mineral extractive process
Source: Author

Transforming such systems would require holistic approaches that cut across different sectors and disciplines that take into consideration causal relationships between the different variables affecting mining activities. The process of establishing what needs to be done and setting priorities on the basis of our knowledge about the system is critical if we are to achieve our desired outcomes. In addition to environmental/ecological challenges, reclamation costs and the availability of funds may also inform policy direction. Furthermore, technological innovation may have a bearing on both the policy changes and land reclamation required which, in turn, could improve the environment. The selected elements and relationships thereof are illustrated in Figure 7.

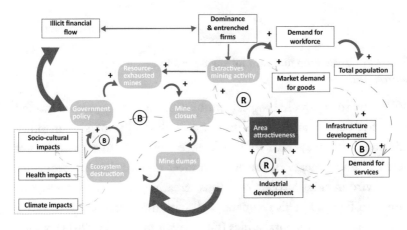

Figure 6: Complex dimensions of mining systems
Source: Author

Overall, government policy may promote the generation of funds for post-mining rehabilitation by enforcing legislative obligations. Legislation can be used to force companies to provide funds for rehabilitation. The government may also provide funding for this purpose through parliamentary grant allocations while private companies may also support this activity through corporate social responsibility funds (Ackerman, 1973). For most countries from the developing world, it is always a challenge to prioritise the environment over the ever-increasing competing priorities generated by different government ministries. Policy on the other hand may foster technological innovation, and the deployment of such technologies thus creates a balancing loop between technological innovation and policy as shown in Figure 7.

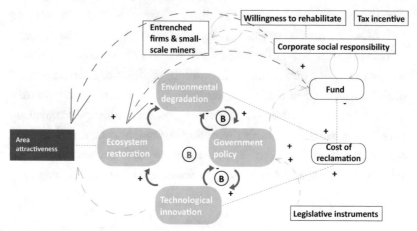

Figure 7: Causal loop diagram for post-mining restoration
Source: Author

With regard to technological innovation, if there are identified gaps that may retard the utilisation of technology for sustainability solutions, this may positively influence the legislative environment through amendments of the Acts to meet the required prerequisites for environmental protection, thus creating a balancing loop as shown in Figure 7. This resonates well with the theories of restorative justice and environmental ethics (Callicott, 1989; Taylor, 2011), which look at defence of the land and respect for nature.

Through the public trust doctrine, under which the state bears the responsibility for protecting natural resources especially on defunct and ownerless mines, various legislation and regulations have been enacted in South Africa. To deal with water resources, for instance, NEMA, the Water Services Act, NWA and the National Environmental Management Waste Act have been passed. Together, financial, institutional and policy frameworks can lead to land reclamation or restoration, thus reducing the level of environmental degradation. The overall system created is, in turn, self-balancing as illustrated in Figure 7.

Acknowledging the plethora of proposed technologies and strategies for concurrent mining and reclamation, the next section focuses on a selection of reclamation/reconstruction measures for the terrestrial ecosystems of post-mine sites, given the urgency of rehabilitating post-mining areas in the South African context.

LAND RECLAMATION ON MINE DUMPS

The concept of resource-exhausted mine land reclamation advances SDG 15, which focuses on life on the land. Ecological systems theory can provide a good starting point for understanding transitions to sustainability. The ecological definition of reclamation is a process of restoration of the ecosystem and its functions (Bradshaw, 1996; Hüttl and Gerwin, 2005). Other scholars have defined it as the recreation of bio-ecological conditions in which the exchange between the soil and the plants provides for successful plant succession and intensive development of soil-forming processes (Krzaklewski, 1988). Adopted here is the classical definition by Sir Arthur George Tansley from 1935 (cited by Golley, 1993), which underscores that the ecosystem should be an integrated system of biotic and abiotic elements in which all trophic levels contain a set of species ensuring circulation of matter and energy flow.

Several scholars have identified afforestation as one of the best sustainable strategies to reclaim mine lands and restore them to their original state (Parrotta et al., 1997; Filcheva et al., 2000; Pietrzykowski and Krzaklewski, 2007). The major critique of this strategy is that long periods are required to assess the effects of afforestation.

The second generation opportunities in post-mine sites are related to the retreatment of dumps. Often, privately owned and controlled dumps remain an additional source of mineral resources, and this realisation has precipitated the advent of increasing numbers of dump projects over the years. Some of the enablers have largely been economic, in particular the quest for employment, technological breakthroughs, as well as environmental challenges and safety, as witnessed in South Africa's chrome- and diamond-rich dumps.

In a nutshell, from a systems perspective, once mining is complete, there are the '3Es' of mine dump retreatment: economic evaluation, environmental assessment and engineering. Land reclamation thus requires a cost benefit analysis of proposed interventions; an environmental assessment points to issues of sustainability and accountability and engineering aspects address technology and implementation. The process requires the collective effort of an array of specialists including engineering, biological, hydrological and wildlife scientists, other environmental experts, economists, political economists and other social scientists.

LAND RECLAMATION: RESPONSE TO SUBSIDENCE

The temporal and spatial analysis – also referred to as 'time-space evolution analysis' – of mining subsidence is one key approach that can be applied to develop an effective response or allow the state to be better prepared to respond to geological disasters in post-mining sites. As propounded by Xiao et al. (2014), monitoring and assessment of mining subsidence in post-mine areas is key to ascertaining the characteristics of marginal fissures, the genesis and development rules of dynamic fissures and their geological conditions. From an ecological perspective, scholars such as Zhang et al. (2013) have developed an artificial, guided ecological restoration model and method with partition restoration which facilitates the planning and prioritisation of restorative investment systems, damage reduction and plant promotion. This is drawn from restoration ecology, which is emerging as one of the most important disciplines in environmental science (Ormerod, 2003).

Xiao et al. (2011) have proffered the resource-based land use assessment model as a key strategy for attaining sustainable urban cities. The model takes into account the degree of land use, land use structure, reasonable input level, output efficiency and land use intensification for a resource-exhausted city. The results of such a model provide an analysis of land suitability with specified land use recommendations in relation to population growth, controlled areas for construction, proposed artificial landscapes, eco-agricultural zones and open water landscape, as demonstrated in the case of Huaibei in China. The model classified land suitability and identified specific priority intervention areas in Huaibei. Clear recommendations such as the establishment of eco-agricultural zones were proposed, in line with growing population trends.

APPROACHES TO ACIDIFICATION
OF WATER RESOURCES

The main thrust of AMD treatment is to remove suspended solids, neutralise free acidity and remove iron and other metals such as sulphates so that the resulting effluent can meet permissible environmental discharge levels (Dinardo et al., 1991). Sulphates are well known to be major contaminants in mine water. Their reaction with other existing metal ions leads to the formation of a wide range of salts that contribute significantly to salinity in the vicinity of discharge (Bowell, 2000). Treatment processes are aimed at lowering sulphate content in a bid to meet discharge regulations. There are two standard classification AMD treatment processes: active and passive treatment systems.

As described by Mujuru and Mutanga (2016), passive treatment systems depend on natural, physical geochemical and biological processes and are mostly applied after mine closure. This approach needs ample land space and can function in a natural environment to effect acid water detoxification. Typical passive treatment systems include constructed wetlands, diversion wells containing crushed limestone, or open ditches filled with limestone and bioreactors. They are often considered to provide long-lasting solutions to the AMD problem. However, the volumes of iron loadings and acidic values of streams determine the success of passive treatment systems. If the volumes are too high, then they may exceed the optimum threshold of the plant.

One of the most practical passive treatment systems is wetlands, often referred to as 'biological filters' (Greenway and Simpson, 1996; Mthembu et al., 2013). The remediation process is attained through complex interactions between plants, microbes and sediments. The use of wetlands has been identified as a cost-effective way of removing heavy metals from AMD. The usage of this approach has increased over the years and processes have been successfully implemented in several cases around the world, including Parys Mountain mine and the southern Afon Goch in Anglesey in North Wales, UK, as well as South Africa's Western Cape constructed wetlands (Dean et al., 2013; Baker et al., 1990; Fyson et al., 1994; Matagi et al., 1998; Gerba, 2000; Lakay, 2013).

AMD treatment via wetlands involves either natural or constructed wetlands. Natural wetlands are characterised by permanent or temporary water-saturated soils or sediments, which support plants adapted to reducing conditions in their rhizosphere.[4] Constructed wetlands are ecosystems designed by humans to mimic their natural counterparts. Constructed wetlands often comprise shallow excavations containing flooded gravel, organic matter and soil to support wetland plants such as cattails (*Typha* spp.), rushes (*Carex* spp.), reeds (*Phragmites* spp.), sedges (*Juncus* spp.) and bulrushes (*Scirpus* spp.) (Droste, 1997). In addition, the purification process in constructed wetlands occurs in a controlled environment, where plants, microbial populations, soil, sediments, speed of water and other parameters are set to maximise the efficiency of the water treatment process. They are well known to be self-sustaining once established. In terms of cost, they become cheaper in the long-run (White, 2008).

There are some complexities associated with wetland restoration, all of which are relevant in the South African context. Rooney et al. (2012) identify four major impediments to wetland restoration, namely:

Lack of policy framework that provides an enabling environment for wetlands usage. (This is in view of the defunct ownerless mines as well as the concurrent mine owners.)

Sometimes the volume of tailings (dump size) and upgrading by-products exceeds the size of the mine pits. Closure of such mine dumps thus results in hills instead of level ground, which inhibit the purpose and design of the wetland.

Geotechnical disasters, such as subsidence, are often associated

---

4   A rhizosphere is a narrow region of soil that is directly influenced by root secretions and associated soil micro-organisms.

with landscape closure.

Often precipitation is less than potential evapotranspiration. Water availability will therefore limit the establishment of wetland areas in the reclaimed landscape.

Active treatment systems, according to Jennings (2008), are engineered treatment facilities that apply a chemical amendment approach to the treatment of AMD in order to achieve a specified water quality standard in a discharge area. Active treatment systems are frequently used at mining sites that are still in operation. This is because active mines often have limited space for remediation systems and drainage chemistry and flow rates can change as mining proceeds. Active treatment methods, for instance sulphur-reducing bioreactor plants, require less space than wetlands. In terms of cost, studies have shown that the passive treatment approach is more economic as a long-term solution (Ndlovu, 2016). Especially with regard to defunct ownerless mines, passive treatments are ideal where available land, chemistry and flow rates are not expected to change. In South Africa, most of these procedures are still laboratory scale or being carried out at a pilot plant, as described in Mujuru and Mutanga (2016). There is great potential for full-scale deployment of these treatment systems as most of the proven techniques have not yet been deployed across South Africa's thousands of abandoned mines.

Commonly used active treatment procedures for the remediation of acidic mine water streams include most abiotic methods such as chemical precipitation, ion exchange, solvent extraction, electrolytic techniques, membrane technologies and adsorption processes (see Table 1). Active treatment is characterised by the ongoing, high intensity flow of chemicals into and out of a treatment plant, which requires continuous monitoring and maintenance by trained personnel. These systems generally require the construction of a treatment plant that is equipped with a variety of systems, such as agitated reactors, precipitators, clarifiers and thickeners and extraction columns (Ndlovu, 2016).

Table 1: Summarised AMD treatment processes

Process	Process technology	Pre-treatment?	Advantages	Disadvantages	Possible improvement
Lime/ limestone	Chemical precipitation	No	• Inexpensive • Low maintenance • Removes trace metals • Produces water quality suitable for irrigation or reuse in the mine • Can be used as a cost effective pre-treatment method to other processes • HDS process proven technology	• Limited sulphate removal • High amount of sludge produced • Costs associated with handling and safe disposal of potential unstable sludge	Reduction in the production of waste or recycling of sludge
CSIR-ABC	Chemical precipitation	No	• A portion of the sludge produced is processed as re-usable or saleable by-products, thereby reducing the total waste sludge volumes to be disposed of. • The pre-treatment steps (neutralisation and gypsum precipitation) are widely applied and mature technologies. The risks associated with these processes are well understood.	• High Costs associated with the thermal reduction of waste • The feasibility of this process relies on the recovery of BaCO3 and Ca(OH)2, and the production of elemental sulphur.	High energy demands, alternative and cost effective routes for recovering the reagents and saleable by-products
SAVMIN	Chemical precipitation	No	• Reduce sulphate to very low levels • High-quality water even at fluctuating feed sulphate levels	• High amount of sludge produced • Success depends on high level of gypsum crystallisation. • Process can be complicated to control	Reduction in the production of waste or recycling of sludge
CESR	Chemical precipitation	No	• High-quality water obtained • Low levels of residual sulphate • Trace metals removal	• High amount of sludge produced. Relatively expensive	Reduction in the production of waste or recycling of sludge

Source: Mujuru and Mutanga, 2016

Abbreviations used in the table:

CSIR-ABC: Council for Scientific and Industrial Research Alkali-Barium-Calcium desalination process

SAVMIN: Sulphate removal though precipitation – Mintek

CESR: Cost-effective suplhate removal process

FINANCING ABANDONED MINES REHABILITATION

Financing abandoned mines rehabilitation has long been a serious challenge for many countries given that few parties are prepared to take responsibility for historic, disused mines. In the South African context, this can be linked to colonial and apartheid laws, which exhibited the lack of willingness by government to hold mining companies accountable. The question of who should provide funds for such an undertaking has been an intractable problem. Fortunately, the state has been able to set aside R45 billion in financial provisions for rehabilitation in 2015 (Olalde, 2017). Despite the requisite finances being available, Olalde found that since at least 2011 no large coal mines operating in South Africa were granted mine closure. This implies that the mines have not been rehabilitated but have rather left behind a legacy of severe pollution. How this can be addressed remains unresolved.

Mhlongo and Amponsah-Dacosta (2015) have described two models that can be used to estimate mine closure, namely a mine closure model and a rule-based model. The mine closure model utilises project management principles, while the rule-based model applies a closure costing approach to determine the importance of financial provision (Du Plessis and Brent, 2006).

Financing abandoned post-mine sites is often impeded by uncertainty around rehabilitation strategies and the level of standards to pursue. The goals and nature of rehabilitation have always made it a challenge to get an accurate estimate of the cost of rehabilitation (Van Zyl et al., 2002). In the context of AMD treatment, Dyosi and Banda (2016) underscore the need for both long-term and sustainable funding of AMD treatment. What emerges from their discussion is that long-term and sustainable funding of AMD treatment requires multi-pronged public, private and public-private funding models. In South Africa, for example, based on conservative estimates of the costs of the rehabilitation of asbestos mine projects, R30 billion is required for the rehabilitation of abandoned mine sites (WWF, 2012).

Looking at the economic benefits of rehabilitation, Mujuru and Mutanga (2016) provide an array of opportunities offered by resource-exhausted mines. Upscaling of second generation by-products can help to generate the funds required to mitigate the legacy of mining. For example, while popular technologies such as reverse osmosis and chemical precipitation result in the generation of secondary waste materials that are environmental pollutants in their own right, the focus could be on recoverable and saleable by-products. The CSIR also reported on the recovery of drinking water from gold mine effluent when using the Alkali-Barium-Calcium (ABC) desalination process. Du Plessis (1983) hints at the possibility of utilising lime-treated AMD for agricultural crops. Other AMD by-products could include the recovery of metals, for example, calcium or barium carbonate and sulphur through processes such as the ABC process (Motaung et al., 2008). Inorganic components in AMD treatment sludge can be used for the production of building materials, such as cement and bricks. Interestingly, the technical feasibility of the microbial fuel cell-based technology for electricity generation has also been envisaged (Mujuru and Mutanga, 2016), meaning that defunct, ownerless mines may be viewed as possible enterprise components. This could reduce the operational cost of treatment but could also allow for the generation and re-use of recovered products, in some cases. Masukume et al. (2013) underscore this by noting that when the value of treated water and by-products exceeds the treatment cost, it is feasible to create enterprises that could provide economic benefits while solving environmental problems. Experimental evidence has shown that AMD treatment can result in water recovery which can be used for agricultural purposes (Mujuru and Mutanga, 2016).

CONCLUSION

Ecological and socio-economic systems are complex, adaptive systems, integrating phenomena across scales of space, time and organisational complexity. This chapter provides a myriad of complex dimensions associated with post-mining sites. The chapter reflects on the complex dimensions of environmental injustice against the backdrop of SDGs 6 and 15 which focus on water and land respectively. On the question of liabilities apportionment, especially for ownerless mines, the chapter posits the need to advance collective or shared responsibility, between government and the private sector, to attain restorative environmental justice. The chapter points to the multiple dimensions and impacts of abandoned mine sites, underlining that countries with mining as a cornerstone of their economic development have the huge task of addressing post-mining reclamation. The chapter contends that integrated holistic approaches are required to face multiple challenges from mineral-exhausted, historical mines.

The chapter acknowledges the need for new ways of looking at development, including the notion of 'wellbeing', with sensitivity to environmental and social dynamics and taking externalities into full account, as described in Fioramonti's chapter 12 in this volume. Innovative solutions for the future of South Africa's mining sector include re-imagining business models such as converting waste to meaningful, eco-friendly, income-generating products, services and funding for reclamation purposes. Essentially, there is great potential to recover valuable products from the huge quantities of mine waste such as AMD and mine dumps which are features of the mining sector in South Africa. A comprehensive cost and benefits analysis of mining *prior* to extraction is the route to sustainability so that we don't create destruction that can't be addressed to begin with.

REFERENCES

Abubakary, S. (2014). 'Haulage system optimization for underground mines: A discrete event simulation and mixed integer programming approach'. PhD thesis. Luleå University of Technology. Available at: https://www.diva-portal.org/smash/get/diva2:989916/ Accessed 14 July 2018.

Ackerman, R.W. (1973). 'How companies respond to social demands'. *Harvard University Review*, Vol. 51(4), pp. 88–98.

Adler, R. and Rascher, J. (2007). 'A strategy for the management of acid mine drainage from gold mines in Gauteng'. CSIR report No. CSIR/NRE/PW/ER/2007/0053/C. CSIR, Pretoria.

Alaoma, A. and Voulvoulis, N. (2018). 'Mineral resource active regions: The need for systems thinking in management'. *AIMS Environmental Science*, Vol. 5(2), pp. 78–95.

Amponsah-Dacosta, F. and Mhlongo, S. E. (2016). 'A review of problems and solutions of abandoned mines in South Africa'. *International Journal of*

Mining, Reclamation and Environment, Vol. 30(4), pp. 279–294.

Auditor-General. (2009). 'Report of the auditor-general to Parliament on a performance audit of the rehabilitation of abandoned mines at the Department of Minerals and Energy'. Pretoria. Available at: https://cer.org.za/wp-content/uploads/2011/10/AG_Report_on_abandoned_mines-Oct-2009.pdf. Accessed 15 July 2017.

Azapagic, A. (2004). 'Developing a framework for sustainable development indicators for the mining and minerals industry'. *Journal of Cleaner Production*, Vol. 12(6), pp. 639–662.

Baker, A.J.M. and Walker, P.L. (1990). 'Ecophysiology of metal uptake by tolerant plants'. In Shaw, A.J. (ed). *Heavy Metal Tolerance in Plants: Evolutionary Aspects*. CRC Press, Boca Raton, FL.

Blaine, S. (18 November 2013). 'Acid water problem "neglected by government"'. *Business Day*. Available at: http://www.bdlive.co.za/national/science/2013/11/18/acid-water-problem-neglected-by-government. Accessed 15 October 2015.

Bowell, R.J. (2000). 'Sulphate and salt minerals: The problem of treating mine waste'. *Mining and Environmental Managemen,*Vol. 8, pp. 11–14.

Bradshaw, A.D. (1996). 'Underlying principles of restoration'. *Canadian Journal of Fisheries. Aquatic. Sciences,* Vol. 53(1), pp. 3–9.

Cabrera, D., Colosi, L. and Lobdell, C. (2008). 'Systems thinking'. *Evaluation and Program Plannin,*. Vol. 31(3), pp. 299–310.

Cairns, J. and Crawford, T. V. (1991). *Integrated Environmental Management: A Handbook for Integrated Environmental Management*. US Bureau of Land Management, USA.

Callicott, J.B. (1989). *In Defence of the Land Ethic: Essays in Environmental Philosophy*. State University of New York Press, Albany.

Claassen, M. (2006). 'Water resources in support of socio-economic development'. *In Vaalco Supplement: Water for a Sustainable Future*. Vaal River Catchment Association, Johannesburg, p. 21.

Clark J.S., Bell, D., Chu, C., Courbaud. B., Dietze, M., Hersh, M., HilleRisLambers, J., Ibanez, I., LaDeau, S., McMahon, S., Metcale, J., Mohan, J., Moran, E., Pangle, l., Pearson, S., Salk, C., Shen, Z., Valle, D., and Wyckoff, P. (2010). 'High dimensional coexistence based on individual variation: A synthesis of evidence'. *Ecological Monographs*, Vol. 80(4), 569–608.

Clark, W.C. and Dickson, N.M. (2003). 'Sustainability science: The emerging research program'. *Proceedings of the National Academy of Sciences of the United States of America*, Vol. 100(14), pp. 8059–8061.

Constitution of the Republic of South Africa. (1996). Available at: https://www.gov.za/documents/constitution/Constitution-Republic-South-Africa-1996-1. Accessed 11 September 2018.

Convention on Biological Diversity (CBD). (2004). 'The ecosystem approach

(CBD Guidelines)'. Secretariat of the Convention on Biological Diversity, Montreal, Canada.

Council for Geoscience. (2011). 'Sinkholes and subsidence in South Africa'. Council for Geoscience Report number 2011-0010. Western Cape Unit, Cape Town.

CSIR. (2009). 'Acid mine drainage in South Africa'. Briefing Note.

Dean, A.P., Lynch, S., Rowland, P., Toft, B.D., Pittman, J.K. and White, K. N. (2013). 'Natural wetlands are efficient at providing long-term metal remediation of freshwater systems polluted by acid mine drainage'. *Environmental Science and Technology*, Vol. 47(21), pp. 12029–12036.

Department of Water Affairs (DWA). (2013). 'Feasibility study for a long-term solution to address the acid mine drainage associated with the East, Central and West Rand underground mining basins: Assessment of the water quantity and quality of the witwatersrand mine voids'. DWA Report No. P RSA 000/00/16512/2. Department of Water Affairs, Pretoria.

Dinardo, O., Kondos, P.D., Mcreaady, R.G.L., Riveros, P.A. and Skaff, M. (1991). 'Study on metals recovery/recycling from acid mine drainage Phase 1A: Literature survey'. Mend Treatment Committee, CANMET, Energy, Mines and Resources Canada and WTC, Environment Canada, Canada.

Droste, R.L. (1997). *Theory and Practice of Water and Wastewater Management Systems*. Wiley, New York.

Du Plessis, A. and Brent, A.C. (2006). 'Development of a risk-based mine closure cost calculation'. *Journal of the Southern African Institute of Mining and Metallurgy*, Vol. 106, pp. 443–540.

Dyosi, Z. and Banda, G. (2016). 'Financing acid mine drainage treatment in South Africa'. In Mujuru, M. and Mutanga, S. (eds). *Management and Mitigation of Acid Mine Drainage in South Africa*. AISA, Pretoria.

Earthlife Africa. (13 June 2010). 'Stream of acid, the Tweelopiespruit'. Available at: http://earthlife.org.za/2010/06/13/stream-of-acid-the-tweelopiespruit/. Accessed 24 July 2018.

EcoPartners. (2016). 'South African mine dump'. Briefing Note Proposal. Eco-Partners, Johannesburg.

Feris, L. and Kotzé, L.J. (2014). 'The regulation of acid mine drainage in South Africa: Law and governance perspectives'. *Potchefstroom Electronic Law Journal*, Vol. 17(5), pp. 2105–2163.

Filcheva, E., Noustorova, M., Gentcheva-Kostadinova, S.V, and Haigh, M. J. (2000). 'Organic accumulation and microbial action in surface coal mine spoils'. *Pernik Bulgaria Ecological Engineering*, Vol. 15, pp. 1–15.

Forrester, J.W. (1961). *Industrial Dynamics*. Productivity Press, Portland, OR.

Forrester, J.W. (1969). *Urban Dynamics*. MIT Press, Cambridge, MA.

Franklin, M.M. (2005). 'Applying modelling 1956 and simulation as part of business process improvement for complex mining logistics'. A Pen State and Colorado School of Mines Conference on Business Process

Improvement in the Extractive Industry. Denver, CO.

Fyson, A., Kalin, M. and Adrian, L.W. (1994). 'Arsenic and nickel removal by wetland sediments'. *Proceedings of the International Land Reclamation and Mine Drainage Conference and Third International Conference on the Abatement of Acidic Drainage*, Vol. 1, pp. 109–118.

Gao, F., Zhou K.P., Chen X.Y. and Luo, X.W. (2012). 'Disaster chains induced by mining and chain–cutting disaster mitigation technology'. *Journal of Disaster Advances*, Vol. 10, pp. 5–15.

Gerba, C.P. (2000). 'Domestic waste and waste treatment'. In Maier, R.M., Pepper, I.L. and Gerba, C.P. (eds). *Environmental Microbiology*. Academic Press, California.

Golley, F.B. (1993). *History of the Ecosystem Concept in Ecology*. Yale University Press, New Haven.

Greenway, M. and Simpson, J. (1996). 'Artificial wetlands for wastewater treatment, water reuse and wildlife in Queensland, Australia'. *Water Science and Technology*. Vol. 33, pp. 221–229.

Gu, S.T., Wang, C.Q., Jiang, B.Y., Tan, Y.L. and Li, N.N. (2012). 'Field test of rock burst danger based on drilling pulverized coal parameters'. *Journal of Disaster Advances,* Vol. 5(4), pp. 237–240.

Hu, Z.Q. (1996). *Coal Mining Subsidence of the Land Resource Management and Land Reclamation*. Coal Industry Press, Beijing.

Hüttl, R. and Gerwin, W. (2005). 'Landscape and ecosystem development after disturbance by mining'. *Ecological Engineering*, Vol 24(1–2), pp. 1–3.

Jennings, S.R., Neuman, D.R. and Blicker, P.S. (2008). *Acid Mine Drainage and Effects on Fish Health and Ecology: A Review*. Reclamation Research Group Publication, Bozeman.

Kaggwa, M., Mutanga, S. and Simelane, T. (2017). 'Systems dynamics application in the management of sugarcane production under known constraints'. In Brent, A. and Simelane, T. (eds). *System Dynamics Models for Africa's Developmental Planning*. AISA, Pretoria.

Kitula, A. (2006). 'The environmental and socio-economic impacts of mining on local livelihoods in Tanzania: A case study of Geita District'. *Journal of Clean Production,* Vol. 14, pp. 405–414.

KPMG. (2017). 'The role of mining in the South African economy'. Available at: https://www.sablog.kpmg.co.za/2013/12/role-mining-south-african-economy/. Accessed 25 August 2017.

Krzaklewski, W. (1988). 'Some methodological aspects of planning and implementation of forest reclamation on the example of open cast mining'. *Sci. B. AGH Academy Mineral Metallurgy Krakow, Sozology*. Vol. 1222, pp. 331–338.

Lakay, V. (2013). 'An analysis of the performance of constructed wetlands in the treatment of domestic wastewater in the Western Cape, South Africa'. PhD thesis. University of Cape Town, Cape Town.

Lwabukuna, O. (2016). 'Interrogating and reviewing legal and policy frameworks governing acid mine drainage (AMD) in South Africa'. In Mujuru, M. and Mutanga, S. (eds). *Management and Mitigation of Acid Mine Drainage in South Africa*. AISA, Pretoria.

Maluleke, G.T. and Pretorius, L. (2016). 'Modelling the impact of mining on socio-economic infrastructure development — A system dynamics approach'. *South African Journal of Industrial Engineering*, Vol. 27(4), pp. 66–76.

Manson, S.M. (2001). 'Simplifying complexity: A review of complexity theory'. *Geoforum*, Vol. 32(3), pp. 405–414.

Masukume, M., Maree, J.P., Ruto, S. and Joubert, H. (2013). 'Processing of Barium Sulphide to Barium Carbonate and Sulphur'. *Journal of Chemical Engineering and Process Technology*, Vol. 4(4).

Matagi, S.V., Swai, D., and Mugabe, R. (1998). 'A review of heavy metal removal mechanisms in wetlands'. *African Journal for Tropical Hydrobiology and Fisheries*, Vol. 8, pp. 23–35.

Mhlongo, S.E. and Amponsah-Dacosta, F. (2015). 'A review of problems and solutions of abandoned mines in South Africa'. *International Journal of Mining, Reclamation and Environment*, Vol. 30(4), pp. 279–294.

Millennium Ecosystem Assessment (MEA). (2005). *Ecosystems and Human Well-being: Synthesis*. Island Press, Washington DC.

Mingers, J. and White, L. (2010). 'A review of the recent contribution of systems thinking to operational research and management science'. *European Journal of Operational Research*, Vol. 207(3), pp. 1147–1161.

Motaung, S., Maree, J., De Beer, M. and Bolongo, L. (2008). 'Recovery of drinking water and by-products from gold mine effluents'. *International Journal of Water Resources Development*, Vol. 24(3), pp. 433–450.

Mthembu, M.S., Odinga, C.A., Swalaha, F.M. and Bux, F. (2013). 'Constructed wetlands: A future alternative wastewater treatment technology'. *African Journal of Biotechnology*, Vol. 12(29), pp. 4542–4553.

Mujuru, M. and Mutanga, S. (2016). *Management and Mitigation of Acid Mine Drainage in South Africa*. AISA, Pretoria.

Mutanga, S.S., De Vries, M., Mbohwa, C., Holger, H., Kumar, D. and Rogner, H. (2016). 'An integrated approach for modelling the Electricity Value of Sugarcane Production System'. *Applied Energy*, Vol. 177, pp. 823–838.

Ndlovu, S. (2016). 'Acid mine drainage treatment technologies'. In Mujuru, M. and Mutanga, S. (eds). *Management and Mitigation of Acid Mine Drainage in South Africa*. AISA, Pretoria.

Olalde, M. (2017). 'Mining and people: The impact of mining on the South African economy and living standards'. The South African Human Rights Commission. Available at http://www.fse.org.za/index.php/item/. Accessed 15 March 2018.

Oosthuizen, A.C. and Richardson, S. (2011). 'Sinkholes and subsidence in South

Africa'. Council for Geoscience. Available at: http://www.geoscience.org. za/images/geohazard/Sinkholes.pdf. Accessed 10 June 2017.

Ormerod, D. (2003). 'Improving the disclosure regime'. *The International Journal of Evidence and Proof*, Vol. 7(2), pp. 102–129.

Parrotta, J.A., Knowles, O.H. and Wunderle, J.M. (1997). 'Development of floristic diversity in 10-year-old restoration forests on a bauxite mined site in Amazonia'. *Forest Ecology and Management*, Vol. 99(1–2), pp. 21–42.

Pietrzykowski, M. and Krzaklewski, W. (2007). 'Soil organic matter, C and N accumulation during natural succession and reclamation in an opencast sand quarry (Southern Poland)'. *Archives of Agronomy and Soil Science.*,Vol. 53(5), 473–483.

Roberts, N., Andersen, D., Deal, R., Garet, M. and Shaffer, W. (1983). *Introduction to Computer Simulation: A System Dynamics Modeling Approach*. Productivity Press, Portland, OR.

Rooney, R., Bayley, S. and Schindler, D. W. (2012). 'Oil sands mining and reclamation cause massive loss of peatland and stored carbon'. *Proceedings of the National Academy of Sciences of the United States of America*, Vol. 109(13), pp. 4933–4937.

Sahu, H.B and Dash, S. (2014). 'Land degradation due to mining in India and its mitigation measures'. *Proceedings of Second International Conference on Environmental Science and Technology*. Singapore. National Institute of Technology Rourkela Sundargarh, Orissa, India.

Salama, A. (2014). 'Haulage system optimization for underground mines: A discrete event simulation and mixed integer programming approach'. PhD thesis. Luleå University of Technology, Sweden. Available at: https:// www.diva-portal.org/smash/get/diva2:989916/. Accessed 14 July 2018.

South African Human Rights Commission (SAHRC). (2016). 'Acid mine drainage and human rights'. The South African Human Rights Commission. Available at: https://www.sahrc.org.za/home/21/files/ AMD%20Booklet.pdf. Accessed 15 March 2018.

Sterman, J.D. (2000). *Business Dynamics: Systems Thinking and Modeling for a Complex World*. Irwin/McGraw-Hill, Boston.

Sterman, J.D. (2002). 'All models are wrong: Reflections on becoming a systems scientist'. *System Dynamics Review*, Vol. 18(4), pp. 501–531.

Swart, E. (2003). 'The South African legislative framework for mine closure'. *The Journal of the South African Institute of Mining and Metallurgy*, pp. 489–492.

Taylor, P.W. (2011). *Respect for Nature: A Theory of Environmental Ethics*. Princeton University Press, Princeton.

United Nations Development Plan (UNDP). (2016a). 'Sustainable Development Goal 6: Ensure access to water and sanitation for all'. Available at: http://www.un.org/sustainabledevelopment/water-and-sanitation/. Accessed 31 July 2017.

United Nations Development Plan (UNDP). (2016b). 'Sustainable Development Goal 15'. Available at: http://www.un.org/sustainabledevelopment/water-and-sanitation/. Accessed 31 July 2017.

United Nations Economic Commission for Africa (UNECA). (2011). 'Minerals and Africa's development: The International Study Group report on Africa's mineral regimes'. Available at: https://uneca.org/publications/minerals-and-africas-development. Accessed 11 September 2018.

United Nations Environment Programme (UNEP). (2011). 'Decoupling natural resources use impact and environmental impacts from economic growth'. International Resource Panel. United Nations Environment Programme, Nairobi, Kenya.

Van Wyk, D. (2012). 'A review of platinum mining in the Bojanela District of the North West Province'. Bench Marks Foundation, Johannesburg. Available at: http://www.bench-marks.org.za/. Accessed 27 January 2013.

Van Zyl, D., Sassoon, M., Digby, C., Fleury, A.M., and Kyeyune, S. (2002). 'Mining for the future'. Main Report, Mining, Minerals and Sustainable Development (MMSD) of the Institute of Environment and Development (IIED), England.

Voulvoulis, N., Skolout, J., Oates, C. and Plant, J. (2013). 'From chemical risk assessment to environmental resources management: The challenge for mining'. *Environ Science Pollution Research,* Vol. 20, pp. 7815–7826.

White, S. (2008). 'Wetland use in acid mine drainage remediation'. Available at: http://home.eng.iastate.edu/~tge/ce421-521/Steven%20White.pdf. Accessed 30 June 2017.

World Wildlife Fund (WWF). (2012). 'R30 billion is just the start of South Africa's Mining hangover.' Available at: http://www.wwf.org.za/?6600/acid-mine-draining. Accessed 15 June 2017.

Xiao, W., Hu, Z. and Fu, Y. (2014). 'Zoning of land reclamation in coal mining area and new progresses for the past 10 years [sic]'. *International Journal of Coal Science and Technology*, Vol. 1(2), pp.177–183.

Xiao, W., Hu, Z., Li, J., Zhang, H. and Hu, J. (2011). 'A study of land reclamation and ecological restoration in a resource-exhausted city – A case study of Huaibei in China'. *International Journal of Mining, Reclamation and Environment*, Vol. 25(4), pp. 332–341.

Zhang, Y., Zhao, T., Shi, C., Wu, H. and Li, D. (2013). 'Dynamic research on vegetation recovering in abandoned mine slopes in Beijing mountainous areas'. *Journal of Arid Land Resources and Environment*. Vol. 27, pp. 61–66.

NINE

A feminist perspective on women and mining in South Africa

Salimah Valiani and Nester Ndebele[1]

In a 2009 article of the World Bank Extractive Industries and Development Series, based on consultations with mining communities around the world, it is argued that the benefits and risks of extractive industries are distributed differentially among segments of affected communities. In greater detail, men have the greatest access to the benefits of extractive industries, in the form of employment and income, while women and the families they support are far more vulnerable to harmful social and environmental impacts (Eftimie et al., 2009).

More recently, through a synthesis of transdisciplinary academic

---

1   The authors acknowledge the female artisanal miners who agreed to participate in this study; the contributions of Patience Salane, who assisted in facilitating the focus groups, transcribing and translating the data and the contribution of Temoso Mashile, who provided general research assistance as well as transcribed and translated fieldwork data.

and policy-oriented research, Katy Jenkins (2014: 330) argues that women of the Global South are disproportionately affected by the negative aspects of mining and that this is not well recognised or understood. Drawing from seven country studies of the impact of extractive industries on women in Africa, WoMin-African Gender and Extractives Alliance (Valiani, 2015) argues that mining in the continent cannot constitute sustainable development because, in most instances, mineral and oil-based development involves the misuse or/and destruction of crucial but typically undervalued resources: community wealth, food production systems and women's labour. In a study of the disjuncture between corporate social responsibility discourse and the impacts of gold extraction on women in Tanzania, Lauwo (2018) underlines that the impact of environmental and social risks has fallen disproportionately on women through the loss of productive land, marginalisation and increased health risks.

Building on these arguments, this chapter focuses on South Africa – with its century-long history of mineral-based wealth generation and the recent, formal inclusion of women in the mining industry. It is argued that from a feminist perspective, mining thus far has not constituted a positive development experience for women. This is due to the myriad of occupational challenges faced by the still relatively small numbers of female mine workers, and the yet larger numbers of women negatively affected by the mining industry since its inception in the last quarter of the 19th century. Lahiri-Dutt and Macintyre (2006) have identified that women in mining areas in developing countries are typically not seen as active participants in the economy. Attempting to help reverse this misconception, it is further argued here that suboptimal use of female labour, destruction of community wealth and stunted social reproduction are the overall outcomes of mining for women in South Africa. The argument is demonstrated through discussion of the experiences of various groups of women examined by both academic researchers and advocate-researchers: female asbestos mine workers, female underground mine workers, female agricultural producers affected by mining and female artisanal mine workers.

Academic studies from the fields of history, health science, sociology, anthropology and others are drawn on here, as well as research and surveys conducted by non-governmental organisations.

Given the paucity of research on female artisanal miners in South Africa, fieldwork was conducted with female artisanal diamond miners in Kimberley, Northern Cape. Though not representative of all female artisanal miners in South Africa, a sub-argument of the chapter is that the artisanal workers who were engaged constitute the one exception whereby mining is a relatively more positive experience for females. The working experience of female artisanal diamond miners in Kimberley provides some insight into the possibilities for a more inclusive future of production in South Africa, though not explicitly tied to mining.

CONCEPTUAL APPROACH

Drawing from ecofeminism, the feminist perspective applied here puts a lens on the majority of women in and affected by mining, the majority of communities affected by mining, as well as the ecosystems of which all human societies are a part.[2] As Lozeva and Marinova (2010) elaborate, patriarchal tradition exacerbates male dominance over the environment, similar to the ways in which it supports male dominance over women. Mies and Shiva underline in the 2014 edition of their classic volume, *Ecofeminism*, that ecological disasters have a greater impact on women than on men, and relatedly, that the maintenance of everyday life within the context of ecological destruction and industrial catastrophes has been made the particular responsibility of women (Mies and Shiva, 2014). Hence Lauwo (2018) argues for the need for feminist approaches, which formulate questions that place women's lives, and those of the marginalised, at the centre of inquiry.

While these notions are especially pertinent to women and communities affected by mining, ecofeminism also offers theoretical insight relevant to an analysis of the experience of female mine workers – particularly through its questioning of the concepts of 'productivity' and 'development'. Shiva (1988: 44) argues that rather

---

2   As such, this paper does not focus on the experiences of female administrative and managerial workers in the mining industry. Far less particular to the mining industry, the experience of these women is much like that of female administrators and managers in other male-dominated industries.

than being 'benign and gender-neutral', dominant models of thought and development strategies are rooted in patriarchal assumptions of homogeneity, domination and centralisation. In turn, Shiva (1988: 42) rejects the notion of 'productivity of labour' as defined in mainstream economics as 'man producing commodities using some of nature's wealth and women's work as raw materials and dispensing with the rest'. As will be shown, such narrow notions of productivity and development underlie the undervaluing and suboptimal use of female mine workers in South Africa, in the past as well as the present.

FEMALE ASBESTOS MINE WORKERS

Crocidolite asbestos, one of three types of asbestos mined in South Africa until 1980, was mined in the north-west of current day Northern Cape province from 1893. The asbestos industry was dependent on female labour, which comprised close to half of the mining workforce in north-west Northern Cape province and north-eastern Transvaal (current day Limpopo) from 1893 to 1980, though records of mining companies and the Department of Mines as well as accounts of most historians date the employment of women from 1893 to 1955 (McCulloch, 2003: 414).

Taking the work not wanted by men, women worked largely in cobbing and sorting asbestos. Small numbers of females were employed in sorting asbestos and pressing it into bags to be transported (Davies et al., 2001: 87; McCulloch, 2003: 421). At some mines, such as Westerberg and Pomfret, long fibres were hand-sorted from the mill feed, trimmed using four-pound hammers and then bagged separately, largely because long fibres fetched better prices in the world market (McCulloch, 2003: 428). But the large majority, specifically three-quarters of female asbestos mine workers, cobbed asbestos or pounded the asbestos ore with hammers weighing two kilograms – the first stage of extracting asbestos fibres from the rocky ore (McCulloch, 2003: 418). Done on the surface, this work was attractive to women, who worked in groups and shared food preparation and child-minding tasks on the mining sites.[3]

---

3   This differed from domestic work, the other employment option for colonised female South Africans of the region, which obliged women to leave their families to earn an income.

The cobbing method was preferred by mining companies because feeding cobbs into processing machines prevented damage to the ducting and other parts of the machinery. In greater detail, when ores were fed directly into the machines, large amounts of host rock and dust also entered, causing damage and lengthening the refining stage (McCulloch, 2003: 417). In addition, the meals prepared by female mine workers for male underground mine workers saved companies the expense of providing meals and running canteens. Despite these gains from having female workers on site, mining companies paid less for cobbing and other female work, in many instances subsuming the wage of female mine workers into the wages paid to males (McCulloch, 2003: 418, 422).

For all of these reasons, the employment of female cobbers and sorters persisted, even following World War II when parts of asbestos mining were industrialised. McCulloch provides extensive evidence of how, up to 1980, asbestos mines were regularly granted exemptions under the 1956 Mines Act (which prohibited the employment of female mine workers) in order to employ women in cobbing, break-sorting asbestos, ore sorting, re-sorting fibre and other 'surface risk work' (McCulloch, 2003: 424).

Though unrevealed to workers for much of the 1893–1980 period, all of the work performed by women in asbestos mines involved far greater health risks than the underground mining of asbestos ore given the close, direct contact that female workers had with asbestos fibre (McCulloch, 2003: 416). A study of 770 female asbestos mine workers employed in the Northern Cape province between 1929 and 1992 found that a clinical diagnosis of asbestosis was made in 96.2 per cent of the women (Davies et al., 2001: 87).[4] The average length of employment was 10.8 years, with 21 per cent of the 770 women first employed in the mines between the ages of 10 and 14, 49 per cent between the

---

4   Asbestosis is a non-cancerous yet deadly lung disease characterised by severe scarring and inflammation of lung tissue. The disease is caused by asbestos fibres lodged in the lungs, leading to symptoms like coughing, shortness of breath and tightness in the chest. Radiological evidence of tuberculosis, the illness more often recorded as cause of death of female asbestos mine workers, or a history of previous treatment for tuberculosis, was found in only 7.5 per cent of these women (Davies et al., 2001: 87).

ages of 15 and 24 and 3.4 per cent before the age of 10 (Davies et al., 2001: 89). Needless to say, the costs of this particularly female work in asbestos mines are immeasurable and most likely undervalued in payments made as a result of court claims brought against asbestos mining companies by former asbestos mine workers.[5]

Valiani (2015: 13) defines stunted social reproduction as the inability of families affected by mineral and oil extraction to reproduce the workforce, nurture talents and potential talents of youth and to grow culturally and socially. Given the contract system used in asbestos and other mines, whereby ill workers were simply terminated and replaced, it is unknown how many asbestos mine workers died over time as a result of asbestos-related illnesses and conditions. In 2001, Davies et al. (2001: 1) argued that there continued to be a 'large, unrecognised burden of asbestos-related lung diseases among rural women' who had worked on asbestos mines. Overall, given the extremity of the known diagnoses – and the imagined but undiagnosed health effects of asbestos mining on women, men and their children who played among cobbs and asbestos fibres – stunted social reproduction can be seen as the defining outcome of asbestos mining in the medium to long run.

FEMALE UNDERGROUND MINE WORKERS[6]

Reversing the banning of female underground mine workers in the Minerals Act No. 50 of 1991, the Mine and Health Safety Act No. 29 of 1996 and the Mineral and Petroleum Resources Development

---

5   The largest of these is probably the £21 million settlement reached by Cape PLC (formerly Cape Asbestos Pty) and former asbestos miners and their families in 2001 (see Hodge, 2002).

6   The analysis in this section draws heavily on data from Asanda Benya's 2009 master's thesis on female underground mine workers at Impala Platinum in Rustenburg. Benya's is the only ethnographic study of female underground mine workers in South Africa. Based on in-depth interviews and participant observation over a period of two and a half months, Benya's study is rich in description, leaving room for analysis. The data is arguably quite representative in that according to the latest data, the largest number of women employed in mining (across occupations) in South Africa today are working in platinum mining, at 21,278. They constitute a large proportion, therefore, of the total 57,472 women employed in mining (Statistics South Africa, 2017).

Act No. 28 of 2002 permitted the employment of female underground mine workers. This change was affected in the name of gender equality (Benya, 2009: 4).

Beyond the legalisation of female underground mine workers, more instrumental in bringing women into the mining industry is the 2007 Mining Charter requirement that females constitute at least 10 per cent of each mining company workforce by 2009. The Department of Mineral Resources (DMR) subsequently increased these targets such that by April 2016 targets for female employment were set, in a new draft Charter, at 50 per cent for top management, 60 per cent for senior management, 75 per cent for middle management, 88 per cent for junior management, 40 per cent for skilled and semi-skilled occupations and 3 per cent in other occupations. In response to this targeting, most companies instituted elaborate programmes to assist women's entry and advancement in the industry. In the case of Impala Platinum, these include career-development plans for women with mentoring and accelerated training to achieve benchmarks such as blasting certificates, pregnancy and sexual harassment policies and separate female toilets and changing rooms. In terms of male mine workers, Impala Platinum has endeavoured to raise awareness around equality and female mine workers' rights through videos and call centres where men can put their questions anonymously (Benya, 2009: 6).

Despite formal inclusion in the industry – in the form of legislative and company measures – retention of women in the industry remains the key challenge, according to both Impala Platinum and the network of professionals Women in Mining South Africa (WiMSA) (Benya, 2009: 5; WiMSA, 2016). According to the Statistician-General, female employees constituted 14.7 per cent of the total platinum mining workforce in 2015 (Statistics SA, 2017). For the mining industry as a whole in 2015, female employees amounted to 57,472 of 353,344 total employees, or 16 per cent of the workforce (Statistics SA, 2017). This low percentage of female employees – well under half and far below DMR targets – is due to persistent structural inequality making for the suboptimal use of female labour in this historically masculine industry. Various instances of structural inequality are elaborated below in support of this argument.

Cockburn (1981) traces the construction of capability and the gender bias of technology through the history of compositors in the British printing trade. Theoretically, Cockburn (1981: 56) argues that bodily strength and capability are socially constructed and deployed politically. This is a process involving several converging practices: the accumulation of physical capabilities by males who are taught, trained and encouraged from childhood to develop the body in certain ways, the definition of tasks that match these capabilities and following from these, the selective design of tools and machines (Cockburn, 1981: 44).

The design of tools, tests and protective gear which are biased toward male physique and capabilities can also be traced in the mining industry. The first test encountered by aspiring underground mine workers is the heat tolerance screening (HTS) test. Designed to ensure endurance while working under high temperatures and in humid conditions typical inside mine chambers, the test is extremely difficult for women to pass. According to an HTS supervisor at Impala Platinum, only 6 per cent of women pass on the first attempt, with most passing on the second or third attempt after following dietary and exercise regimes recommended by the Impala Platinum HTS Centre (Benya, 2009: 62). This is largely due to women's higher body temperature level (relative to men's) rooted in different phases of the menstrual cycle. That women do tend to eventually pass the exam after following dietary and exercise programmes is an instance of Cockburn's social construction of bodily strength and capability; when trained a certain way, women's bodies perform as men's in the hot temperature and humidity underground.

Perhaps even more important to stress in terms of structural inequality within the mining industry, however, is that the standard of the test is set by what has been deemed normal for males. A worker with a body temperature exceeding 37 degrees Celsius at the start of the test is not admitted into the test. This immediately bars many women from taking the test. On completion of 30 minutes of light exercise (i.e. climbing up and down stairs) underground, a body temperature of 37.6 degrees Celsius or less is considered a pass, meaning that the worker is deemed able to perform under heat levels rising to 32 degrees (Benya, 2009: 57–59). If the industry invested in developing HTS science particular to the norms of female bodies, both the starting and passing body temperatures of the HTS test would likely be set differently for females.

In terms of tools, headlamp batteries and rescue packs must be carried while working underground at all times. These tools are designed for men and are heavy for female workers to carry. Given that the majority of females are allocated work involving the transport of heavy materials within the mine – to be discussed in greater detail below – the weight of the batteries and packs adds to the weight of materials which are either carried by female workers or pushed in a cart.[7] Benya gives the example of a female transport worker who carries three packets of explosive material from one work station in the mine to another. The packets weigh 23.5 kilograms each and the route involves climbing a rope (Benya, 2009: 95). Other material typically transported by female workers are roof bolts, pipes for water and compressed air and ore (Benya, 2009: 67, 95). If headlamp batteries and rescue packs were designed with consideration for the needs of female workers, the load for female workers would be lessened. Paradoxically, despite regular handling of these heavy loads, female mine workers are seen as 'weak' and only 'fit for household work' by male mine workers (Benya, 2009: 72).[8]

Male bias can also be seen in the protective gear provided to underground mine workers. Most shafts have one-piece overalls, which are a challenge for female workers when using the toilet. Toilets are very small and not cleaned regularly. Female workers must therefore remove headlamp batteries, rescue packs and overalls outside the toilet before use (Benya, 2009: 97). In addition to complete lack of privacy and increased sense of vulnerability from removing the overall and

---

7   Females are by far not the only workers doing this type of work. According to Impala Platinum data, by June 2008, 978 female and 20,549 male mine workers were employed as 'unskilled and defined decision-making personnel' of a total 21,527 in this category.

8   While providing ample evidence, Benya (2009) does not identify this paradox. Similarly, she does not link the design of tools, tests and protective gear to gender bias. In this sense her argument is limited to describing the various forms of discrimination faced by female mine workers. In turn, Benya does not have a structural explanation for this discrimination. She attributes it to biological difference and individual prejudice.

exposing the body, this takes time and contributes to male workers' perceptions of female workers as slow (Benya, 2009: 97).[9]

Beyond tools, tests and protective gear, the standards by which bonuses are awarded to workers are set in accordance with male worker abilities, making bonuses unobtainable for female workers in a variety of ways. Given that male wages are low – another structural inequality inherent in the South African mining industry – bonuses are regarded by male workers as crucial. In greater detail, between 2000 and 2008 for instance, labour (i.e. workers as opposed to management) at the three large platinum producers only received 29 per cent of the value added produced (Bowman and Isaacs, 2014: 1). As Bowman and Isaacs point out, this is much lower than the national average whereby 50 per cent of value added was distributed to workers in the South African economy over the same period. On the flip side, 61 per cent of value added produced in the platinum industry went to profit (Bowman and Isaacs, 2014: 1).

Hence it is understandable that male workers and several human resource managers interviewed by Benya (2009: 78) reported that bonuses could exceed monthly wages, amounting at times to more than R10,000 per worker per month. Bonuses are awarded by team for elevated production levels with targets set by dominant (i.e. male) standards of production. Even one female worker in the team is thus seen by male workers as a hindrance. Verbal harassment of female workers follows from this (Tau, 2015). As one shift supervisor remarked to Benya:

> *For a team to take on a woman it means forgoing a man, which means less production and compromising bonus. Workers and myself, I do not want my team to not get bonus because they all have slow-working women that are always complaining and arguing with other team members* (Benya, 2009: 79).

---

9   In shafts where two-piece overalls are available, appropriate sizes for females are often missing. Female workers then tend to revert to using one-piece overalls (Benya, 2009).

Although the annual base wages of platinum miners have since increased due to the 2014 Association of Mineworkers and Construction Union strike, the average 8 per cent increase (just above inflation) means that the relative largesse of bonus schemes remains the same. In turn, the structural constraint for female workers to be accepted and treated equally in teams by male workers also remains.

Moreover, the highest bonuses are received by teams working in 'stoping', or the actual extracting of ore containing the mineral, where the vast majority of female miners do not work (Benya, 2009: 77). This is due to formal and informal job allocation by male and female managers, as well as male underground workers. In addition to being seen as slow and weak, female mine workers are perceived as lazy and less willing to take risks (Benya, 2009: 80). Female workers are thus allocated jobs in 'development' sections responsible for opening up new spaces to mine (i.e. installing pipes for water and compressed air), and in the transport of materials as elaborated above (Benya, 2009: 67). Other jobs done by females include cleaning the travelling way between sections, loading ore onto hoppers and taking ore from various underground sections to the surface, installing and uninstalling railway lines, issuing supplies to other workers, carrying the shift supervisor's bag and measuring work stations (Benya, 2009: 73).

The few women who are formally allocated positions at higher skill levels in stoping sections – for instance as panel operators, winch operators or rock drill operators – though paid the corresponding wages, rarely work in these positions because they are considered 'too difficult for women' by male colleagues (Benya, 2009: 73). Male team members thus operate the machines for female workers while females are given 'small easy tasks' (Benya, 2009: 82). Without stoping experience, promotions are difficult.

For all of these reasons, female mine workers encountered by Benya frequently expressed the desire to leave mining (in contrast to males) and common complaints were around unequal wages and the inability to support family and extended family on mining wages (Benya, 2009: 84). This is consistent with findings of a survey carried out by WiMSA (2016) in which only 34 per cent of female professional mine workers expressed that their employer pays females and males equally, and only 38 per cent of whom said that disparities were performance related. Given that male underground workers generally have more years of service and are socially, culturally and physically more able to accumulate skills and experience in the industry, male workers generally earn more than female workers. While some female underground workers envision moving to jobs on the surface, this is only possible for those with matriculation passes or those able to study while working underground (Benya, 2009: 83).

It is not hard to understand, then, why taking a *nyatsi*, or supplementary sexual partner, underground is a widespread practice among female mine workers. The *nyatsi* typically provides money, transport to work (a particular need for morning shifts, when the use of local taxis and buses is particularly unsafe for women), promotions and dinners and holidays in exchange for sex. This is an open practice among underground workers. Most female workers and the *nyatsi* have permanent partners, such that these underground relationships are limited to transactional sex (Benya, 2009: 84).[10] Transactional sex allows the female worker to transfer most of her salary to the family members she supports while her own daily financial needs are taken care of by the *nyatsi* (Benya, 2009: 85).

Sexual violence underground is the final major structural inequality preventing greater participation of women in underground mining. In Benya's findings, female workers in stoping teams report avoiding night shifts due to fears of sexual violence by male team members, and several females report experiencing groping and sexual rubbing in the cage descending to the mine and while waiting for chairlifts (Benya, 2009: 91, 93). Occasional media reports have featured female mine workers and their female union representatives speaking of rapes in mine changing rooms and forced sex in exchange for assistance with strenuous tasks underground (Tau, 2015; *News24*, 2015). Perhaps the most publicised case of rape and murder underground is that of

---

10 It should be noted that because of company policy to not perpetuate the migrant labour system in the hiring of female mine workers, females are employed within a 60-kilometre radius of Impala Platinum mines. Female workers thus live with families and commute daily to mines while more male workers live at the mines.

Pinky Moisane in 2012 (*News24*, 2013).

Given the high rates of male sexual violence against women in South Africa, what is particular to the mining industry is the approach taken by employers to sexual violence in mines. According to the Centre for Applied Legal Studies (CALS), mines maintain that male sexual violence underground is a criminal justice matter and not a responsibility of mining companies (CALS, 2015: 7). Though CALS views employer protection of female mine workers as falling within occupational health and safety standards under labour legislation, the fact that such protection is not conceived of, by mining employers, as under their jurisdiction is further evidence of the structural barriers preventing significant female presence in the industry.

In sum, gender-biased tests, tools, protective gear, performance standards and formal and informal work allocation prevent the optimal use of female labour in underground mining in South Africa. Female mine workers' dependence on transactional sex – to compensate for the lack of work experience, meaningful skills development, occupational advancement, and in turn, adequate wages – sexually objectifies female mine workers and reconfirms the suboptimal use of female labour in the industry. The line between transactional sex and sexual violence in mines is, not surprisingly, thin, and the threat of such violence, in addition to the gender biases elaborated above, makes the low retention rate of female workers in the industry equally unsurprising.

FEMALE AGRICULTURAL WORKERS AFFECTED BY MINING

As with the instance of female mine workers, there is a paucity of studies on rural women affected by mining in South Africa. To the best knowledge of the authors, only one study on the subject exists: a report by the advocacy organisation, WoMin-African Gender and Mining Alliance (2017), entitled *No longer a life worth living – Mining impacted women speak through participatory action research in the Somkhele and Fuleni communities, northern Kwazulu-Natal, South Africa*. The report tells the story of water pollution, largely caused by coal mining in northern KwaZulu-Natal, and focuses on how women living in the area cope with the resulting crisis. Further or future research with a focus on women affected by mining in other parts of the country – including Mpumalanga and Limpopo (coal), North West, Gauteng and Free State (gold) and Limpopo, Gauteng and Mpumalanga (platinum group metals) – would likely tell similar stories of the destruction of community wealth and the resulting burden of crisis management carried by women.

In greater detail, the participatory research was carried out over a period of eight months in Somkhele and Fuleni, two communities (of several villages each) in the uMfolozi River catchment area. With a total population of 180,000 in Somkhele and 16,000 in Fuleni, just over half of the households in the communities are female-headed (WoMin, 2017: iii). Despite this, due to the gendered nature of 'communal land tenure' under the Ingonyama Trust, of which Somkhele and Fuleni are a part, a hierarchy of traditional leaders makes all major decisions related to land use. Precisely through this relation of gendered control and exclusion – as in several other parts of South Africa – the Petmin Group-owned Tendele coal mine was established in Somkhele.

Compounding this inequality, since 2015 drought has severely affected water supply in Somkhele and Fuleni. In addition to the drought, the Petmin Group-owned Tendele coal mine in Somkhele is largely responsible for depleting the water supply through its underground and surface exploitation of the uMfolozi since 2007 (WoMin, 2017: iii). According to the women of Somkhele, this has been done primarily through the pumping of water from the uMfolozi River, the main water supply which had run dry by the time of the research, and the fencing off of communal water sources. In addition to water grabs, the Tendele mine has caused air, water and soil pollution, which all research participants link to the destruction of crop and livestock farming in the area (WoMin, 2017: iii).

Water available to the communities has become unclean water collected from dams, giving rise to gastrointestinal problems such as vomiting, nausea and diarrhoea (WoMin, 2017: 22). As women are responsible for water and food provision, desperate attempts to purify the dirty water include using Jik bleaching liquid, cement and ash (WoMin, 2017: 22). According to the affected women, the use of this water is causing skin irritations and vaginal infections (WoMin, 2017: 22). The WoMin study highlights the case of a Macibini (Somkhele) family admitted to hospital after consuming water 'purified' with cement. Given the toxic compounds in cement – lime, crystalline silica and chromium – cement-laced water damages lungs, corrodes human tissue and causes allergic reactions.

Mine-related destruction of community wealth, in the form of water, thus leads to stunted social reproduction in Somkhele, whereby the community is unable to reproduce itself and grow culturally and socially. With regard to Somkhele women in particular, the inability to farm and raise livestock due to mine-related air, soil and water pollution further amounts to the suboptimal use of their labour. All of this occurs within a historical structure of gendered decision making around land which puts not only the majority of women, but the majority of the communities of Somkhele and Fuleni at the receiving end of – typically uncalculated – human and environmental costs of mining operations. For these reasons, female leaders of Fuleni have been among those at the forefront of opposing Petmin's intended expansion of coal mining into Fuleni (see Centre for Civil Society, 2016).

FEMALE ARTISANAL MINE WORKERS

In her survey of academic and practitioner-based literature on women and mining in the Global South, Jenkins (2014: 332) underlines the importance of recognising the continued resilience, entrepreneurship and tenacity of female artisanal miners, with the proviso that these should also not be romanticised. Focusing on South Africa, Thornton (2014: 127) asserts that *zamazama* are widely misrepresented by the public press and misunderstood by the South African government, for example, as the stereotype of gang-run work teams engaged in dangerous turf wars underground and perceived as 'stealing from the nation and legitimate mines'. In an anthropological study of gold-mining *zamazamas* in Roodepoort, Johannesburg, Nyoni (2017: 144) underlines that there is 'scarce information' on the roles and experiences of women in these 'spaces'.

All of these broad points are consistent with the findings of the fieldwork with female artisanal diamond miners undertaken for this chapter. As the only study focusing solely on female artisanal miners in South Africa, this chapter cites direct quotes from the participants as much as possible below. It is argued that of the various groups of women in and affected by mining examined here, the experience of female artisanal diamond miners demonstrates a relatively more positive mining experience for women and provides some broad strokes of a future vision of inclusive development favourable to women.

The fieldwork undertaken was with a group of female artisanal diamond miners on the outskirts of Kimberley, Northern Cape. A total of 35 women gathered voluntarily over two days to speak of their living and work experiences in focus groups. In order to gather more in-depth personal and work histories from those willing, five of the 35 women were also interviewed. All of the women who participated work at Beefmaster and/or Samaria camps, an artisanal mining area and informal settlement stretching over some 5 kilometres on the southeast edge of Kimberley, at the tail end of Samaria Road.[11] Some of the women live in Samaria camp and others in townships surrounding Kimberley. The miners search for diamonds left behind in a mine dump sold by De Beers to Ekapa Mining in 2015.

Living conditions
Not unlike other informal settlements surrounding mines, Samaria and Beefmaster are completely lacking in water sources, electricity and other basic infrastructure. An exposed, broken water pipe about two kilometres from Samaria camp is used as a water source which women access using buckets. Running water, toilets, lights and a health clinic were therefore stressed by participants as key, immediate needs. Given the particular health needs of women as physical reproducers of the community, as well as the fact that women carry the heaviest burden when basic resources and infrastructure for social reproduction are absent, these negative aspects of female miners' lives are significant.

---

11 There are six camps in total with an estimated 5,000 artisanal miners: Samaria, Collville, Beefmaster, Gumtree, Green Point and Buffalo.

In the words of Lindiwe,[12] one of the first female miners at Samaria camp and ANC ward committee member:

> *I drew up and submitted a proposal to the Premier's Office for assistance in building a water pipe system. The councillors from both the ANC and the DA came and viewed our camp site and asked what would happen in the case of a fire. I answered that if one shack burns all of them would burn down. Luckily for us that has never happened. They then asked us why we don't build our own water system since we make enough money. We told them that we want the water to be funded from the government. After many back and forths, I spoke to the Speaker who is ANC, we agreed that they would deliver jojo tanks in the interim but we are still waiting for them* (Lindiwe, Focus Group 2, 27 May 2017).

Participants also drew connections between the need for infrastructure and the particular dangers of informal settlement living for women:

> *We have to fetch water from far and most women don't have the strength to do so because the buckets are heavy. It's also not always safe as a woman to walk that distance alone, sometimes at night, carrying heavy water. So its toilets, water and lights because at night it gets really dark* (Akhona, Focus Group 4, 28 May 2017).

> *We have a mobile clinic that comes once every three months. We need a permanent clinic since working with the sand does have its disadvantages. I particularly get shortness of breath because of constant contact with the sand. The clinic could help with regular TB testing and other special tests specifically for women since we use the open veld as our toilet and that can bring with it many kinds of diseases* (Silvia, Focus Group 3, 28 May 2017).

---

12 To protect identities, all names of artisanal miners have been changed.

Working conditions

There are six informal mining camps in total, with some 5,000 artisanal miners – both men and women – searching the top soil for diamonds left behind by De Beers industrial mining operations. A mining permit for the area was previously held by De Beers and has now expired. Another company now claims to hold a permit to mine the area. The artisanal miners were served an eviction notice by Ekapa Mining and an interdict by the municipality of Kimberley, both of which they were challenging in court when the fieldwork was undertaken (Christiana, interview, 26 May 2017).

As with workers in underground mines in South Africa, the artisanal miners around Kimberley are of multiple origins. Unlike in underground mines, where ethnic divisions are prominent, at Samaria and Beefmaster miners reported amicable coexistence. As two of the female miners put it:

> *We all respect each other, it's as if we all come from one place or town but we're all different: There's Xhosas, there's Sothos and all kinds of people. It's a normal neighbourhood for us where we borrow things to each other when the other one needs something* (Lindiwe, Focus Group 2, 27 May 2017).

> *Remember the majority of the people here do not originate in Kimberley. You find people from Lesotho, from Mozambique, and then there are South Africans from different provinces. There is a common understanding that everyone comes from somewhere so there are no conflicts around that issue* (Christiana, interview, 26 May 2017).

Female artisanal miners reported a wide range of previous occupations, including domestic worker, underground mine worker, security guard, care worker and soldier. Most of the occupational stories encountered reflect the observation of Janet Munakamwe (2017, 155) that non-formal mining activity is the result of economic exclusion and inequality. About one third of the participants reported not working outside the home prior to undertaking either artisanal mining or accompanying male partners to the informal settlement. The role of the latter recalls that of female asbestos mine workers in the 20th century:

I was home when he came to fetch me to come and stay with him here. The reason was that it was difficult for him because when he finishes at the mines there isn't anyone at home helping him to clean, cook and heat the water (Thandi, Focus Group 4, 28 May 2017).

Two artisanal miners reported being former underground mine workers and both recounted experiences resonating with Benya's findings:

I used to work at Kumba Resource as a miner. I would say I prefer working here because here I am my own boss; when you work with a supervisor they don't always treat you well, so here I am able to navigate everything at my own time and pace. It's much better here (Khosi, Focus Group 2, 27 May 2017).

I used to work in the mine underground for about five years, it wasn't always safe because we always had a fear that when we are travelling in the cages that the ropes would come undone. I decided to take a break from underground to try my luck with the informal sector, I haven't found anything yet in the month that I've been there but I am hopeful (Gabi, Focus Group 2, 27 May 2017).

Several domestic and care workers reported taking up artisanal mining to replace or supplement traditionally undervalued, female work:

I used to work as a domestic worker, I came here once and in two days I found a diamond. After selling it I immediately realised that this was something I could do full-time; it was much better than waiting for a small salary at the end of the month. So I quit and came to work here (Reena, Focus Group 2, 27 May 2017).

I also worked as a domestic worker. But the salary was so terrible that we used to run out of food in the house in the middle of the month. When I came here our livelihoods really improved, there was a difference from before. I am even able to save a little bit of money in the bank (Nokuthula, Focus Group 2, 27 May 2017).

I am a care worker. I want a good life for my children. Only one of my children has a good life. I work here in the evenings because I only earn a stipend. I have been coming here for five years but I haven't really been able to supplement my income yet (Sonia, Focus Group 3, 28 May 2017).

I used to work at Mediclinic as a hostess. The difference now is that I don't have to deal with the stress of patients and supervisors screaming at me all day. And the money here is also much better (Buni, Focus Group 4, 28 May 2017).

Not unlike the underground mine workers and caregivers, the one former soldier encountered in the focus groups opted for artisanal mining as a means to escape discrimination and achieve a degree of freedom and respect in the workplace. As a professional, unlike other participants, her assets allow her to excel in artisanal mining:

I was 24 years in the air force. We were the first women to be integrated: two ladies and 600 men. We weren't welcomed. So we shifted, started behaving like men. In 1999 more women started entering. We trained the incoming women, and then watched them get promotions while we didn't. Our inferiors were then our bosses and they were horrible to us. I had a certificate to work with ammunition and did the job of four people but didn't get paid for it. I became taken for granted, so I thought, I know of another job. I came to the mine in 2013 and saw only men. I thought to myself, no, I have worked with men. I am comfortable with men. I can do this job. I interviewed the men and they told me you can find small stones a couple in a month. So I went back to the air force for three years. I wrapped up some commitments, and then I started mining. I got my first luck on my third day. Two weeks later I found another rock. Then my fiancé and I started driving to other camps to search. Because we can travel far we find good rocks every month, me and my fiancé. When we find small stones we don't sell immediately. We sell in parcels and get better prices from better buyers. Christiana helps to connect to the buyer who pays higher. Then she gets a commission (Zandi, interview, 28 May 2017).

In addition to better financial reward, which most participants agreed was the major attraction of artisanal mining, many talked about the freedom of determining one's work and break times and the pace of work. Several participants highlighted the enjoyment of working with the soil, the challenge of sharpening their skills to detect soils likely to contain diamonds and a near-spiritual hope that the searching would be fruitful. This combination of materiality and belief without excessive materialism among female artisanal mine workers is unique as compared to other 21st century workers of South Africa and elsewhere. Though most female artisanal miners do not find much of value on a regular basis, they remain driven by the space to develop and the autonomy offered by the work. In the words of a former store manager and a former security guard:

My father died in 2009 so I have been responsible for my mum, sister, two brothers, four children and one grandchild. They all live in Charlene (a village) on my mother's grant. No one has a job. I do piecework with the school lunch programme to buy food, so I can spend most of my time working hard in mining. I have hope when I do this work. And I have patience and trust that something will come together. I don't feel this in the piecework. I have been here since 2014, working alone. I haven't found anything yet (Karabo, interview, 28 May 2017).

The advantages are impressive because you manage your time, you push your own production. You have it in your mind that you are running your own company, so you are able to wake up at a time that suits you. The other advantage is the money. I was working as a security guard and my salary was about 3,500 Rands a month for 21 days of work a month, 12 hours a day. That's 12 hours that you spend away from home, away from your blankets. But if I find say, a three-carat diamond, I would have to work 6 to 8 months as a security guard to achieve the same money. In my year here, I have found one 68 point diamond, with colour. I got 70,000 rand, which I and my two kids are still surviving on. We of course don't get the big diamonds that the mining companies get so the sizes are small compared to what's really out there. So really, we are working on their leftovers. When they have cleaned the soil, there are the small leftovers (Botshelo, large group discussion and interview, 27 May 2017).

In terms of access to the mining camps, virtually all participants reported that artisanal mining was accessible only to those who had tools. Tools involve one shovel, a pick, two buckets, some bags for sorting and two sieves. Tools thus become a barrier to entry for those who can't afford them but want to mine. Theft of tools is common, particularly given the temporary nature of the informal structures in which people live in the camp. In large part, the materials used are recycled heavy plastics. Most huts are therefore not lockable, making for easy entry.

As in other cases of female workers of lesser means, tools can and do become a currency for abuse. The spread of the problem was not clear through the focus groups conducted:

If I as a man have tools, I lure women without tools and that is how they end up staying together. It's a very abusive system because the men end up physically abusing the women. We have many cases of domestic abuse here (Lindiwe, Focus Group 2, 27 May 2017).

That's something that needs to be highlighted; some women come here, and they do not have tools. They then have to rely on their male colleagues for assistance, the males would use that to their advantage by taking advantage of the women (Kananelo, Focus Group 2, 27 May 2017).

When asked specifically about transactional sex, one former underground mine worker and one artisanal miner shared experiences.

> *I worked with a lot of men underground. I had my own group of guys that I would work with, so I never felt unsafe. I hated working the night shift, I was scared and asked that they give me a straight shift, but they refused. I was hired in the right way, although I know of many women who weren't. Some would sleep with their bosses to either get hired or get promoted. I have heard cases of when the women do get these promotions, the bosses keep going back to them and requesting sex. When the boss is having problems at home, you have to 'substitute' those problems with your body because he did you a favour with the promotion. If you refuse he threatens to fire you or rescind the promotion* (Gabi, Focus Group 2, 27 May 2017).

> *Even some women who arrived here arrived in that way. They had to pay a protection fee to have access to the soil* (Lindiwe, Focus Group 2, 27 May 2017).

Most maintained that this was not typical:

> *There is a rough sieve and fine sieve; there are different soils. We came with tools, my daughter and I, and a guy showed us on our first day how to mine with what instruments. We worked together, three people, for three weeks and we found a 68-point diamond. And we shared what we earned. So, I am unfamiliar with the need to use the vagina as a means to access all this* (Botshelo, large group discussion, 27 May 2017).

What was much clearer is that what one woman termed 'financial abuse' was a common experience for female miners working with their male partners. In the words of Lindiwe, one of the first women in the camp and in whom many women seemed to confide:

> *We also need to discuss the financial abuse that all of us women*

deal with on the camp. You'll find that you've been living with a man for over six months, and both of you are in the business of mining diamonds, and you don't always go together to the site. Then there will be an instance where he just disappears with no reason, only to find that he had discovered a diamond that day. It happens with other women as well, like clockwork. All the while you wash his clothes, cook for him, give him sex, and that is how you are repaid. It's really painful for women; some of them when the women have made discoveries steal them in the middle of the night and never come back. Either that or they come back from the field claiming they have not found anything then the next day you hear that they have sold a diamond. It really is painful (Lindiwe, Focus Group 2, 27 May 2017).

Our partners are not honest. They get the diamond and then they disappear. They come back only when they are broke. This is the reason why we make sure we are part of every step of the mining process – even digging (Buhle, Focus Group 1, 27 May 2017).

Far from all participants work with their male partners. Some work alone, some work in teams of women, others in teams of women and men. Most agreed that working relations with males were generally harmonious, as this exchange among four female miners in Focus Group 2 (27 May 2017) demonstrates:

We work very well with the men.

They are our colleagues. They're our brothers, our fathers. And they're very loving.

Very loving; we are close to most of them.

In terms of tasks, unlike the gendered tasks in artisanal gold mining in Roodepoort noted by Munakamwe (2017) and Nyoni (2017), female artisanal diamond miner participants reported working in all aspects of the mining work. In artisanal gold mining, much like in the asbestos mining of the last century, females work in crushing or grinding the gold ore from chunks of rock while males extract the rock from underground. Female artisanal diamond miners reported working with shovels and picks, as well as in sieving and sorting. Some reported only sieving and sorting. Some reported mixing the tasks with male team mates in different ways depending on the season:

> *We spend the day working. He is going to do the digging and I'll do the sieving, then he will help me after he has finished digging, because we sieve twice. After that we sit together, then we sort together. When we are done and if we have found something we go to Makgomosha, where we sell. There is work there also, we have to negotiate, we don't just give in and go, or accept then just go. We do negotiate with them. If you say no, you can go to another buyer, there are too many here, so you can go to another one if you are not satisfied with the price, no hard feelings. My son is six now, he is in his first year of school, so after selling, I will send my son and mother money, so they can at least have something in the morning when they wake up. I can sleep hungry, I don't care, as long as they are fed, I am satisfied. That's why I work, for my son and my mom, because they are the most important people in my life (Reena, Focus Group 2, 27 May 2017).*

> *She and I work together with three males here in Samaria. We all dig, sieve and sort. When the day is hot, in the summer, our male team mates bring bags of soil home and we sort through the night (Jody, Focus Group 3, 28 May 2017).*

As with the large majority of working women, female artisanal miners in all focus groups reported carrying the full burden of care work in addition to mining work:

> *We do everything as women. We go to work and work the same as men. Yes, we know there are some jobs we can't do, that they do. Like, when it comes to this, you have to take a big shovel, you use a shovel to pick up soil and put it in a bucket. But you can't take a*

week or the whole day. A man can do it in an hour or 30 minutes, he can handle that, but you can't. When you get home, he can't handle that also, you have to cook, he's tired and you're tired, but you have to cook. Both of you come from work but you have to do something, you can't sit like him. He's going to get there and sit because he is the man and you are the woman, you have to do the work (Brenda, Focus Group 3, 28 May 2017).

I did speak to my partner. I told him that sometimes he has to help, when he comes from work he needs to make the fire and I will wash the pots then we can sit together. He helps me with something else and I wash the dishes because we have to take food to work. So, I'll wash the dishes while I put the pots, so he must help to get wood. I also have to clean the house because I can't sleep for two days without cleaning the house, I have to, no matter what, and this will happen after I bath, after we have eaten and then I start cleaning the house, and he'll be sleeping. So, it's the same, it is always women, no man (Anathi, Focus Group 1, 27 May 2017).

Enlisting responses on the topic of selling and earning was the most challenging element of the fieldwork. Participants were generally hesitant to reveal the number and frequency of finds as well as the value of diamonds discovered. This is not only because most diamonds are sold on the black market, but also because theft, as in the case of tools, is a major threat in the camp:

It is dangerous to talk about diamonds here in the camp. A guy in 2015 got 50,000 rand for a parcel. He put it in his old fridge. One day he left his shack. He hadn't been paid yet. His neighbour went searching for the money and couldn't find it. Some days later the man returned and when he was sleeping, his neighbour burnt him to death in anger (Christiane, Focus Group 3, 28 May 2017).

Participants were quite aware of the drawbacks of using the black market, including undercutting by buyers and the lack of confidentiality.

As poor workers and miners without permits, however, there is little other choice:

> *Because we are hungry, and we are moneyless and foodless. They stay just across the road, we call them, he/she look at it and will say let's go to the ATM, ask for a price, you say six thousand, he/she will agree. The buyers will get about fifteen thousand for that diamond* (Lerato, Focus Group 3, 28 May 2017).

> *The other day, a guy got an 8 or 9-point diamond. It had many blisters, spider nets inside and dots. They were supposed to get at least five million, but they were paid 900,000 Rand. That's daylight robbery. That's how the black market works. If that guy went with Baba Lucky through the Board, he would probably have 20 million rand now* (Christiane, Focus Group 3, 28 May 2017).

Something of a hierarchy was evident in terms of accessing better prices in the black market. Zandi, the former soldier, mentioned going through Christiane and paying her a commission. This was confirmed by Christiane and others:

> *Christiane helps to connect to the buyer who pays higher. The she gets a commission* (Zandi, Focus Group 3, 28 May 2017).

> *Sometimes we share buyers, some of the ladies would come to me and I would connect them with a buyer and the buyer would compensate me. We meet at a remote place and make the exchange. If the money is too much they deposit it into our accounts* (Christiane, Group 3, 28 May 2017).

Similarly, selling diamonds to the South African Diamond Exchange and Export Centre (referred to by participants as 'the Board'), where prices are far more favourable, was mediated through one male artisanal miner who had a mining permit at the time of fieldwork:

Baba Lucky has a licence, he can buy from us and go sell it at the Board. When you get to the Board, you put your diamond inside, they write it in and they give you a slip, you have to wait a couple of days because there are other people who will come and bid on your diamond, the higher price is your price. For instance, you get a seven-carat diamond, which is a lot of money, we are talking about millions here. If your diamond is at the Board, buyers will come and bid on your diamond, one will say 50 thousand, another 80 thousand, someone else will say 3 million, another 20 million, and that will be your price. The Board is fair, you get a quotation, if you have a bank account, they will put the money directly into your account and nobody can arrest you (Christiane, Focus Group 3, 28 May 2017).

From the buying and selling issues it is clear that 'legality' is one of the main obstacles for Kimberley female and male artisanal miners alike. In order to attempt to grapple with this, at the time of research the miners were in the process of forming co-operatives with a view to applying for mining permits. As Munakamwe (2017: 155) observes of artisanal gold miners, artisanal diamond miners also lack formal political voice and 'devise strategies of passive resistance'. Though only constituting about half of the artisanal mining population, women were in the lead of this initiative and spoke enthusiastically of the strategy:

We have registered 330 co-operatives of ten people each so far. We have a long way to go though, with 4,000 plus people in all six camps. We registered in Pretoria because the Small Enterprise Development Agency here turned us away. We are waiting now to open bank accounts and then apply for mining permits. With the permit, we can sell directly to the Board which will then give us proof, so we can bank our earnings. We want people to know that we are working women in the camps' (Christiane, interview, 28 May 2017).

In early June 2018, the artisanal miners of Kimberley had successfully negotiated – with Ekapa Mining – formal access to 600 hectares of mine dump land to continue mining for diamond remnants. In tandem with this, the DMR granted two mining permits for shared use by the miners (DMR, 2018; *Solomon Star*, 2018).

Finally, with regard to their visions for the future, participants largely talked about their children. Participants were unequivocal about the fact that they did not envision their children succeeding them as artisanal miners. In the words of one of the most successful miners, Zandi, the former soldier:

> *I want more than just the basics here in Samaria. I have a vision for our children and their children. I work here full time and am here for the future and future generations. I don't want my kids and theirs to come and do the pick and shovel. I want them to be our future leaders* (Zandi, Focus Group 3, 28 May 2017).

CONCLUSION

Based on the existing, rather sparse body of literature on women in and women affected by mining in South Africa, supplemented by primary research undertaken, this chapter demonstrates how suboptimal use of female labour, destruction of community wealth and stunted social reproduction are the predominant experiences of women in mining in South Africa. The experience of female underground miners is argued to be an instance of the suboptimal use of female labour, the experience of female rural agricultural producers in northern KwaZulu-Natal an instance of both the destruction of community wealth and suboptimal use of female labour and the history of female asbestos mine workers an instance of stunted social reproduction. Artisanal diamond mining seems to be the one exception whereby mining is a relatively more positive experience for female artisanal miners, though these women do not see a future for their children in artisanal mining.

Broadly, underground mining and artisanal mining today, like asbestos mining in the past, provide an exit for women from unemployment and undervalued, typical female work. Though employed and earning relatively better than many female workers, female mine workers have endured and continue to endure particularly intense hardships, from the direct contact with asbestos for female asbestos mine workers, to the gender biased technologies and standards confronting underground miners, to the harsh living conditions faced by artisanal miners. Gendered power imbalance, discrimination, and violence against women are persistent in all these instances but to a far less extent in artisanal diamond mining. Damage to health and harm to the reproduction of communities are the manifestations of mining-related destruction of community wealth, the major burden of which is carried by women.

If ample supply of land and relatively open access to it are the key structural conditions for the positive experience of female artisanal diamond miners in Kimberley, these provide some insight into the possibilities for future production inclusive of women and peoples of multiple ethnicities and nationalities. Similarly, worker-selected production teams, worker-determined pace of production, team-controlled earnings and co-operative structures for legal and administrative purposes are important features to retain from the fieldwork presented here. Beyond mineral extraction, these structures and features can be adapted to the production of various ecologically and socially sound goods, including the mining of waste discussed in other chapters of this volume.

REFERENCES

Benya, A. (2009). 'Women in mining: A challenge to occupational culture in mines'. Master of Arts dissertation. University of the Witwatersrand, Johannesburg

Bowman, A. and Isaacs, G. (2014). 'Demanding the impossible? Platinum mining profits and wage demands in context'. Interim Report. Corporate Strategy and Industrial Development, University of the Witwatersrand, Johannesburg.

Centre for Applied Legal Studies (CALS). (2015). Submissions to Special Rapporteur on Violence against Women, its Causes and Consequences [by] Dr. Dubravka Šimonovic . Available at: https://www.wits.ac.za/media/wits-university/faculties-and-schools/commerce-law-and-management/research entities/cals/documents/Submisison%20Rapporteur%20on%20violence%20against%20women%20CALS%20submission.pdf. Accessed 5 July 2017.

Centre for Civil Society. (2016). 'Fighting petmin coal in Somkhele, KZN, activists under attack'. Available at: http://ccs.ukzn.ac.za/default. asp?2,68,3,3673. Accessed 14 June 2018.

Cockburn, C. (1981). 'The material of male power'. *Feminist Review*, Vol. 9, pp. 41–58.

Davies, J., Williams, B., Debeila, M. and Davies, D. (2001). 'Asbestos-related lung disease among women in the Northern Province of South Africa'. *South African Journal of Science*, Vol. 97, pp. 1–6.

Department of Mineral Resources (DMR). (2018). 'Deputy Minister Oliphant to hand over mining permits to former Zama Zamas in Kimberley'. Available at http://www.dmr.gov.za/news-room/post/1716. Accessed 6 June 2018.

Eftimie, A., Heller, K. and Strongman, J. (2009). 'Gender dimensions of the extractive industries: Mining for equity'. *World Bank Extractive Industries and Development Series*, No. 8, pp. 1–56.

Hodge, N. (2002). 'South African asbestos victims win compensation but claim halved.' Available at https://www.wsws.org/en/articles/2002/01/asb-j09.html. Accessed 3 July 2017.

Jenkins, K. (2014). 'Women, mining and development: an emerging research agenda'. *The Extractive Industries and Society*, Vol. 1, pp. 329–229.

Lahiri-Dutt, K. and Macintyre, M. (2006). 'Where life is in the pits (and elsewhere) and gendered'. In Lahiri-Dutt, K. and Macintyre, M. (eds). *Women miners in developing countries: Pit women and others*. Ashgate, London.

Lauwo, S. (2018). 'Challenging masculinity in CSR disclosures: Silencing of women's voices in Tanzania's mining industry'. *Journal of Business Ethics*, No. 149, pp. 689–706.

Lozeva, S. and Marinova, D. (2010). 'Negotiating gender: Experience from Western Australian mining industry'. *Journal of Economic and Social Policy*, Vol. 13(2), Article 7.

McCulloch, J. (2003). 'Women mining asbestos in South Africa, 1893-1980'. *Journal of Southern African Studies*, Vol. 29(2), pp. 414–432.

Mies, M. and Shiva, V. (2014). *Ecofeminism*. Zed Books, London.

Munakamwe, J. (2017). 'Zamazama – Livelihood strategies, mobilisation and resistance in Johannesburg, South Africa.' In Nhemachena, A. and Warikandwa, T. (eds). *Mining Africa: Law, Environment, Society and Politics in Historical and Multidisciplinary Perspectives*. Langaa Research and Publishing Common Initiative Group, Bamenda.

News24. (1 September 2013). 'Death in the pit of sexism'. Available at https://www.news24.com/Archives/City-Press/Death-in-the-pit-of-sexism-20150430. Accessed 1 August 2018.

News24. (10 March 2015). 'Woman raped in Amplats changing room'. Available at: http://www.news24.com/SouthAfrica/News/Woman-raped-in-Amplats-mine-changing-room-20150310. Accessed 5 July 2017.

Nyoni, P. (2017). 'Unsung heroes? An anthropological approach into the experiences of "zamazamas" in Johannesburg, South Africa'. In Nhemachena, A. and Warikandwa, T. (eds). *Mining Africa: Law, Environment, Society and Politics in Historical and Multidisciplinary Perspectives.* pp. 155–185. Langaa Research and Publishing Common Initiative Group, Bamenda.

Shiva, V. (1988). *Staying Alive: Women, Ecology and Survival in India.* Kali for Women, New Delhi.

Solomon Star. (11 June 2018). 'Artisanal miners get more land'. Available at: https://www.facebook.com/photo.php?fbid=2167247903289 338andset=gm.1493674730742959andtype=3andtheaterandifg=1. Accessed 13 June 2018.

Statistics South Africa. (2017). 'Mining industry 2015'. Report No. 20-01-02 (2015). Available at: www.statssa.gov.za/?page_id=1854andPPN=Report-20-01-02. Accessed 4 July 2017.

Tau, P. (19 December 2015). 'Mines of rape, fear.' *City Press.* Available at http://city-press.news24.com/News/mines-of-rape-fear-20151212. Accessed 5 July 2017.

Thornton, R. (2014). 'Zamazama, "illegal" artisanal miners, misrepresented by the South African press and government'. *The Extractive Industries and Society*, Vol. 1, pp. 127–129.

Valiani, S. (2015). 'The Africa Mining Vision – A long overdue ecofeminist critique'. Analytical Paper. WoMin-African Gender and Extractives Alliance, Johannesburg. Available at: https://womin.org.za/images/docs/analytical-paper.pdf. Accessed 4 July 2017.

WiMSA. (2016). 'Industry survey results 2016'. Available at: wimsa.org.za/wp-content/uploads/2016/08/WiMSA-Survey-2016.pdf. Accessed 4 July 2017.

WoMin. (2017). 'No longer a life worth living report – Mining impacted women speak through participatory action research in the Somkhele and Fuleni communities, Northern Kwazulu-Natal, South Africa'. WoMin-African Gender and Mining Alliance, Johannesburg. Available at: https://womin.org.za/resource-library/women-building-power/no-longer-a-life-worth-living-report.html. Accessed 1 August 2018.

TEN

Trade union organising in the mining sector

A structural perspective on worker insurgency and shifting union strategies

KHWEZI MABASA AND
CRISPEN CHINGUNO

INTRODUCTION

This chapter explores the evolution of trade union organising in the mining sector. It argues that the forms of union power and activism that emerged have been shaped by structural evolution. Therefore, debates about organising must explore the interplay between socio-political structure and worker agency in specific historical eras. Changes in the fundamental make-up of South Africa's mining sector influence union organising and politics. This dialectic is at the core of organisations' strength, renewal and decline. We develop this argument by using a political economy approach, which focuses on evolving mining capital, state and labour relations.

The theoretical framework used is based on various schools of neo-Marxism, with an emphasis on scholarship that specifically examines racialised capitalism in South Africa. Mine workers' resistance and organising can only be understood through a pluralistic Marxist framework that transcends the analytical tools of the classic tradition. These scholars — like Magubane, Turok and Wolpe — relate class to broader dimensions such as race, culture, anti-colonial resistance and uneven spatial development. This approach enriches Marxist analyses of class formation and evolution in various epochs of capitalist development. Our chapter draws from published labour movement literature, as well as new and original ethnographic studies recently conducted in the platinum belt.[1] The chapter begins with a discussion on colonial and apartheid mining labour regimes. We then proceed to examine early forms of union resistance in the 20th century and link these to the emergence of the National Union of Mineworkers in the 1980s. The last sections of the chapter analyse the impact of structural reform in the mining sector on worker agency and union organising in post-apartheid South Africa.

LABOUR AND THE MINERALS REVOLUTION

South Africa's political economy was predominantly rural and agrarian prior to the discovery of minerals in the 19th century. The transition to a mining-based political economy shaped the emerging system of racialised capitalism. It structured power relations in the political, economic and social spheres. Magubane captures the industry's prominence in these words: 'The discovery of diamonds in 1864 and more particularly gold on the Witwatersrand in 1884 brought about economic changes in South African life tantamount to a full-fledged revolution' (Magubane, 1979: 102).

This minerals-led revolution (dominated by gold) was evident in the structural composition of South Africa's political economy. The gold

---

1   Chinguno (2015) and Ntswana (2014) ethnographic studies track the trajectory of labour relations regimes and workers' struggles for collective power, drawing on the experience of platinum mine workers in South Africa. Crispen Chinguno is the co-author of this chapter.

mining industry accounted for 80 per cent of the country's exports by 1910 (Magubane, 1979: 103). This dominance continued throughout the 20th century, with mining contributing 18 per cent towards South Africa's gross domestic product and making up 80 per cent of export revenue in 1985. The industry had a strong workforce of over 700,000 by the mid-1980s, which was primarily comprised of black workers from various southern African states (82 per cent) (Kraak, 1993: 91).

The last trend is related to a central feature of South African mining: an evolving relationship between capital and labour (Maloka, 2014; Terreblanche, 2012; Turok, 2011). The availability of undervalued black labour allowed for the development of the industry and propelled South Africa to become one of the leading exporters of minerals. However, labour's relationship with mining capital has been characterised by contradictions, compromises and contestation. The following section discusses capital-labour relations by focusing on the mining labour regimes instituted during the colonial and apartheid eras. It elucidates the innate connection between worker activism and structural features of the political economy.

COLONIAL AND APARTHEID MINING LABOUR REGIME

The term 'labour regimes' is drawn from Beckman and Sachikonye, who define it as 'the complex of institutions, rules, and practices through which relations between labour and capital are regulated both at the work-place as well as in society at large' (2001: 9). Buhlungu (2010: 2) uses this definition to periodise South African capital-labour relations into three specific eras: colonial, apartheid and market or neoliberal labour regimes. These categories of regimes are intertwined with industrial relations governance, which operates on a continuum with two extremes. The first labour regime is a *non-hegemonic system* based on violent coercion; whilst the second *hegemonic system* is centred on manufacturing some consent through collective bargaining (Burawoy, 1979; Von Holdt, 2003). Colonial and apartheid industrial relations structures relied on domination and discrimination for most of the 20th century. The discussion explores these non-hegemonic systems by examining the following three structural features of colonial and apartheid mining labour regimes.

The first feature is the relationship between mining and racialised authoritarian political governance. State power shaped relations between labour and capital during these epochs. Both colonial and apartheid governments introduced repressive legislation that marginalised black labour and the broader African populace (Maloka, 2014; Terreblanche, 2012). The early exploitation of mine workers can only be understood within the context of a broader political economy analysis. Burawoy's conception of 'production politics' explains this inherent link between labour processes and what he describes as the 'political apparatus' (1983: 587). The latter shapes the labour process by regulating contestations between capital and labour. Thus, mining labour relations must be connected to deployment of state power in a colonial socio-political context.

Most scholars concur that labour shortages inspired the political and economic subjugation of African mine employees in the 19th and 20th centuries (Magubane, 1996; Terreblanche, 2012; Turok, 2011; Wolpe, 1972). This labour scarcity was attributed to three factors: a resilient black peasant economy, low sector wages and white mine worker insurgency (Bundy, 1988; Bezuidenhout and Buhlungu, 2011; Webster, 2017). The state played a significant role in resolving the crisis by using a coercive 'political apparatus' (Buhlungu, 2010; Burawoy, 193).

The black population was systematically dispossessed and exploited. This process included measures such as the despotic regulation of labour movement, violence, segregation policies and land appropriation. The last intervention was particularly important, because black agrarian production ensured that 'rural communities were relatively self-sufficient through subsistence agriculture' (Bezuidenhout and Buhlungu, 2011: 244). This destruction of non-capitalist agrarian livelihoods was motivated by the need to create cheap black mining and agricultural labour (Atkinson, 2007; Cousins, 2011; Neocosmos, 1993; Wolpe, 1972). It epitomised what Marx (1867) termed 'primitive accumulation'. He describes this as 'the process which divorces the worker from the ownership of the conditions of his own labour' (Marx, 1867: 874).

However, there are two crucial differences relating to South

Africa's historical development path. One, dispossession did not lead to the total destruction of the peasantry or of agrarian livelihoods, as envisioned by Marx. It established dual agrarian systems that co-existed simultaneously: largely white-owned commercial agriculture and subsistence agriculture on the generally unproductive land assigned to black people under apartheid. Both systems supported a racialised political economy dominated by an interdependent agro-mining economic structure described later (Kariuki, 2009; Mafeje, 2003; Neocosmos, 1993). Two, social differentiation was not only driven by capital and class relations. Hierarchical race and gender power dynamics were constitutive elements of this system.

The second characteristic of South Africa's mining labour regime up to 1994 is controlled migration and geography (Bezuidenhout and Buhlungu, 2011: 237). This observation challenges traditional analyses that focus primarily on agency and dynamics within the workplace. These conventional studies overlook how social relations in other institutions shape worker agency. This purist understanding of labour fails to appreciate the influence of household dynamics, local struggles and spatial governance on labour relations (Von Holdt, 2003: 286).

Magubane (1979: 106) reminds us that workers were drawn from many countries in the early stages of mining. He cites examples of indentured Chinese labour and workers 'recruited' from various Portuguese colonies before and after the Anglo-Boer War. This continued throughout the 20th century, with mine employees coming from all over southern Africa (Forrest, 2015: 511). However, the use of southern African mine labour expanded rapidly between 1960 and 1970. According to Massey (1983: 440), this segment made up 80 per cent of mine workers by 1973. At the heart of this system was an institutionalised recruitment structure. The Chamber of Mines (renamed the Minerals Council of South Africa in 2018) led this process when it established two powerful recruiting agencies in the first decade of the 20th century: The Employment Bureau of Africa and the Witwatersrand Native Labour Association (Bezuidenhout and Buhlungu, 2011; Forrest, 2015). These institutes were supported by state-run bodies in segregated homelands and neighbouring southern African states. According to Kraak, the so-called 'bantustans' established under apartheid set up 'tribal labour bureaus' that facilitated mine worker recruitment (1993: 4). Access to sufficient income for black workers was minimalised through such collective recruitment, as this system allowed mining companies to pay 'uniform low wages' (Kraak, 1990: 91). Mining companies did not compete for labour; rather, they imposed wages on mine workers (Massey, 1983: 430).

This migrant regime depended on two centres of physical control: rural areas governed by tribal authorities under apartheid legislation and male-dominated mine compounds located in the apartheid state's segregated urban localities (Bezuidenhout and Buhlungu, 2011: 238). The state fortified this system through heavily policed pass laws, influx control and prohibitions on freedom of association. Social interactions within compounds were governed by authoritarian policies, as employers attempted to control all aspects of mine workers' social reproduction (Bezuidenhout and Buhlungu, 2011: 245). Control over mine worker migration was not only driven by political objectives. It also facilitated mining capital accumulation. The recruitment nexus was based on the super exploitation of black labour.

International political economy dynamics also helped keep mining wages down. Global gold prices were determined in external markets, and mining houses had limited price-setting power. Thus, surplus value could only be guaranteed by exploiting and extracting additional cheap labour (Capps, 2015; Kraak, 1993; Magubane, 1979; Massey, 1983). There was minimal employment security because mine workers were contracted for short periods (mostly six months) and had to return home afterwards in order to reapply for jobs (Bezuidenhout and Buhlungu, 2011; Kraak, 1993). These contracts were not negotiated, and workers faced criminal prosecution for contractual non-compliance. Mining employers, working through recruitment agencies, exacerbated the exploitation by channelling black workers into the 'least skilled, worst paid and most arduous categories of work' (Kraak, 1993:4). Their white counterparts occupied highly skilled categories of work with better conditions. This precarious employment status undermined all attempts to organise workers and increased the capacity of mining capital to dispose of labour at any time.

Moreover, the costs of mine workers' social security and reproduction were externalised to areas designated for black people under successive colonial and apartheid governments: native reserves and subsequently bantustans (Massey, 1983: 434). Wolpe (1972: 434–435) explains that reserves functioned as a form of social security for super-exploited migrant mine workers. Women in rural areas performed household duties to support male workers who lost their jobs either through illness or dismissal. More importantly, rural subsistence production supplemented low mine wages, which could not generate socio-economic security for households. Reserve female household labour subsidised the social reproduction of black male mine workers. This dualism constituted a fundamental element of South Africa's core and peripheral agro-mining complex.

Thirdly, both colonial and apartheid despotic mining labour regimes produced social differentiation. The division of mine workers along racial lines was crucial to this phenomenon. Racism shaped the nature of employment, income and organisational rights (Burawoy, 1976; Magubane, 1996; Moodie, 2015; Webster, 2017). Mining companies adopted job reservation policies, which secured certain categories of work for white employees. This was codified in various pieces of legislation, including the Mines and Works Act No. 12 of 1911 which barred black mine employees from performing certain tasks e.g., driving locomotives and occupying management roles (Kraak, 1993: 93).

The practice of job reservation based on race was largely influenced by the insurgency of white mine workers, who were motivated by racism and narrow class interest (Burawoy, 1976: 1054). Their rebellion culminated in the 1922 Rand Strike, by white mine workers. According to Webster, this formalised the class alliance between a nascent Afrikaner nationalist movement and white workers (Webster, 2017: 142). Bezuidenhout and Buhlungu's (2011: 246) study on mining compounds adds another dimension to the social differentiation analysis employed to explain fragmentation amongst the labour force. Their study highlights the importance of ethnicity in shaping divisions amongst workers. This phenomenon is related to state and mining company's tight control over the physical spaces in which the workers lived and worked.

Racialised capitalism was anchored on a segregationist development model, which politicised and institutionalised ethnic identity (Mamdani,

1996; Mamdani, 2012). Mamdani explains this accurately by arguing that 'cultural difference was reinforced, exaggerated, and built up into different legal systems, each enforced by a separate administrative and political authority' (2012: 48). Mining employers played an important role in shaping this political framework, which facilitated South Africa's racialised capitalist system, including the organisation of work and living spaces along ethnic lines.

The main aim of employers and the state was to weaken worker solidarity using institutionalised racism. In addition, as Mamdani (1996) has shown, tribal hierarchies were reinforced in accommodation structures and job allocation processes. The elevation of sectarian and regional political interests was a central element of this strategy. However, 20th century African mining labour resistance exposed the limitations of this strategy. The following sections link this agency to wider structural developments in the apartheid industrial order.

AGENCY, RESISTANCE AND VIOLENCE

The preceding discussion highlights the intertwined nature of state and mining capital power, and of forced patterns of migration and social differentiation in the industry. These structural phenomena shaped worker agency and insurgency over various epochs. Struggles between mine workers and employers centred on attempts to re-order power relations produced by the structural dynamics elaborated above. Worker resistance was primarily grounded in a social movement unionism model, which links shopfloor issues to broader political, international and community struggles (Pillay, 2015; Saul, 1986; Von Holdt, 2003). This tactic was an essential catalyst for mine worker activism in the early 20th century (Crush, 1989; Kraak, 1993; MISTRA, 2015; Moodie, 2015).

Black employee struggles against colonial and apartheid mining labour regimes were highly politicised. Government's role in reproducing exploitative capital-labour relations elevated the necessity (from the mid-1920s) of combining workplace and political freedom (O'Meara, 1975). Worker resistance was primarily grounded in a social movement unionism model, which links to leading figures in the Communist Party of South Africa and the Transvaal African National Congress (ANC). The latter organisation played a pivotal role in establishing the African Mine Workers' Union (AMWU). It established a committee in 1941 that called for the formation of a black-led mine workers union. This initiative, led by J.B. Marks, was mandated to the Communist Party of South Africa and the Transvaal African National Congress (ANC). Previous attempts to organise black mine workers, in the 1920s and 1930s, were undermined by state repression and the alliance between an emerging racist regime and white trade unions. The Pact government of 1924, for example, introduced various pieces of legislation which entrenched the marginalisation of black worker organising and rights. Some leaders of the white labour movement attempted to represent black working class interests during the 1930s but their success was minimal, and they struggled to obtain political legitimacy. Legislation in the first decades of the 20th century outlawed black workers' freedom of association and right to organise.

Strategies for worker organising had to take into account the need to challenge systemic, structural authoritarianism which maintained an exploitative labour process. This could not be achieved without adopting various forms of social movement unionism. This mode of organising continued throughout the 20th century and strengthened the drive for black union recognition in the 1980s (Crush, 1989; Moodie, 2015). However, this strategy was constrained by state authoritarianism. A prime example is the 1946 strike that led to 12 African deaths and over 1,000 injuries (O'Meara, 1975; Webster, 2017). State-led violence was essential for repressing early black mine worker insurgency. This method of control also relied on private security forces within the mines. The literature on mining labour regimes prior to 1994 highlights high levels of militarisation in the mines and surrounding compounds (Bezuidenhout and Buhlungu, 2011; Crush, 1989; Kraak, 1993; Massey, 1983).

Informal networks were essential organising instruments in the aftermath of the strike, as the state responded with increasing repression from the 1940s onwards. This continued throughout the 1950s and 1960s, an epoch characterised by black trade union organisational weakness. Mining employers believed that informal formations, however, were non-political, and in some instances allowed them to develop without employer control or interference. Some prominent examples of these networks include: religious groups, sports societies, compound marshals, cultural organisations and worker committees (Bezuidenhout and Buhlungu, 2011; Moodie, 2015). The employers thought that the growth and support of these organisations would dampen worker militancy and organising. However, these formations provided a platform for collective discussions of grievances. Members of these networks brought organising experience into the revival of black mining unions in the late 1970s (Bezuidenhout and Buhlungu, 2011; Buhlungu, 2010; Moodie, 2015). Worker agency had to be built through informal networks, because of the authoritarian industrial relations structure. This exemplifies the structure-agency dialectic in mine worker organising argued in this chapter.

Black mine workers also developed strategies to challenge the despotic migrant labour system. These were included in early forms of resistance and continued until official union recognition in the 1980s. According to O'Meara (1975: 14), AMWU's 1944 demands included a statement on eradicating the oppressive mine worker migrant regime. Black employees carried out other tactics of subversion in the following decades. These included non-compliance with mining contracts, circumventing segregation laws and challenging compound governance (Crush, 1989; Forrest, 2015; Massey, 1983). Some employees bypassed laws about urban segregated spaces by residing in townships. This increased contact between mine workers and other members of the urban African proletariat accelerated black union political activism during the 1970s. Mining companies and state authorities attempted to curb the insurgency of this period by introducing stringent penalties. But this produced minimal results, as mine workers continued to subvert the migrant system (Crush, 1989: 7).

Violence was viewed as a legitimate instrument for managing contestations between capital and labour throughout the 20th century. During this epoch, state and private mining authorities used coercion to marginalise trade union organisations. This repression played a central role in maintaining hierarchical racial and tribal relations in the workforce. This violence was not only exercised by employers and state security forces. It was also prevalent in the informal networks and worker organisations that have always characterised the mining labour force. Thus, we argue for a distinction between violence from above and below. The former is exercised by state or mining officials whilst the latter is deployed by workers themselves. Both forms of violence were exercised in struggles to organise workers and advance campaigns for official union recognition. Some mine workers used violence in recruitment or to intimidate scab labour (Chinguno, 2013; Chinguno, 2015).

This resistance culminated in the unco-ordinated and fragmented mine strikes of the 1970s. This industrial action was sporadic and not based on unified mandates or campaigns. Strikes tended to be around company-specific or sectarian black worker grievances rather than about common employee struggles. The 1970s was a significant decade in the history of the South African labour movement, with increased political and labour militancy. The volatile political milieu of the 1970s created favourable conditions for mass mobilisation of mine workers. Black trade union revival had been inspired by the 1973 Durban strikes, which strengthened the campaign for formal union recognition (Buhlungu, 2010: 22). Mining companies struggled to use traditional forms of control, like policing compounds, to stop the worker activism. This was largely due to the effective worker resistance and shifting power balances explained in later sections of the chapter. Crush describes this era as the 'proto-organisational phase' characterised by large-scale instability in mining and the broader political economy (1989: 7). He concludes that this phase demonstrated 'how defective existing structures of control were in defusing, much less managing, worker protest' (Crush, 1989: 7).

Both state and mining officials attempted to counter massive resistance with new forms of structural domination. These methods produced minimal results and inevitably led to internal employer debates on establishing an institutionalised *hegemonic system*. The historic rise of the National Union of Mineworkers (NUM) must be located within this broader political economy context. The discussion below illustrates the intersecting relationship between NUM's ascendency and reconfigurations in the political economy of mining. This strengthens the chapter's main contention, namely that the dialectic between structure and agency shapes trade union power. In other words, labour organising takes place within a contested structural context, characterised by various macro-social features. Worker agency is moulded by institutions and regimes that underpin a particular social order. However, these structures of power are in turn altered and remade by the same trade union insurgency.

NUM AND THE APARTHEID CORPORATIST MODEL

The establishment and early successes of NUM in the 1980s are attributed to several salient structural factors. The primary one is the evolving mining migrant labour regime. Mining experienced a huge crisis in the 1970s when the governments of various southern African states withdrew their citizens' access to South Africa's mining labour markets. Leaders of various states terminated their employee supply agreements with the apartheid regime. This move was motivated by solidarity with liberation movements opposing the apartheid regime (Massey, 1983: 440). Regional migrant worker networks were used to promote political activism and recruitment into liberation forces. Furthermore, internal resistance in South Africa and subversion of the migrant system was growing. Some prominent examples include black workers violating employment contracts, pass laws and influx control legislation.

The findings of the Mining Riots Commission (1975) and Riekert Commission (1977, reporting in 1979) informed alternative approaches to labour migration. The two commissions were established by state authorities, in response to employers' concerns about the negative economic impact of industrial and political unrest across the country. The two reports argued that creating a stable black urban working class was necessary for industrial stability. They also cited the limitations of being over-dependent on southern African labour.

In the aftermath of the commissions, mining companies and state authorities worked together to put in place polices that allowed for the creation of a mostly South African black mine workforce. These measures include a review of migrant policies and decreasing influx control. By 1980, 60 per cent of black mine workers were

South African (Massey, 1983: 441). Employers also introduced a host of incentives to stabilise the domestic supply of migrant labour and to retain employees (Crush, 1992: 63). These structural shifts created conditions conducive to strengthening worker organising. The development of a stable proletarian workforce created suitable conditions for building trade union strength and worker consciousness (Crush, 1992; Forrest, 2015).

These changes, and increased resistance from black mine workers, produced a structural shift in employers' labour relations strategies. One of South Africa's largest mining conglomerates, Anglo American, advocated the rationalisation of industrial relations. This proposal culminated in the establishment of the Wiehahn Commission (1977, reporting in 1979). Anglo American viewed the unionisation of black workers as an essential step towards establishing a corporatist industrial relations regime (Crush, 1989; Butler, 2013). The reforms were also informed by efforts to remove the colour bar, and in turn, reduce the costs of white skilled labour. Anglo American drew from its experience in post-independence Zambia, which illustrated that harmonious industrial relations could only be guaranteed through formal black union recognition. It became imperative to incorporate black unions into a hegemonic industrial system based on mutual consent (Burawoy, 1979).

However, South African capital has historically been divided into two main dominant factions: Afrikaner and Anglo-Saxon capitalists. The latter group, represented by Anglo American in mining, was strongly opposed by other major mining houses at what was then called the Chamber of Mines. The chief critics were Gencor and Goldfields, which defended the maintenance of a coercive and non-hegemonic labour relations regime. Both these enterprises were aligned to the Afrikaner faction of capital. These different approaches to workplace management shaped the emergence, expansion and organising strategies of trade unions in mining. Though limited by repressive legislation, collective bargaining between employers and black workers' unions in other industries had already begun, following the gains of the 1973 Durban strikes.

This intra-capital contestation was resolved by the outcomes of the Wiehahn Commission. It called for unionisation of black workers and the establishment of a non-racist industrial relations regime. But the entrenched systems of control in mining continued to present challenges for black employee mobilisation. Industrial conflict in the industry intensified after 1980, with black workers experiencing victimisation (Chinguno, 2015). Nonetheless, the NUM emerged in 1982 and grew into one of the largest trade unions in South Africa. It was established by the Black Consciousness dominated Council of Unions of South Africa (CUSA). Black mine workers at Kloof Goldfields approached CUSA for assistance in forming a black mine workers' union (Butler, 2013). Butler argues that CUSA received support from Anglo American. He argues that Anglo American had an interest in the formation of a black workers union, which it saw as a vehicle to reduce overall costs by foiling the growing demands of white unions. Further, the company viewed this formal union recognition as a key instrument for controlling black worker resistance. The union terminated its CUSA affiliation in the mid-1980s and played an important role in the Congress of South African Trade Unions' (COSATU) formation. In the following sections we explore how NUM used these structural shifts to increase membership and to build associational power in the 1980s. This term refers to the ability to increase worker solidarity by organising employees 'collectively into trade unions' (Webster, 2017: 149). Structural power, which is derived from workers' capacity to withdraw labour and halt production, complements the exercise of associational power (Wright, 2000; Webster, 2017).

NUM'S STRATEGY OF THE 1980S

The structural factors discussed in the previous section had a direct impact on NUM's early successes in the 1980s. Our discussion of the organisation's strategies commences with changing power relations in the compounds. There was consensus among state and mining house officials that the socio-economic conditions in the compounds sparked resistance and violence. Moreover, militarised policing of the compounds exacerbated instability on the mines. This induced some companies to introduce reforms in their governance of accommodation.

Anglo American took the lead by demilitarising spatial planning and housing skilled black workers in township accommodation (Crush, 1992: 394). This intervention facilitated union politicisation and solidarity amongst various segments of the urbanised black working class. Mine workers engaged employees from other industries and drew lessons on organising. Bezuidenhout and Buhlungu (2011: 248) point out that NUM established a strong presence in the mine compounds throughout the 1980s. The organisation democratised the structures of governance by usurping the authority of tribal leaders (*izinduna*) (MISTRA, 2015: 21). Democratically elected worker committees took over the internal management of compound life. The union also used informal networks, communication platforms within the compounds and recreational activities to organise workers (Bezuidenhout and Buhlungu, 2011: 248–250).

The subversion of compound authority also had a substantial impact on political subjectivity. Previous sections highlighted how management entrenched tribal hierarchies and activities through control over accommodation. This enforced social differentiation was crucial for maintaining ethnic, factional rivalries within the workforce. But NUM managed to overcome this in the promotion of collective worker consciousness within the compounds. The organisation used astute methods such as incorporating cultural practices and leisure activities into union work (Bezuidenhout and Buhlungu, 2011: 248–250). It did not attempt to erase cultural memory or subjectivity. Rather, it channelled these identities and practices so that they would support unionisation (Buhlungu, 2010: 29). Political factors also played an essential role in diffusing ethnic tensions. Both the Black Consciousness and Charterist movements elevated the significance of challenging tribal and racial divisions (Bezuidenhout and Buhlungu, 2008: 267). Even more importantly, NUM challenged the material bases for ethnic divisions by questioning tribal-based allocation of jobs and status within the mining accommodation system.

This resistance to authoritarian social differentiation was not limited to ethnic or tribal hierarchies. It extended to challenging what Von Holdt (2003: 5) describes as the 'apartheid workplace regime'. The manifestations of racialised work organisation in mining were outlined in previous sections. In this discussion we explore strategies used by NUM in subverting workplace racism. One of the early tactics employed by the organisation was invoking black masculinity (*ubudoda*), to challenge racist micro-violence in the workplace. Patriarchal identity motivated black workers to violently resist racist punishment or abuse. These employees also defied white supremacist norms, which governed social interactions in the production space (Bezuidenhout and Buhlungu, 2011; Moodie, 2009).

On a larger and collective scale, NUM argued for equal remuneration and the termination of the colour bar. In some cases, African and white employees performed similar tasks. But they received different pay rates, because job reservation policy barred black workers from obtaining certification. The union elevated this matter and by the mid-1980s the issue formed part of major central bargaining demands (Kraak, 1993: 93–94). Moodie (2009: 51) views the focus on institutionalised racism as a slight shift in organising strategy. In its early years, NUM emphasised safety, wage campaigns and solidarity. However, 'since 1984 shaft stewards had urged the union to embark on more general resistance to white control both at the point of production and in the migrant hostels' (Moodie, 2009: 51).

The shift relates to another salient aspect of NUM's organising: strengthening the social movement union tradition. This phenomenon should not be reduced to political alliance formation (Von Holdt, 2003: 9). It was also about remaking the authoritarian social order, which shaped unequal capital and labour relations in the first place (Von Holdt, 2003: 9). More importantly, social movement unionism incorporates fundamental principles such as deepening democracy, popular representation and worker control within unions (Von Holdt, 2003: 9). The last concept refers 'to the entrenchment of a political ethos characterised by decentralised, participatory and bottom-up decision-making in a union' (Mabasa, 2017: 7). Worker control can only be fully realised when an organisation strengthens political education, leadership development, internal democracy and broad participation (Byrne et al., 2015; Mabasa, 2017).

NUM implemented most of these organising principles in the 1980s. According to Crush (1989: 10), membership growth in the early years was enhanced by the following important factors: responsive leadership, strengthening of internal democratic practices, building strong regional structures and popular representation. His analysis is echoed in other accounts, which highlight how the union ensured accountability at all levels. Mass meetings and decentralised decision-making built the union's legitimacy amongst workers (Bezuidenhout and Buhlungu, 2008; MISTRA, 2015).

Furthermore, participatory democracy and worker control empowered members, who received intensive political and occupational education. This produced a pool of competent leaders at all levels, with some being appointed as permanent officials in the union. Another salient factor was the ability of the union to maintain resilience and solidarity during industrial action, especially in the context of management attempts to divide workers. The NUM also ensured that other socio-economic issues, not related to wages, such as health and safety, social security and housing became central pillars of union activism. The organisation built internal infrastructure and capacity to support workers' access to these basic rights. This included developing an effective bureaucracy made up of researchers, organisers and specific sub-committees (Bezuidenhout and Buhlungu, 2007; Bezuidenhout and Buhlungu, 2008; MISTRA, 2015; Moodie, 2009)

Lastly, and most importantly, was NUM's ability to take advantage of the structural crisis in the nation's economy that also affected mining. The organisation built internal infrastructure and capacity through a series of crises from the 1970s until the democratic transition. International sanctions, heightened union insurgency, social unrest, skilled labour shortages and the global crisis drove this development impasse (Nattrass and Seekings, 2011: 344).

It is within this context that NUM managed to obtain formal recognition and expand membership. Crush (1989: 8) explains that union numbers leaped from 14,000 to 360,000 between 1982 and 1988. These achievements were shaped by both structural factors and the accompanying organised mine worker agency. This culminated in what we describe as the 'apartheid mining corporatist industrial relations order'. The term is drawn from political economy literature, which explores the global phenomenon of class compromise (Wright,

2000; Wright, 2012). This takes place when 'contending forces find a way to actively cooperate in ways that open some space for non-zero-sum gains. Active forms of mutual cooperation help both workers and capitalists to better realise their interests than is possible by simply extracting concessions through confrontation' (Wright, 2012: 2–3). Employers, the state and NUM began establishing a new corporatist industrial order from the 1980s. Mining capital had a strategic interest in the existence of a strong black trade union as part of its arsenal of control and manufacturing consent, while NUM required access to institutional power in order to consolidate its organisational and structural gains. The institutionalised power dimension refers to the establishment of legislation and supporting structures (i.e. collective bargaining, labour courts, etc.), which allow unions to exercise their associational or structural strength. This development highlights the shift from a non-hegemonic to hegemonic industrial order described earlier. However, this was taking place in an authoritarian political context created by a repressive state apparatus.

The acceptance of NUM as a voice for black workers, following its recognition in 1983 by the then Chamber of Mines, culminated in the institutionalisation of the union-management relationship. The union made the choice to adopt institutionalisation as part of a strategy to defend the interests of black workers. This was a gradual process which started in the 1980s and crystallised in the mid-1990s. However, this institutionalisation of industrial relations presented a paradox to NUM. It transformed its strategy from mobilisation to servicing; i.e. from structural and associational strength to institutional power.[2] This transition is associated with centralised decision-making and bureaucratisation in organisations, which is conducive to oligarchical

---

2 We draw the distinction between structural and organisational power from Silver (2003), Wright (2000) and Dörre, Holst and Nachtwey (2009). Silver and Wright argue that structural power relates to the power workers have because of their position in the production process and organisational power is derived from workers' collective organisations such as trade unions. They further argue that it usually substitutes structural power. Dörre, Holst and Nachtwey (2009) identify institutional power as a third form of power developed collectively by workers, deriving initially from organisational power before transforming to institutional power.

rule (Michels, 1962). We explore this last phenomenon in the discussion on NUM's post-apartheid organising challenges.

The following sections discuss the evolution of South Africa's mining industrial relations system from the early 1990s. We review the literature on the political economy of mining, which describes major structural changes within the sector since 1994. The transition from apartheid to democracy produced numerous structural reforms, which transformed mining. These present some challenges and opportunities for trade unions organising in the industry.

THE TRANSITION AND MINING

The early transformation towards an institutionalised industrial relations regime was not aligned with changes in the political climate. Rather, it took place within an authoritarian context and reflected the power balance associated with the apartheid mining corporatist industrial relations order. As noted by Chinguno (2015: 580), 'a typical trade union operates within a continuum: contesting employer hegemony, challenging exploitation and striking compromises (co-operation)'. These concessions emerge from institutionalising the relationship between labour and capital. The transformation of industrial relations was only fully accepted by workers following the integration of industrial and political transformation. This was strengthened by COSATU's decision to formalise its alliance with the dominant liberation political parties: the ANC and the South African Communist Party in 1993 (Buhlungu, 2010: 58). NUM was one of the leading affiliates supporting this decision, which played a crucial part in ensuring the ANC's electoral victory in 1994. Many workers in the mining sector still associate NUM with the ruling party (Buhlungu and Bezuidenhout, 2008: 272).

South Africa's post-apartheid industrial relations regime has sought to democratise and institutionalise various means of resolving industrial conflict. Much of the shift is due to the political agency of COSATU within the alliance. The democratic government adopted a new constitution, which defended freedom of association, collective bargaining and the right to strike. This included passing a new Labour

Relations Act No.66 of 1995 that contained provisions for dispute resolution. Moreover, the post-apartheid government introduced comprehensive labour laws and institutions to strengthen the transition towards a fully democratic industrial order (Buhlungu, 2010: 76). Contemporary mining trade union organising, and agency, has taken place in a new democratic hegemonic industrial relations system.

Political economy researchers identify the shift to platinum mining, rather than gold mining, becoming dominant within the industry as the most defining feature of mining capital and labour relations in post-apartheid South Africa (Capps, 2015; Forrest, 2015; Mnwana, 2015a). As elaborated in other chapters of this volume, South Africa is endowed with the world's largest platinum reserves (87 per cent), and platinum production expanded by 70 per cent from 1994 to 2009. By 2010 the platinum employment figures were higher than those of gold and coal (Capps, 2015: 500). According to Stats SA (2015: 20), 198,952 workers were employed in the platinum group metals sub-sector by 2015. These workers constitute just over 40% of total mine employment (Stats SA, 2015: 20)

South Africa's mining labour market has also been transformed by the transition to a 'neo-liberal labour regime' (Buhlungu, 2010: 2). The literature identifies several important developments that have ruptured traditional concepts of employment. Mining unions have been confronted by the increased sub-contracting of mining operations, atypical employment, mechanisation and externalisation of labour costs (Buhlungu, 2010; Kenny and Bezuidenhout, 1999). This fragmentation of labour using complex institutions and actors is not peculiar to the mining sector. It is prevalent throughout the country's economy (Benya and Ncube, 2015; Theron, 2016). These developments require a more nuanced conception of 'working class'. As Standing (2016) explains: 'There has been class fragmentation, so that the old nomenclature is no longer fit for understanding the dynamics of class struggle' (Standing, 2016: 192).

Moreover, the labour force has growing numbers of female workers; but unequal gender relations persist in mining (Benya, 2009; see Valiani and Ndebele in this volume). The gendered pay gap and the fact that - increasingly - many of the workers in the most precarious positions are female, epitomise this inequality. Another important development is the increased job mobility of black, male mine workers since the transition. The introduction of democratic labour legislation has created opportunities for advancement into skilled positions, which come with better remuneration and benefits.

Forrest (2015: 514) provides an insightful analysis into the restructuring of the migrant labour system. She develops her argument in a discussion on changes in the mining labour recruitment regime. It has become more diverse, with labour sourced through different agents within a complex nexus. This trend is particularly prevalent in the platinum sector, which has several autonomous recruiting agencies and institutions (Capps, 2015; Forrest, 2015). This network includes labour brokers, contractors, informal networks, and, sometimes, traditional authorities (Kenny and Bezuidenhout, 1999; Forrest, 2015; Mnwana, 2015b). Democratisation has also developed competition within the labour market, as companies are regularly in competition with each other for access to skilled labour (Forrest, 2015: 516). There is also a new order in relation to workers' living and working spaces. This has profound implications for trade union organising and solidarity, which depended heavily on compound structures to strengthen mobilisation (Buhlungu and Bezuidenhout, 2008: 267). Mine employees are no longer restricted by authoritarian laws about where and how workers can live. Furthermore, some have access to living or housing allowances that provide the means to gain access to various forms of accommodation.

Mnwana (2015a: 500–501) elucidates another significant change in the spatial connection between mining and the rural, former apartheid 'homelands'. He explains: 'In South Africa, gold has fuelled the development of large urban industrial centres. But the increased attention paid to platinum post-apartheid has shifted the geographical focus to rural areas' (Mnwana 2015a: 500). (This is because South Africa's major platinum reserves are to be found in the former homeland areas of North West and Limpopo provinces.) Platinum mining expansion has reconfigured the agro-mining complex discussed in earlier sections. The traditional demarcation of rural and urban centres in mine workers' livelihoods is inadequate. It does not explore the social, political and economic implications of this 'geographical shift' accurately (Mnwana, 2015a: 501). This structural change has huge implications for conceptions of trade union organising and broader working-class solidarity in mining communities. We unpack this phenomenon within the context of the platinum case study later. These structural changes in mining have shaped both worker and trade union agency. The next discussions link these features to the organisational experiences of NUM in post-apartheid South Africa. They highlight the connection between organisational strength and structural change.

NUM AND THE PARADOX OF INSTITUTIONALISATION: FROM ORGANISATIONAL TO INSTITUTIONAL POWER

Institutionalisation of industrial relations is paradoxical; it creates institutional security for the union, but on the other hand it displaces the requirement for continuous organising on the shop floor (Purcell, 1993). Moreover, it is tied to bureaucratisation and alienation from branch workplace issues (Fantasia and Voss, 2004). Bureaucratisation is exemplified in the increased power of full-time union technocrats. Some NUM members question the political orientation and legitimacy of these officials. The general concern is that worker control and union culture is undermined by these professional employees (MISTRA, 2015: 34). It is also reflected in hierarchical power relations within unions, between various leadership structures or leaders and members. This 'subverts democratic participation by members in decision-making' (Buhlungu, 2010: 118; MISTRA, 2015).

Moreover, union power and survival are arguably more dependent on the unions' capacities to organise and empower workers on the ground than on institutional security. Institutionalisation enhances the risks of unions becoming demobilised and alienated from their broad membership base. A mine worker cited in Chinguno (2015: 181) during the 2012 platinum belt strikes captured this succinctly: 'the union is no longer there for us anymore as it was in the past. It is there just to guarantee management that we won't revolt against the exploitation'. This alienation is also captured in an unpublished research report MISTRA undertook for NUM, entitled 'The National Union of Mineworkers: Exploring pathways to organisational renewal'. The report found that a 'common complaint amongst members is that branch meetings are hardly convened' (MISTRA, 2015: 24). The decline of grassroots membership participation is a systemic problem, which is reflected in the NALEDI (2015) publication based on COSATU worker surveys conducted in 2006 and 2012. Only 37 per cent of workers confirmed that their unions held meetings once a month (Byrne et al., 2015: 68).

The accelerated fragmentation of the labour market has caused additional political contradictions and organisational challenges. Neoliberal globalisation is associated with the proliferation of flexible labour regimes, which create new forms of precarious, atypical employment. South African trade unions have been grappling with this challenge for the past two decades. Von Holdt (2003: 242) views this as 'authoritarian restoration', which is introduced by employers in order to reconstruct coercive control over labour in new conditions (Von Holdt, 2003: 242 and 2010). He argues that outsourcing, labour brokering and subcontracting are designed to enhance capital's control of the labour process and further maximise profit. More importantly, these trends impede union organising and power (Kenny and Bezuidenhout, 1999; Forrest, 2015). As Forrest explains in her recent study on mining labour recruitment: 'one of outsourcing's defining features has been that it bypasses organised labour and state regulations through the short-term hire of labour for both core and non-core activities' (Forrest, 2015: 514).

Forrest captures the effect of labour fragmentation on mine worker organising. Buhlungu and Bezuidenhout's (2008) extensive case study on the state of NUM reveals that the organisation has had only minimal success in responding to this challenge. It shows how NUM's membership continues to be drawn largely from employees in standard employment. Only a small segment of the precarious atypical workforce is represented by the organisation. The authors conclude that: 'most members in our survey (93 percent) were employed on permanent contracts of employment' (Buhlungu and Bezuidenhout, 2008: 278).

The MISTRA (2015) report explores this shortcoming and reveals union organisers' inability to develop adequate adaptive strategies. It points out that labour fragmentation erodes associational power, and 'the changing nature of work in the industry limits membership growth' (MISTRA, 2015: 58). This representation gap is prevalent throughout COSATU (NALEDI, 2015). Ninety per cent of the federation's membership is drawn from employees in secure, standard employment (Benya and Ncube, 2015: 23). This is alarming when one considers that non-standard employment has expanded rapidly, especially in sectors like mining. Atypical employment accounted for 28 per cent of the mine workforce by 2013 (Mining News, 2013). Most of these employees with non-standard contracts work in the platinum sector. This 'crisis of representation' presents huge political challenges as some workers are not unionised and are excluded from the institutionalised industrial relations system (Buhlungu and Bezuidenhout, 2008; Webster, 2005). Chinguno's (2015: 134) study illustrates how labour fragmentation erodes mine worker solidarity and inevitably leads to non-conventional forms of worker insurgency. He demonstrates that during the platinum strikes of 2012 and 2015, precarious mine workers formed informal worker committees because they believed unions failed to represent their interests.

Intra-union contestation also depletes worker solidarity. This is mainly driven by careerism and individualism. A mine worker cited in the MISTRA (2015) report explains this in the following words: 'People are fighting for positions now. We are fighting for positions to get benefits and not to represent the union' (MISTRA, 2015: 28). There are various forms of benefits associated with the phenomenon of careerism: occupational mobility, political deployment and access to union resources. More worryingly, these material battles have reignited ethnic divisions in some cases (Buhlungu and Bezuidenhout, 2008: 284). This ultimately leads to heightened factionalism within the organisation, as current and prospective office bearers 'resort to extra-ordinary means' to either retain or gain access to positions (MISTRA, 2015: 30). The structural transformations discussed earlier have exacerbated this challenge particularly given that democratisation has meant increased opportunities for black worker mobility. The result of factional intra-union contestation is patronage, which erodes the union's legitimacy amongst employees.

The erosion of NUM's legitimacy is exacerbated by its loss of social movement unionism. Labour and community solidarity in mining areas has declined, and in some instances, members of mining-affected communities view migrant mine workers as foreigners who cause socio-economic upheaval (Benya, 2015; Mnwana, 2015b; see Mnwana chapter 7 in this publication). This contestation also finds prominence in formal, local institutions such as clinics and government offices. Migrant workers and their partners are confronted by ethnic discrimination in platinum belt locations (Benya, 2015; Mnwana, 2015b). Moreover, class fragmentation and mobility separate some workers from local community struggles in townships or rural areas (in the case of the platinum belt).

The close proximity of NUM to the ANC also comes with substantive costs to the union's legitimacy. As one platinum worker explains: 'COSATU and therefore union affiliates like NUM have become a branch of the ANC and the mine bosses; that is why when we ask for more money and better conditions the unions won't hear us – because it's like you are creating confusion in their household' (Ntswana, 2014: 46). Rival unions have exploited this perception by invoking their official status of political independence. But this too is questionable, as Ntswana's (2014) study of the Association of Mineworkers and Construction Union (AMCU) reveals that some political groupings influence workers within these organisations. The issue of union legitimacy amongst mine workers is explored further in the following analysis of platinum strike waves.

LESSONS FROM THE PLATINUM BELT STRIKES

Previous sections illustrated that the expansion of platinum mining occurred within a context of structural change on both micro and macro levels. The shift happened during the political transition and the period of accelerated neoliberal globalisation. As a result, platinum mining capital adopted a different organising strategy to manage labour relations. It was not keen on adopting centralised collective bargaining processes because keeping labour costs down was one of the key drivers of competitiveness among different mining houses. Thus, the major industrial relations shift accompanying platinum production growth is decentralised collective bargaining at company level (Capps, 2015: 499). This strategy is the primary cause of socio-economic disparities among employees in the platinum sub-sector. Labour fragmentation based on income, the nature of employment, poor representation and access to benefits generated instability in industrial relations, which culminated in the strike waves of 2012 (Chinguno, 2015).

Moreover, expanded platinum extraction coincided with the demise of the compound system. As said earlier, workers increasingly have access to different forms of accommodation and are exempt from authoritarian spatial control (Bezuidenhout and Buhlungu, 2011: 251). However, platinum mine workers have minimal access to benefits such as housing or living out allowances, because this sub-sector accounts for most of the precarious workers in the mining sector. The few workers with such benefits opt to remit the funds home. This explains the elevation of accommodation challenges during the platinum strikes of 2012 to 2014.

The reconfigured spatial regime has presented significant organisational challenges for NUM. It has exacerbated social differentiation amongst mine workers in the post-apartheid era who now have 'fragmented residential patterns' (Bezuidenhout and Buhlungu, 2011: 256). The union cannot rely on the traditional compound as a site of political mobilisation and organising. Developments from the platinum belt indicate that most exploited workers reside in informal settlements. These exist within a complex set of accommodation options: Reconstruction and Development Programme houses, family units, township backroom renting and suburbs. Thus, contemporary union organising strategy needs to incorporate this new spatial order and challenge workers' uneven access to affordable, safe and decent accommodation. NUM has not been very successful in organising many of the mine workers in the informal settlements who are usually in temporary, precarious positions.

The platinum insurgency was initiated by rock drill operators (RDOs) dissatisfied with their remuneration in comparison with amounts received by RDOs at other mining houses. These workers questioned NUM's approach and strategy in the wage negotiation dispute with Impala Platinum in 2012. The strike gained overwhelming support from the more precarious segments of the workforce. It was characterised by insurgent unionism organised through informal workers' committees. These structures presented demands directly to management through informal representatives. Leaders were selected in popular democratic processes such as mass meetings. The workers were not only demanding better conditions of service, but also rejecting representation by NUM, which was the most recognised union in the region. One worker (cited in Chinguno, 2015: 196) said: 'We did not want the NUM to be involved because it no longer represented our interests. It was compromised by management. It only served the interests of management …. It was only there to legitimise our exploitation.'

The fallout with NUM stems from claims that the union had been captured by management and was incapable of representing workers' collective interests. These workers argue that NUM has lost touch with ordinary workers and their daily socio-economic realities. This is an example of the paradox of institutionalism discussed above and in Chinguno's study (2015: 248). Moreover, mine workers claimed that the organisation was incorporated into an exclusionary industrial relations system. Therefore, they created alternative organising tactics in the form of informal structures and attempted to resolve their issues outside conventional bargaining systems (Chinguno, 2013; Chinguno, 2015).

The striking workers later invited a new union, AMCU, to address their concerns. It engaged the informal workers' committees and promised to be a new voice for the mine workers. A NUM official cited in Chinguno (2015: 216) recounted how this happened:

AMCU came in during the strike but they were quick such that during the strike they were already the ones who were assisting in formulating strategies and advising workers and committees not to listen to management and NUM. We did not realise this initially but were just surprised by the level of arrogance and belligerence exhibited by our members.

AMCU presented NUM with a considerable challenge as it attracted workers from the latter's traditional membership base. It appealed to employees on the basis of knowledge about their daily experiences and shop floor issues. Furthermore, it incorporated informal workers' committees into union structures (Ntswana, 2014: 59). This increased associational power has created contradictions within AMCU. The leadership fears that the committees will be used to undermine their political and organisational power. But members argue that worker committees are essential to avoid the bureaucratisation and centralised decision-making associated with NUM pitfalls. A worker cited in Ntswana's (2014) analysis of platinum belt worker committees explains this concern:

We are very worried about the way the union is acting; this workers council of the madalas (old men) was a good thing as it kept the union in check but now AMCU has come and said all other workers committee structures existing in the mines should be disbanded because they are dividing workers. The truth is that the unions are threatened by the workers council and committees because they know that these are building consciousness among workers to call for accountability (Ntswana, 2014: 61).

Another issue raised by AMCU members is the dominance of national leadership, particularly the president, Joseph Mathunjwa. The general concern is that the organisation is developing a cult of personality, centred around Mathunjwa and other national officials. This decreases participatory democracy, worker control and branch cadre development. An AMCU member captured this concern in the following words: 'AMCU needs to build people in the union structures who can handle things in the absence of Mathunjwa, but it seems, the leaders are threatened by the young men who are shop stewards' (Ntswana, 2014: 62).

This tension between institutionalisation, leadership authority and maintaining workers' control over their own structures was a prominent theme in debates about organising during the platinum strike waves. Trade unions that are engaged in orthodox collective bargaining require some level of hierarchy and technical capacity. However, this must be counterbalanced with continued accountability and participatory decision-making. The workers who migrated from NUM to AMCU accuse NUM of overlooking the latter principle. More importantly, NUM is viewed as an enforcer of the established industrial order, which is controlled by mining companies (Ntswana, 2014: 59–61).

Recent reports suggest that AMCU is also battling with intra-union rivalry. Its leadership is accused of corruption and authoritarian governance. Some of these contestations are related to ethnic tensions within the organisation (Montosho, 2018). This has culminated in court battles, where members accuse the leadership of using organisational machinery to remove political opponents in branches or regions. More worryingly, some employees have resigned from union structures because of death threats and the prevalence of violence within the union (Montosho, 2018).

CONCLUSION: TRADE UNION ORGANISING AND THE FUTURE OF MINING INDUSTRIAL RELATIONS

This chapter explores the organisational challenges experienced by mining trade unions in colonial, apartheid and post-apartheid South Africa. It links structural change with worker agency in an effort to expand the literature on mine workers' power, organisation and resistance. We argue that the intersecting relationship between worker resistance and structural change provides insights into historical as well as contemporary trade union strengths and weaknesses. Furthermore, the chapter shows how these organisational issues have affected the making and re-making of industrial relations in South Africa. Worker resistance and organising impacts on the political legitimacy of labour institutions. The analysis highlights the following five key points about organisational strategies for renewal and related debates.

First, there is a need to overcome the challenge of representation by organising varied segments of the workforce. Union membership is primarily composed of those working under standard employment conditions, whilst the prevalence of atypical labour conditions is increasing. This trend has fragmented worker solidarity and challenged the legitimacy of unions in workers' eyes. The non-conventional forms of insurgency in the platinum belt highlight this shortcoming. Moreover, it questions the sustainability of the corporatist post-apartheid industrial relations system.

Second, the research shows that it is essential to rebuild internal democracy and worker control within union structures. Evidence from union research and ethnographic studies indicates that both have declined within unions. The sense of disempowerment and alienation expressed by workers in various interviews highlights this trend. South Africa's labour movement and many of its accomplishments have been built on a democratic, decentralised organising model. This can be restored through reviving the capacity of sub-national structures: branches and regions. Accountability and participatory democracy are fundamental elements of these primary structures of worker control.

Third, unions must seriously consider returning to the practice of social movement unionism. Solidarity between those waging labour and community struggles has weakened in the post-apartheid era. However, evidence from AMCU-dominated worker committees indicates some attempts to re-establish these links, though limited to certain places and their spatial politics. There are currently debates underway relating to social and labour plans that allow for the identification of issues that could facilitate a rebuilding of the alliance. A key aspect of social movement unionism is the relationship with political parties. The de-legitimisation of NUM, in the view of the workers cited in this study, is exacerbated by its historical link with the ruling party (ANC). Thus, state and party shortcomings are externalised onto the union, with some workers arguing that it cannot exercise independent political agency. The union needs to continue debating the nature of its complimentary and contradictory partnership with the ANC, especially in the post-apartheid context in which segments of the workforce are challenging the policy and class orientation of the party.

Fourth, trade unions need to enhance their strategies for organising by linking them to structural changes. Organising must not be viewed in isolation from broader macro-economic and industry-specific trends. This requires a detailed empirical analysis of how the changes in the industries of coal, platinum group metals and other essential minerals have transformed the nature of work. This analysis must in turn help to generate innovative strategies for recruitment and for servicing workers' needs. Organisational renewal requires strategic, long-term analysis. A good starting point for this analysis would be current international debate about value-chain organising capacity. Both NUM and AMCU organise across value chains in raw minerals extraction, energy and construction. The political economy literature highlights strong linkages between these sectors; there is a wealth of international experience of organising this way from which to draw. Thus, exploring various strategies for strengthening union power through value chain mobilisation may be fruitful.

Finally, collective action in standard industrial relations is often exercised through formal institutions, like trade unions. However, the platinum belt strike wave exhibited the power of collective action outside formal industrial relations institutions (Chinguno, 2015). These unconventional worker protests are politically significant because they raise two pertinent questions:

1. Is South Africa's current industrial relations system suitable for addressing the grievances of all segments of the country's workforce?

2. Do current trade union organising strategies speak to those who make up the restructured, post-apartheid mine worker labour force?

REFERENCES

Atkinson, D. (2007). *Going for Broke: The Fate of Farm Workers in Arid South Africa*. Human Sciences Research Council, Cape Town.

Beckman, B. and Sachikonye, L.M. (2001). *Labour Regimes and Liberalization: The Restructuring of State-society Relations in Africa*. University of Zimbabwe Publication, Harare.

Benya, A. (2009). 'Women in mining: A challenge to occupational culture in mines'. Masters dissertation. Faculty of Social Science and Humanities, University of the Witwatersrand.

Benya, A. (2015). 'The invisible hands: Women in Marikana'. *Review of African Political Economy,* Vol. 42(146), pp. 545can P

Benya, A. and Ncube, P. (2015). 'COSATU's demographic profile'. In *COSATU Workers' Surveys of 2006 and 2012: What Do They Tell Us?* NALEDI, Johannesburg.

Bezuidenhout, A. and Buhlungu, S. (2007). 'Old victories, new struggles: The state of the National Union of Mineworkers'. In Buhlungu, S., Daniel, J., Southall, J. and Lutchman, J. (eds). *State of the Nation: South Africa 2007.* HSRC Press, Pretoria.

Bezuidenhout, A. and Buhlungu, S. (2011). 'From compounded to fragmented labour: Mineworkers and the demise of compounds in South Africa'. *Antipode,* Vol. 43(2), pp. 237.

Buhlungu, S. (2010). *A Paradox of Victory: COSATU and the Democratic Transition in South Africa.* University of KwaZulu-Natal Press, Pietermaritzburg.

Bundy, C. (1988). *The Rise and Fall of the South African Peasantry.* James Currey Publishers, London.

Burawoy, M. (1976). 'The functions and reproduction of migrant labor: Comparative material from southern Africa and the United States'. *American Journal of Sociology,* Vol. 8(5), pp. 1050–1087.

Burawoy, M. (1979). *Manufacturing Consent.* Chicago University Press, Chicago.

Burawoy, M. (1983). *The Politics of Production: Factory Regimes under Capitalism and Socialism.* Verso, London.

Butler, A. (2013). *Cyril Ramaphosa.* Jacana Press, Johannesburg.

Byrne, S., Rees, R. and Orr, L. (2015). 'Internal democracy and worker control in COSATU'. In *COSATU Workers' Surveys of 2006 and 2012: What Do They Tell Us?* NALEDI, Johannesburg.

Capps, G. (2015). 'Labour in the time of platinum'. *Review of African Political Economy,* Vol. 42(146), pp. 497–507.

Chinguno, C. (2013). 'Marikana and the post-apartheid workplace order'. Sociology, Work and Development Institute (SWOP), University of the Witwatersrand, Johannesburg.

Chinguno, C. (2015). 'The shifting dynamics of the relations between institutionalization and strike violence: A case study of Impala Platinum, Rustenburg (1982–2012)'. Doctoral dissertation. Faculty of Humanities, University of the Witwatersrand.

Cousins, B. (2011). 'What is a smallholder'? Class-analytic perspectives on small-scale farming and agrarian reform in South Africa'. In Hebinck, P. and Shackleton, C (eds). *Land Resource Reform in South Africa: Impacts on Livelihoods.* Routledge, Abingdon.

Crush, J. (1989). 'Migrancy and militancy: The case of the National Union of Mineworkers of South Africa'. *African Affairs,* Vol. 88(350), pp. 5–23.

Crush, J. (1992). 'The compound in post-apartheid South Africa'. *Geographical Review,* Vol. 82(4), pp. 388–400.

Dörre, K., Holst, H. and Nachtwey, O. (2009). 'Organizing – A strategic option for trade union renewal?' *International Journal of Action Research,* Vol. 5(1), pp. 33–67.

Fantasia, R. and Voss, K. (2004). *Hard Work: Remaking the American Labour*

Movement. University of California Press, Berkeley.

Forrest, K. (2015). 'Rustenburg's labour recruitment regime: Shifts and new meanings'. *Review of African Political Economy*, Vol. 42(146), pp. 508–525.

Kariuki, S. (2009). 'Agrarian reform, rural development and governance in Eastern and Southern Africa: Policy brief 59'. Centre for Policy Studies, Johannesburg.

Kenny, B. and Bezuidenhout, A. (1999). 'Contracting, complexity and control: An overview of the changing nature of subcontracting in the South African mining industry'. *The South African Mining Institute of Mining and Metallurgy*, pp. 185–191.

Kraak, G. (1993). *Breaking the Chains: Labour in South Africa in the 1970s and 1980s*. Pluto Press, Colorado.

Mabasa, K. (2017). 'An empty pot at the end of the rainbow? Trade unions and twenty years of democracy'. Friedrich-Ebert-Stiftung, Johannesburg.

Mafeje, A. (2003). 'The agrarian question, access to land, and peasant responses in Sub-Saharan Africa'. United Nations Research Institute for Social Development, Geneva.

Magubane, B.M. (1979). *The Political Economy of Race and Class in South Africa*. Monthly Review Press, New York.

Magubane, B.M. (1996). *The Making of a Racist State. British Imperialism and the Union of South Africa 1875–1910*. Africa World Press, Trenton.

Maloka, E. (2014). *Friends of the Natives: The Inconvenient Past of South African Liberalism*. 3rd Millennium Publishing, Durban.

Mamdani, M. (2012). *Define and Rule: Native as Political Identity*. Wits University Press, Johannesburg.

Marx, K. (1867). *Capital. Volume One*. Lawrence and Wishart, London.

Massey, D. (1983). 'Class struggle and migrant labour in South African gold mines'. *Canadian Journal of African Studies/Revue Canadienne Des Études Africaines*, Vol. 17(3), pp. 429–448.

Michels, R. (1962). *Political Parties: A Sociological Study of the Organizational Tendencies in Modern Democracies*. The Free Press, New York.

Mining News. (5 August 2013). 'Current SA contract mining models not sustainable'. Available at: http://www.miningne.ws/2013/08/05/current-sa-contract-mining-models-not-sustainable/. Accessed 30 June 2018.

MISTRA. (2015). *National Union of Mineworkers: Exploring Pathways to Organisational Renewal*. Johannesburg, MISTRA.

Mnwana, S. (2015a). 'Mining and "community" struggles on the platinum belt: A case of Sefikile Village in the North West Province, South Africa'. *The Extractive Industries and Society* (2), pp. 500–508.

Mnwana, S. (2015b). 'Between "locals" and "foreigners": Mining and rural politics of belonging in North West Province, South Africa'. *Labour Capital and Society*, Vol.48(1and2), pp. 500–526.

Montosho, M. (2018). 'Amcu president protects corrupt leaders, court hears'.

IOL. Available at: https://www.iol.co.za/news/south-africa/north-west/amcu-president-protects-corrupt-leaders-court-hears-14153723. Accessed 29 June 2018.

Moodie, T.D. (2009). 'Managing the 1987 mine workers' strike'. *Journal of Southern African Studies*, Vol. 35(1), pp. 45–64

Moodie, T.D. (2015). '"Igneous" means fire from below: The tumultuous history of the National Union of Mineworkers on the South African platinum mines'. *Review of African Political Economy*, Vol. 42(146), pp. 561–576.

NALEDI. (2015). *COSATU Workers' Surveys of 2006 and 2012: What Do They Tell Us?* NALEDI, Johannesburg.

Nattrass, N. and Seekings, J. (2011). 'State-business relations and pro-poor growth in South Africa'. *Journal of International Development*, 23, pp. 338–357.

Neocosmos, M. (1993). 'The agrarian question in Southern Africa and accumulation from below'. Nordic Africa Institute, Sweden.

Ntswana, N. (2014). 'The politics of workers control in South Africa's platinum mines: Do workers' committees in the platinum mining industry represent a practice of renewing worker control?'. Masters dissertation. Faculty of Humanities, University of the Witwatersrand.

O'Meara, P. (1975). 'The 1946 African mine workers' strike and the political economy of South Africa'. *Journal of Commonwealth and Comparative Politics, Vol.* 13(2), pp. 146–173.

Pillay, D. (2015). 'COSATU and the Alliance: Falling apart at the seams'. In Sagtar, V. and Southall, R. (eds). *COSATU in Crisis*. KMM Review Publishing, Johannesburg.

Purcell, J. (1993). 'The end of institutional industrial relations'. *The Political Quarterly*, Vol. 64(1), pp. 6–23

Saul, J. (1986). 'South Africa: The question of strategy'. *New Left Review*, Vol. 160, pp. 3–23.

Standing, G. (2016). 'The precariat, class and progressive Politics: A response'. *Global Labour Journal*, Vol. 7(2), pp. 189–200.

Statistics South Africa. (2017). 'Mining industry 2015'. Statistics South Africa, Pretoria.

Terreblanche, S. (2012). *Lost in Transformation: South Africa's Search for a New Future Since 1986*. KMM Review Publishing, Johannesburg.

Theron, J. (2016). 'What is decent about "decent work"? An argument for a right to decent work in South Africa. In Busby, N., Brodie, D. and Zahn, R. (eds). *The Future Regulation of Work: New Concepts, New Paradigms*. Palgrave Macmillan, London, pp. 167–192.

Turok, B. (2011). *Development in a Divided Country*. Jacana Media, Johannesburg.

Von Holdt, K. (2003). *Transition from Below*. University of Natal Press,

Pietermaritzburg.

Von Holdt, K. (2010). 'Institutionalization, strike violence and local moral orders'. *Transformation: Critical Perspectives on Southern Africa*, Vol. 72(1), pp.127–151.

Webster, E. (2005). 'New forms of work and the representational gap'. In Webster, E. and Von Holdt, K. (eds). *Beyond the Apartheid Workplace: Studies in Transition.* University of KwaZulu-Natal Press, Pietermaritzburg.

Webster E. (2017). 'Marikana and beyond: New dynamics in strikes in South Africa. *Global Labour Journal,* Vol. 8(2), pp. 139.

Wolpe, H. (1972). 'Capitalism and cheap labour power in South Africa: From segregation to apartheid'. *Economy and Society,* Vol. 1(4), pp. 425–456.

Wright, E.O. (2000). 'Working-class power, capitalist-class interests, and class compromise'. *American Journal of Sociology,* Vol. 105(4), pp. 957–1002.

Wright, E.O. (2012). 'Class struggle and class compromise in the era of stagnation and crisis'. Paper is based on a presentation.

Section Three

Beyond Mining: Just Transition and Wellbeing

ELEVEN

The mining-energy nexus, climate change and prospects of just transition

Reflections of a labour educator

HAMEDA DEEDAT

A SIGNIFICANT PART OF THIS BOOK focuses on the mining and energy sectors from historical and political economy angles. Climate change and the future of the mining and energy sectors are deliberated from these perspectives. The potential for renewable energy sources such as biogas, solar and wind energy is also explored. Despite the massive ramifications for labour[i] resulting from change in the mining and energy sectors, apart from the impact on employment as a statistic, labour is seen as an extraneous variable rather than as core to the debate. This chapter is both purposeful and deliberate in placing labour at the centre of the discourse around mining and energy. The point of departure is that labour plays a pivotal role in terms of the means of production and as an influential force in determining the mining and energy trajectory for South Africa. Furthermore, for South Africa to attain a meaningful 'just transition'[ii] in the energy and mining sector, organised labour is central. The mining-energy debate has become topical in South Africa.

Energy took centre stage from 2015 when former President Jacob Zuma (president of South Africa from 2009 to February 2018) began advocating strongly for nuclear sources. In 2017, amidst the nuclear debates, coal emerged as a bone of contention as Eskom announced its move away from coal, citing renewables or the pressure to renew the independent power producers (IPP) contracts as the motivation to close coal-fired power stations. Responses from organised labour included a resounding 'no' to renewables in favour of coal, while others took to the streets and blockaded the roads in and out of Pretoria and at the seat of government, the Union Buildings.

For those of us working as and with labour on the issues of mining, energy and climate change, this response meant that our work was far from over. The developments demonstrated, almost categorically, how far behind we were in terms of labour education, both on these issues and on the question of just transition. As an educator in the labour space, I work with both organised labour across the federations and civil society organisations. This is not to say that trade unions are not part of civil society, but given the strong role South African trade unions played in the struggle against apartheid, trade unions are viewed as distinct and more significant. This conception is emulated at the National Economic Development and Labour Council (NEDLAC), which identifies the social partners as business, government, labour and community. In keeping with this superficial divide, the education work on climate change, energy and mining has been aimed concertedly at driving an education programme for labour.

A secondary goal was to facilitate areas of common struggle between communities, community-based organisations and labour so that they could augment each other. The work undertaken in this area has unfolded over the last few years and points to several complexities. For some, the issues are quite clinical and clear-cut, with renewables being clearly 'the way to go'. For others, and for labour in particular, it is not quite so simple. The complexity is compounded by the current state of trade unions, which has evolved drastically since the start of South

Africa's democracy. Labour is constantly juggling its role as a partner in the Tripartite Alliance, and its mandate towards workers and the working class, with the latter proving to be much more of a challenge.

Fundamentally exacerbating this contradiction, has been the shift by trade unions away from the reliance on worker subscriptions for sustainability towards gains from investment companies, controversially described by Bezuidenhout et al. (2017) and others as 'business unionism'. And yet, as elucidated in the previous chapter in this book, the decline in union density as a result of increased informalisation, lowering labour standards and technological changes in the mining, security, finance, retail and energy sectors, have all elicited discontent and insecurity amongst trade unions and their members. This confluence of change cannot be more visible than in the mining-energy nexus. It is this reality, in my view, that explains the inertia we see in the labour movement, particularly from the quarters that give voice to the movement's concerns.

Like many constituencies, labour is not homogeneous. Trade union members, amongst one another, and unions of different federations may hold a range of positions on a particular issue. Similarly, workers and trade union affiliates may respect congress resolutions of their federations, but this will not preclude them from critically debating and discussing the same issues at hand at affiliate level. Two relevant and recent examples come to mind: nuclear energy and coal versus renewables. These examples are central to this chapter's reflection; I shall discuss them at length.

It is important to underline that this chapter's reflection does not purport to speak on behalf of labour (see also footnote 1). Similarly, the discussion does not offer a definitive account of the positions of organised labour in South Africa vis-à-vis the mining-energy nexus and just transition. Rather, it offers an analysis of the dilemmas and challenges (based on several interactions and engagements with trade unions in the mining and energy sectors) on the subject of climate change. These engagements began between 2010 and 2011 in preparation for the United Nations Framework Convention on Climate Change Conference of the Parties (COP17) hosted in Durban in 2011. Subsequently, the National Labour and Economic Development Institute (NALEDI), COSATU and organisations like the World Wildlife Fund (WWF) worked with various COSATU affiliates, including NUMSA (a COSATU affiliate at the time), FAWU and the National Union of Mineworkers (NUM). Discussion on climate change was located at the leadership level within affiliates and at the COSATU level. In fact, the NUM leadership and senior officials were quite advanced in their own engagements on climate change and were reviewing the option of nuclear energy as an alternative to coal as an energy source.

Their engagements took the form of an international study tour to various nuclear sites to learn and explore nuclear energy as a possibility for South Africa. The official recommendations stemming from the international exchange came to an abrupt end when the COSATU Congress of 2012 adopted an anti-nuclear resolution (i.e. nuclear energy is not an energy option for South Africa). The decision was taken in light of the havoc and devastation wreaked by the Fukushima nuclear disaster of March 2011. The NUM recommendations did not even make it to the table.

After 2011, work on climate change with trade union affiliates continued. It was, however, on a much smaller scale and by 2013 the importance of climate change work for trade unions seemed to have lost momentum. There was a need to reactivate interest and propel labour to engage in the debates about climate change at national and international level. It became imperative for labour to intervene in the energy discussions in order to define and advocate for a just transition for labour, without diminishing the challenges for labour.

NALEDI in collaboration with COSATU, through the COSATU climate change reference group (constituted by representatives from COSATU, its affiliates and civil society organisations) championed this cause. Over the last three years (2015–2018), NALEDI has made significant inroads as a result of the provincial workshops on climate change that we hosted, across the affiliates and federations. While not limited to the energy sector, energy took centre stage with prevailing climate change engagements and contributions by and large being around the just transition and green jobs. South Africa's energy mix, energy democracy and nuclear and renewable energy were areas therefore in which labour was engaging at both national and international levels and within formal high-level union processes such as central executive committees of both affiliates and federations.

The chapter's primary contribution is the insights into labour's engagement with the issues of energy. It also offers insight for environmental organisations that have carved out the 'energy future' and 'energy democracy' for South Africa – without grasping the standpoint of labour. Finally, this chapter sheds light on eco-socialism as recourse for labour rather than labour's current fall-back of settling for a fraught system rigged against it.

THE LABOUR CONTEXT

Between 2008 and 2015, South Africans experienced widespread load shedding of electricity. This had quite a negative impact on the economy and adversely affected businesses, including coal mines and Eskom, and the workers they employed. For some, the impact was so detrimental that firms closed down, while others chose to divest and relocate outside South Africa. As early as 2008 speculation on the 'health' of Eskom began, and yet the parastatal prevailed. Several tariff hikes were instituted and in 2011 the United Nations Climate Change Conference (COP17) took place in Durban. Labour and civil society were present, and coal and Eskom featured centre stage in public discourse. Nevertheless, it was largely business as usual. While talk of renewable sources of energy gained momentum, the coal monopoly (whose hegemony in the energy and mining sector was entrenched) had diffused political and financial responsibility for its part in global warming.

It was not until 2015 that I started working with organised labour on climate change. However, the groundwork had already been established as a result of key events, one of which was COSATU's 2012 publication of a booklet on climate change and the just transition. This was part of the lead up to the COP17 and symbolised the moment at which climate change was placed on the agenda of trade unions. It coincided with COSATU's 2012 congress pronouncing resolutions on climate change and nuclear energy.

By 2015, despite NUM's legwork, it was NUMSA and FAWU

that were at the forefront of climate change issues, together with COSATU (national). The energy debate and the move towards energy democracy were high on NUMSA's agenda. In 2015, the Department of Environmental Affairs (DEA) developed its intended nationally determined contributions (INDCs) and began dialogues in preparation for COP17. Stakeholders, including labour, were asked to comment – but labour was almost invisible.

Global context

It was at this same time that keeping the earth's temperature increase to below two degrees Celsius became imperative; the aim was for below 1.5 degrees. For all intents and purposes coal was identified as one of the worst emitters contributing to greenhouse gases and the global call, particularly by environmentalists like the WWF, was to 'keep the coal in the hole'. NUM, as a result of its engagements on the issues of climate change, both locally and globally, supported the transition from a high carbon intensive economy to a low carbon intensive economy. It noted the impact of the carbon intensive mining and energy sectors on the environment as a key contributor to climate change. Nuclear energy had been gaining traction in political circles in South Africa. It also found its way into the mainstream media and was often seen as a viable option, an alternative energy source for the country. Yet, a strong cohort of labour representatives and civil society organisations made forceful arguments against nuclear to the portfolio committee of the DEA in September 2015 during the parliamentary hearings. They advocated strongly for government to take an anti-nuclear stance.

While anti-nuclear, however, was a formal COSATU resolution, the workings of the resolution had not been fully interrogated. As such, I found myself in hot water at a NEDLAC summit on energy in 2015. I was asked to represent the community (portfolio) and raise the issues related to the high cost of energy, inefficient energy supply by Eskom and the rumours around nuclear energy. At the end of my presentation I was challenged publicly by a senior COSATU leader to qualify my statement to 'keep the coal in the hole'; I was asked if this meant job losses and unemployment for coal mine and Eskom workers and whether I had considered the implications that such a decision would have for NUM.

My naivety dawned on me. A demonstration or articulation that reflects an understanding of how coal contributed to climate change as an emitter of greenhouse gases is not synonymous with 'keep the coal in the hole', nor with an acceptance of the consequent job losses such a position implies. I also realised that I was not alone in this naivety. For trade unions and leaders in particular, because of who and what they represent, simply taking the moral high ground was not going to bode well for workers who would potentially lose their jobs. This therefore begged the questions of who and how many workers were to lose their jobs? Was there an alternative? Yet again, labour had the answer: the just transition! But at this juncture it proved to be largely a theoretical exercise.

ENGAGING THE HARD REALITIES

It was not until 2014 that fortune smiled upon some of us as Development Bank of Southern Africa funding, through the Green Fund, enabled NALEDI to run provincial workshops on climate change with the trade union federations and their affiliates. Since the proposal was an outcome of the COSATU climate change reference group,[1] the group participants were central to the content and focus of the workshops. As a result of several engagements of the reference group, the group identified key sectors for each province, the composition of the unions attending, the sectoral relevance for that province and the key debates on climate change. Mining and energy sectors were taken up in all of the provinces.

Allow me to dispel some myths: while NUM is one of the oldest and best-established trade unions in South Africa, the history and political economy of the mining industry, and more specifically, coal mining, is not common knowledge to NUM members. An analysis of the apartheid economy and its persistence was also absent. Specifically

---

1   The COSATU climate change reference group comprised COSATU affiliates and non-governmental organisations like WWF, Greenpeace and Earthlife Africa. The One Million Climate Jobs was also active. Common ground was being forged on several fronts.

lacking was an understanding of how apartheid had established the mining industry structure in which largely migrant, indigenous black workers and, by extension, families in rural areas, were the source of super-profits for mining companies. The insidious nature of how the state further sustained mining company profits through huge subsidies on water and electricity, as well as tax rebates, was also not understood. As labour educators we were confronted with discussions about the importance of mining given its contribution to the country's economy and the fact that coal mining is directly tied into the energy sector and all the upstream industries, including transport and trucking both poor-grade goal to Eskom and good-grade coal to harbours for export. Jobs and employment were constantly juxtaposed with climate change concerns.

The discussions were intense and, perhaps for the first time, grounded in reality, as workers confronted the prospect of job losses as a consequence of demanding a cleaner planet. It is common knowledge that workers in Eskom and the coal-mining sectors often suffer with work-related illnesses and that poor occupational health and safety conditions, as well as the environment of work itself, negatively impact the health of workers. The adverse impact is manifested in the communities surrounding mining operations (historically black, poor, disadvantaged) and among workers, children and babies who are sentenced to a life of suffering and death, often stemming from chronic heart disease. The number of babies born deformed and children suffering from asthma and other health-related illnesses has reached distressing proportions. *The Bliss of Ignorance* (2015), a documentary produced by Groundwork and Friends of the Earth, depicts graphically the plight of these communities and the extent to which Eskom and mining companies were left to their own devices, 'getting away with murder'! The footage of suffering inflicted on these communities, for generations, evoked outrage.

The true cost of coal became evident. These realisations placed coal at the heart of the debate on the need for a just transition in the energy sector – and it had to be without coal. Historically disadvantaged rural communities, including those living in close proximity to the coal-fired power stations, often continued to have no access to potable water and were forced to drink contaminated water, while Eskom used thousands of litres of water per second. Thus the need for labour to vigorously advocate for change in the energy sector, based on the just transition, was evident and was openly engaged. There was concrete deliberation on how the various trade unions (both their individual unions and as a collective) needed to advance the debate. Renewable energy as a new energy source was discussed; it was well received as an emerging sector. The positive impact it had on energy supply in the face of load shedding was noted. By 2015 in South Africa, renewable energy had stemmed the frequency of load shedding; it kept the lights and geysers on in winter, serving as a complementary energy source on the grid. The idea of renewable energy as a cleaner, greener more sustainable source of energy to address the inadequacies in Eskom (that had resulted from poor operation and maintenance) was also well received. An openness around renewables emerged and it was seen as a viable option for an energy-just transition.

ESKOM THROWS A SPANNER IN THE WORKS

Over the period 2015 to 2017, NALEDI (through the provincial climate change workshops) had achieved major legwork on the ground with representatives of trade unions on the issue of climate change. There were also several meetings on energy and the just transition and/or energy democracy[2] that labour was engaged in alongside civil society. However, the synergy was seriously set aback when Eskom announced the closure of its coal-powered power stations. Was it five or was it four? Even the news reports are contradictory, with the figures constantly being interchanged. The number is by no means insignificant, especially for labour, as each power station closed meant the loss of thousands of jobs (1,000 direct and 40,000 indirect, with a direct adverse effect on both NUM and NUMSA). The announcement on 8 March 2017 by Eskom spokesperson Khulu Phasiwe confirmed that Eskom planned to expedite its plan to close four coal-fired power

---

2   Energy democracy is a political, economic, social and cultural concept that merges the technological energy transition with the strengthening of public participation.

stations (see Slabbert, 2017). According to Phasiwe, the 3000MW Kriel, 1000MW Komati, 2000MW Hendrina and 600MW Camden, which are all in Mpumalanga, were to close. He added that the move on the part of Eskom was to accommodate renewable IPPs (Slabbert, 2017). Quoted in this article by Slabbert (2017), he added that the power stations would gradually close down as renewable energy was connected to the grid.

The response was revealing. Truck drivers, many of whom were members of the South African Transport and Allied Workers Union, blockaded the entrance and exit roads of Pretoria and those to the Union Buildings in a massive protest against Eskom. They were clear that the closure of the four power stations spelt unemployment for them. NUM and NUMSA directed their anger at the announcement towards the Renewable Energy Independent Power Producers Programme. Both threatened mass action and NUMSA went so far as to request a section 77[3] at NEDLAC against Eskom (see Van Rensburg, 2017). Both unions felt that while renewables were good, they could not be introduced at the cost of coal jobs (Van Rensburg, 2017).

Similarly, COSATU, which was the only union federation with a climate change agenda and a policy on both climate change and just transition, committed to mobilising workers to oppose the move. COSATU further took the position, joining NUM, that South Africa's climate change obligations and commitment to introduce renewable energy into the grid should not be used as a means to push in privatisation through the back door (Slabbert, 2017).

Eskom had deliberately and categorically stated to the press that the key reason for closing the coal-powered power stations was renewables. Mentioned in passing were the facts that the power stations were nearing the end of their life span, that there were operation and maintenance issues, rather than attributing it to economic factors and losses incurred

---

3   A section 77 is a right given to workers under the Labour Relations Act (No. 66 of 1995; there have also been several amendments to the Act) to partake in protest action to promote or defend their socio-economic interests legally (they are protected against dismissal, victimisation or disciplinary action for participating). Trade unions thus serve notice of their intent, at least 14 days prior, at NEDLAC.

by Eskom due to slow growth rate and low gross domestic product. Eskom further stressed that in the context of an existing surplus in coal-fired energy, renewable energy merely increased the surplus further. The possibility of reducing electricity costs for domestic users, which could have effectively addressed the over-supply, was not raised and likely not contemplated. More significantly, this could have represented quite a fundamental turning point in our energy history. Eskom could have framed this as the positive impact of renewables, providing the opportunity to explore transitioning away from coal, with labour as a key partner and looking towards a just transition. Quite to the contrary and in the interest of 'coal monopolies' whose interest had to be best served, Eskom set the stage for contestation between labour and the new emerging IPPs in renewables (more of a threat to the interests of coal monopolies than they are to labour). The gravity of the situation and the nature of the response from labour are best qualified by David Sipunzi, NUM's general secretary in the NUM press statement of 24 March 2017. He explained that the estimated job losses were to be close to 10,000 direct jobs, and a further 40,000 indirect jobs could be lost around each coal-powered power station. This was reiterated again in 2018 by NUM and NUMSA respectively (Moosa, 2018; Niselow, 2018).

In hindsight, it might have dawned on the trade unions that Eskom, not to mention its capital investors, were not yet ready to relinquish their cash cow, having profited from the coal sector for decades. With climate change on the horizon, and South Africa figuring as the ninth worst emitter in the world as a result of coal-based energy and petroleum (SASOL), at face value it seemed that business had to take the moral high ground and commit to the reduction of carbon emissions. However, the trump card was to play into the unions' hand by stroking the Achilles heel of organised labour: massive job losses. NUM has already seen thousands of jobs lost across the various mineral sub-sectors and was reeling from the rapid pace. Similarly, NUMSA, no longer part of COSATU, lost members to the Liberated Metalworkers Union of South Africa, as well as due to job losses in the auto and steel sectors. Neither unions at this point or in the immediate future needed more job losses.

Rather than challenging Eskom, for the unions the issue became renewable energy versus coal jobs. The crux of the matter is that the two energy sources dialectically oppose each other. The one is carbon intensive and the other is zero carbon, the one has huge health implications and the other has none, the one is very expensive and the other one is relatively cheap and the one is exceptionally bad for the climate and the other one is climate friendly. The key element that sadly got lost in all of this was that one was job shedding and the other one job creating. Renewable energy had and has the potential to change the manufacturing and energy landscape in South Africa, the region and the continent as a whole. It also has the immense potential to bring about energy democracy and improve energy accessibility with its accompanying off-spins. Renewable energy offered the platform for setting the stage for the just transition and carving out a trajectory that other sectors could follow.

The mention of potential job losses and the fear of its ramifications for trade unions, their members and their communities, had such a knee-jerk response from labour that many NGOs and civil society organisations felt that the solidarity forged towards energy democracy and the just transition was a misinterpretation. And for a moment it was!

JUST TRANSITION

The naivety, even on the part of those of us within labour, cannot be understated. The essence of accomplishment was not in chanting the mantra as climate change warriors, but in implementing systems, possibilities and modalities that ensured that *the climate changed*, and here I mean more than the weather. It became abundantly clear that defining and advancing the struggle for a *just transition* was paramount to the struggle against climate change. NALEDI, in collaboration with COSATU, recognised the need for us as labour to regroup and refocus. A three-day meeting was called amongst COSATU and its affiliates to interrogate the announcement by Eskom, the readiness and support where needed by NUM and of course the responses issued by the various quarters of organised labour in response to Eskom. Central to these discussions were energy democracy, just transition and nuclear energy as a viable energy alternative for South Africa as we transition away from the country's overwhelming dependence on coal to a climate-friendly energy mix.

The three days were filled with robust discussion and debate, and issues like energy democracy, nuclear energy, just transition and its implications for NUM were openly placed on the table and equally openly engaged. Nuclear energy, for example, had both staunch proponents and opponents across affiliate representatives, despite COSATU's resolution which was anti-nuclear. The discussions not only allowed for critical debate and discussion, but for learning as well. A positive outcome of this debate was the COSATU resolution on nuclear energy and that the affiliates stand by this position. However, the informative engagements highlighted some gaps and issues of difference around information on nuclear energy. It was proposed that a more in-depth discussion on nuclear energy happens, even at affiliate level, to ensure full support for the resolution or alternatively table the need at the Central Executive Committee of COSATU to review the resolution.

This robust discussion filtered into other discussions too, particularly the one on energy choice and just transition. This had direct implications for NUM that had the space to articulate both the contradictions of being supportive of efforts to avert a climate disaster and at the same time realising that no coal jobs without a plan on how to transition implied unemployment for workers. NUM raised the point that Eskom was premature in its announcements and there was hope that the NEDLAC processes would stem this tide for now. NUM also realised that the moment for labour to engage government on what is meant by a just transition and to use the energy sector as a case in point was most opportune. Of course the nuts and bolts of what this would mean, and if renewable energy could be the vehicle for the transition as opposed to being the catalyst for job losses, needed research and closer interrogation. NUM expressed a genuine openness to exploring the possibilities under renewables. It also came out quite strongly that if South Africa is going to achieve a true energy democracy then the motivators for change need a different modus operandi. Renewable energy was promoted under private ownership, despite the major potential and beneficiation that could be realised if it were publicly owned. This recognition and the decision to advance not just renewable energy but also the shape it has taken were massive strides forward after the preceding setback.

After this insightful meeting our work was cut out for us. The task was to interrogate the role of renewable energy in the South African labour market. What model was advocated by government and what were the possibilities envisioned by labour? Central to this work were questions around costing, traction (questions of whether renewable energy would gain support as an energy choice, and who would support it), and, most significant of all, what was its job-creating or enabling potential? For trade unions, particularly NUM, this was paramount. The burning question was: does the renewable energy sector have the potential to create one renewable energy job for every coal job lost?

This positive reflection process created space yet again for NGOs and labour to re-engage. Civil society organisations that were quite warm to working with labour were taken aback by organised labour's responses to Eskom's announcement to close coal power stations. For the organisations, such closure was a step in the right direction, and as partners in the struggle against climate change the expectation was that labour would perhaps accept (and would even be prepared for) the fait accompli and negotiate a just transition. Many of the organisations we worked with called us immediately after reading the media statement and, amidst the surprise, there were perceptions of labour being reactionary, conservative, knee-jerk and defensive. At one point it even felt as if we in labour were required to be apologetic.

Times of adversity present opportunity and this tension between our friends and comrades in civil society afforded us an opportunity for introspection and also to come to grips with the heart of many of the issues which, on the surface, seemed cohesive. It also demonstrated the need for us, as labour, to engage substantively around just transition and to settle our inconsistencies or differences of opinion (internally). More positively, it allowed us, as labour, to re-engage with our civil society comrades from a position of confidence and be very clear about what our role and vision is as organised labour. The result saw the interchange and nature of the discourse move from an outright 'no' to renewables to the notion that 'fighting for coal jobs doesn't necessarily mean the same as saving jobs in coal'. In other words, it was not renewables but the form and ownership of IPPs that was contested. Privatisation versus publicly owned, centralised versus decentralised, community-owned energy systems with potential to feed into the grid – these were the framings of new discussions and debates. Most critically, we engaged with the questions: Were renewables potentially the catalyst for an innovative industrial policy that would support the creation of new manufacturing industries in South Africa, as opposed to the handful of jobs stemming from wind farms and solar panel installation? Could this be the just transition for a large number of workers otherwise destined to a future of unemployment?

The significance of these discussions, not only for labour but South Africa as a whole, became evident as organised labour became a voice in the energy debate. This proved key in the face of a strong political push towards nuclear energy, giving rise to coalitions between organised labour, energy caucuses and coalitions that were contesting nuclear as an energy choice for South Africa. This collaboration then filtered into the just transition discussion that was emerging among NGOs. This culminated in joint discussions between organised labour and organisations like Project90, WWF, One Million Climate Jobs Campaign, Green Peace and Earthlife Africa around a vision for South Africa to move from a high carbon intensive society to low carbon, with just transition being the conduit to achieving this. As the collaboration and engagements intensified, so too did the synergies as various efforts to advance a common agenda vis-à-vis the just transition took shape.

The commitment to the vision of a just transition for South Africa towards attaining energy democracy had much support; representatives from civil society and organised labour participated in a two-day retreat on the just transition that was held in July 2017. It was a collaborative effort between WWF, NALEDI and the South African Climate Action Network (SACAN). Each partner identified key stakeholders that were central to these discussions and who were senior enough within their organisations to take forward the just transition agenda. SACAN, as the coordinator of the climate change network, ensured quite a large

NGO contingency and NALEDI, having worked with affiliates and all three labour federations, facilitated the organised labour constituency. There were approximately 40 delegates in total for the two days. The timing of the retreat was also most opportune, noting the threat of job losses and the outcomes of our own engagements as COSATU. The retreat created the space and opportunity to raise the concerns around job losses but also to brainstorm the potential for the just transition against this looming threat. It furthermore provided NUM the platform to shape its position on the just transition, while allowing for reflections and input.

More significantly it provided non-labour participants first-hand insight into the complexities of finding the middle ground (as organised labour), which needed to be both reflective of labour's constituencies' concerns and in line with the struggle against climate change. This prospect was daunting and NUM, despite the pressures, remained true to vision. It is important at this juncture to point out that several NGOs, in recognising this difficulty, hosted events in support of a just transition in the energy sector and provided substantial information on the renewable energy sector, its job creation potential and the models for implementation. The information provided several international examples for labour to draw on and also acted as a motivator for labour to undertake its own research. There were several moments in these deliberations that created food for thought, but none more 'provocative' than the reality that while the renewable energy sector holds vast potential for attaining energy democracy, it was not a panacea with regard to job loss. More specifically, while the sector has job creation and absorption potential, it would certainly not be a 'one renewable energy job for one coal job' translation.

The articulation of labour's agenda thus advanced. The discourse saw a shift from 'no to job losses' to 'fighting for coal jobs doesn't necessarily mean saving jobs in the coal sector'. This positive reflection process created space, yet again, for NGOs and labour to re-engage, but this time more substantively around the just transition. The situation around the closure of the coal power stations was, instead of being approached as a threat, now approached as an opportunity to learn and advance the just transition. So, for example, on the issue of the ownership of renewable energy IPPs had been granted licences as the owners and producers of renewable energy – a privately owned model – which was a conscious choice made by government. Labour's response was for ownership to be public. Instead of being based in privately owned IPPs, labour advanced that renewable energy should preside within the ambit of Eskom or another publicly owned entity.

Secondly, in light of the commitments made at the COPs and the nationally determined contributions, there was no need for renewable energy production to be capped. To the contrary, as demonstrated in chapter 5 of this book, renewable energy production should be increased. South Africa has the capacity for renewable energy production on a large scale, which will have positive ramifications for the country and the region and for a key strategic sector which could potentially be the just transition the energy sector is looking for. Labour had become acutely aware of the potential for renewable energy to become a game changer for South Africa through the development of an altogether new manufacturing sector. Renewable energy could be the catalyst for the development of an innovative industrial policy. This could result in South Africa specialising in renewable energy, thereby mitigating job losses.

This emphasis on high-level job creation contrasted with the model in terms of there being a mere handful of low-skilled jobs generated through IPPs. Additionally, renewable energy through decentralised and/or community-owned energy systems could lend itself to households being able to produce energy to feed into the grid, with positive outcomes for households as well as communities.

The NUM reorientation towards renewables saw COSATU obtaining a section 77 against the IPPs at NEDLAC, with the principle objection being the form of the IPPs, namely private ownership. For organised labour, objecting to private ownership was central. However, for some NGOs this strategy was seen as regressing rather than advancing the just transition agenda.

THREE STEPS FORWARD TWO STEPS BACK

The dialogues on just transition within organised labour, civil society and between the two seemed to have gained traction within both labour and the NGO spaces. However, noting that the energy context was quite volatile and spurred on by the looming doom of nuclear, efforts to unite around the just transition in energy were often tested. Tension arose, for instance, when Eskom announced the intention to close four coal-fired power stations with an estimated direct job loss of 40,000. Eskom qualified the decision by citing the renewal of IPP licences as the reason, thereby pitting coal jobs against the renewable energy sector. COSATU's response was one of raised eyebrows, but nevertheless COSATU created a space for dialogue among the federations and across their affiliates. A key milestone in this process was the opportunity for labour to highlight its concerns, as representative workers' bodies, to civil society organisations less concerned with the livelihoods of workers. This exchange eased doubt in certain quarters of civil society around labour's commitment to a just transition.

COSATU's section 77 application put labour in the position of having to explain itself again and specifically that it was not anti-renewables but rather advocating a paradigm shift. For some civil society organisations 'energy democracy and just transition' was measured by the form of clean and green energy with the number of jobs lost or gained, far less important.

The more nuanced position is that energy democracy is the end point, with just transition as the vehicle and labour as a key partner. The main contestation, however, is the paradigm. Labour is arguing for state-owned renewable sectors, while many of the NGOs that support renewable energy as part of the just transition are impartial to the issue of ownership, so long as the renewable energy sector comes on board and a certain number of jobs are created. So, despite the collaboration and engagement, awareness raising highlighting that workers ultimately stand to bear the brunt has unfortunately not translated into the NGOs taking action to defend job losses. The penny that the advancement of a just transition is equally about jobs and about carbon reduction has yet to drop! Events at the NALEDI climate change conference and Eskom's reiteration that job losses are pending, yet again demonstrated the complexity and sensitivities that organised labour has to grapple with.

CONCLUSION

The commitment to a just transition, the possibilities under renewables and the need for South Africa to move away from a high carbon based economy are a given, even for labour. However, the realities of high levels of unemployment, informalisation and a growth rate of 2 per cent generally, present a gloomy picture for unions like NUMSA and NUM and those of their members set to join the ranks of the unemployed. For both these unions this reality is daunting and, in the absence of a contingency plan on the part of government, Eskom and coal mines, important questions arise. Can these unions genuinely be expected to advocate for green jobs when 'green jobs' spell 'no jobs' for their members? This chapter does not purport to provide an answer, nor advocate what must be done, and least of all take the moral high ground and advocate what is right. To the contrary, this reflection presents the realities and difficulties that confront the labour movement, whose engagement in the just transition towards attaining an energy democracy goes beyond rhetoric, policy engagement, lobbying and advocacy. This reflection underlines how and why the labour movement is forced to address the practical realities of workers' lives and livelihoods, where staunch principles, no matter how morally correct, bear little relevance if they do not put bread on the table. In other words, in the current socio-economic climate – where there is little to no commitment from government to mitigate job losses (both direct and indirect jobs) and the impact of job losses on workers – and where there is no plan for re-absorbing, reskilling and redeploying workers back into the labour market (no mitigation at all!) – the burning question then becomes what do these unions do?

Herein lies the bone of contention! Trade unions have unfortunately not been proactive enough on the issues of a just transition, despite COSATU having taken the lead by developing principles for a just transition. As a result they are on the back foot. However, every challenge presents an opportunity and, as such, this situation has allowed for NGOs and labour to unite and to push collectively for a common agenda in spaces like the IRP process, the 'no to nuclear' campaigns and INDC processes, etc. – all of which have been positive.

Tensions arise, when workers' livelihoods are at stake, and trade unions become defensive in the face of job losses; they then prioritise jobs above all else. Interestingly, NGOs and organised civil society in the energy sector in South Africa are not able to grasp this complexity. In certain instances where NUM and NUMSA and even COSATU took positions to defend jobs, NGOs and civil society organisations become very critical and almost dismissive of the challenges. In the struggle for an energy democracy for South Africa, with a united front between labour and organised civil society at the forefront, actions on the part of labour were seen as 'three steps forward and two steps back'. So, despite various attempts to reconcile differences, organised labour and organised civil society have not found common ground on all of the issues. There have been and continue to be various engagements and platforms on the just transition and on energy democracy, with the most notable one anticipated, at the time of writing, for the second half of 2018.

There has, however, been strong agreement on renewable energy, nuclear energy and a just transition. This contrasts with the differences that persist on what the drivers of energy choice are, the paradigm in which energy is delivered, public versus private models of implementation, job creation versus job losses and decent work versus green jobs. For some civil society representatives the move from a high to low carbon economy is the only concern. For others, exploring the potential for new jobs in sectors like renewable energy is an opportunity which labour must embrace with the knowledge that a simple exchange of one renewable job for every coal job lost will likely not ensue. The question of ownership and the trajectory for achieving energy democracy arguably remains strategic and a means to fundamentally restructure the system of energy delivery.

During the NALEDI climate change conference of February 2018, we tried to facilitate a process to take the discussions a step further. Vishwas Satgar and Jackie Cock, both activists and comrades at heart who have been working extensively on energy issues, shared a perspective on how labour could move this issue forward. The key point they brought home was more than the challenges; labour is presented with an opportunity to advocate for fundamental systems change, which must be anti-capitalist. This, according to Satgar, is the only way energy democracy can be attained. Cock (see also Cock, 2015), building on the conceptual framework and ideas of Satgar, advocated for labour to be far more assertive. She further agitated that a 'true just transition for labour cannot be realised outside of eco-socialism'[4] – anything less could be described as minimalist. She concluded her presentation by encouraging unions and labour as a constituency to become more vocal and active and added that without such action energy democracy for South Africa would remain a 'visionary ambition'. Both the inputs were warmly received by labour representatives and space was afforded for deliberation and reflection. It will indeed take many further discussions and engagement before eco-socialism is advocated as the paradigm for just transition.

In sum, this chapter has aimed to share reflections on the role of labour in major early-21st century energy debates. Key to this reflection has been the highlighting of partnerships and challenges of partnership for labour in this arena. While in principle supporting 'just transition' and 'energy democracy', unions are conflicted as their members are the bearers of the costs of transition – whether it be workers employed directly in coal mining and energy generation, or those employed in related sectors. Given the lack of a constituency base and various ideological tendencies, civil society positions have collided with those of organised labour, and, at the worst, have been oblivious to the consequences for workers. Rather than current notions of a minimalist just transition, labour's centring of the issues of ownership and decentralisation are a sound beginning to the framing of a just transition that is meaningful and true to the goal of energy democracy. With its broad membership base and as an alliance partner, organised labour has an opportunity to take the lead in a moment of global climatic urgency.

---

4   According to Cock, an alternative growth path that offers the opportunity to do things differently.

REFERENCES

Bezuidenhout, A., Tshoaedi, M. and Bischoff, C. (eds). (2017). *Labour Beyond COSATU: Mapping the Rupture in the South Africa's Labour Landscape.* Wits University Press, Johannesburg.

Cock, J. (2015). *Alternative Conceptions of a 'Just Transition' from Fossil Fuel Capitalism.* Rosa Luxemburg Stiftung, Johannesburg.

Friends of the Earth and Groundwork. (2015). *The Bliss of Ignorance.* Documentary. Available at: https://www.imdb.com/title/tt4529628/. Accessed 10 August 2018.

Moosa, F. (5 April 2018). 'Why NUMSA and NUM hate the new independent power producers deal'. *The Daily Vox.* Available at: http://www.thedailyvox.co.za/why-numsa-and-num-hate-the-new-independent-power-producers-deal-fatima-moosa/. Accessed 2 August 2018.

Niselow, T. (5 April 2018). 'NUM hits out at Radebe on IPP deal, threatens to end ANC support'. *Fin24.* Available at: https://www.fin24.com/Economy/num-hits-out-at-radebe-on-ipp-deal-threatens-to-end-anc-support-20180405. Accessed 10 August 2018.

Slabbert, A. (8 March 2017). 'ESKOM to close four power stations'. *Moneyweb.* Available at: https://www.moneyweb.co.za/news/south-africa/eskom-to-close-four-power-stations/. Accessed 30 July 2018.

Van Rensburg, D. (9 April 2017). 'New twist in IPP battle as Eskom may close station early'. *Fin24.* Available at: https://www.fin24.com/Economy/Eskom/new-twist-in-ipp-battle-as-eskom-may-close-stations-early-20170407. Accessed 30 July 2018.

NOTES

i   Labour is an integral part of South African society and general reference to it implies workers and the working class, the poor or proletariat, the historically disadvantaged. As a constituency and as an alliance partner, labour refers to organised labour, and more specifically, the Congress of South African Trade Unions (COSATU). This use of the term labour in this paper imbues all of these, including trade union officials, educators and even officials in leadership positions. In the context of climate change and the just transition it extends beyond COSATU and includes the National Council of Trade Unions, the Federation of Unions of South Africa, the Democratic Municipal and Allied Workers Union of South Africa, the National Union of Metalworkers of South Africa (NUMSA) and the Food and Allied Workers Union (FAWU). It is important to point out that labour like any other collective group under a single term does not imply homogeneity. There are various voices, perspectives, positions, perceptions, orientations, beliefs and practices and some are fundamentally opposed to the other. As such, the author puts forward a perspective from within labour, shared by many, but by no means representative of labour. At best the views presented here are reflective of a labour perspective.

ii   The just transition is used to describe a process of transition which is based on fairness and gives due consideration to workers, communities, the poor and disadvantaged, the vulnerable and the environment in the transition process. The just transition was adopted by labour internationally and more specifically in the context of climate change, in recognising the necessity of moving or transitioning from a high carbon intensive economy to a low carbon intensive economy. However, this transition needs to take cognisance of the particular impact for workers/labour, jobs and communities that will be affected adversely. The just transition is thus not ad hoc; it is a well-planned process that mitigates the negative impact through redeployment, reskilling and redirecting employment to new sectors. In this way, the consequence of transitioning as a result of the capitalist mode of production is not a consequence borne by labour. Moreover, it is an opportunity to challenge the capitalist system and the modes of production, which are responsible for the climate crisis and offer an opportunity to change the system. The just transition also creates the impetus and opportunity to rethink jobs and recreate jobs as green jobs and meet the Sustainable Development Goal 8 objective of decent jobs. However, for most civil society organisations in the 'green space', the just transition by and large means transitioning from a high carbon to low carbon economy – with 'just' implying due consideration is given to those impacted by the transition, without challenging the system or mode of production that is responsible. Herein lies the contestation.

TWELVE

The future of mineral mining

Pathways for a wellbeing economy approach

LORENZO FIORAMONTI

THE CRISES THE WORLD FACES require a fundamental economic transformation. From climate change to rampant inequality, contemporary challenges point to the need to move away from the extractive development model that dominated the 20th century, towards a different approach to prosperity that goes beyond simply increasing production and consumption. We need a balanced economy, which reconnects human beings with each other and with the natural ecosystems underpinning their very existence.

The term introduced to describe this different development model is 'wellbeing economy' (Fioramonti, 2017b). Unlike conventional growth-based industrialisation, an economy that pursues wellbeing must recognise that 'value' can be created in many different ways, not only through conventional industrial production. In particular, it must appreciate the importance of both natural and social systems in value creation, which have been traditionally neglected by conventional approaches to growth. This means breaking the vicious cycle of consumption, which generates enormous negative externalities for nature and society. Besides reconnecting human beings with natural ecosystems, a wellbeing economy must empower citizens to become active change makers and value creators, rejecting the passive notion of consumers.

Africa has a legitimate aspiration to develop. For too long, it has been pillaged by colonialism and by foreign and national corporations extracting non-renewable resources and exploiting human beings, leaving behind a wake of destruction, both in the social and natural realms. But if Africa's new development journey is to be successful, it needs to start by rethinking what development really means in the 21st century. The 'Africa Rising' discourse (Fioramonti, 2014), which became popular after 2010, has assumed that mass-scale production and consumption is exactly what the continent needs in order to prosper. It has touted more consumerism and more extraction as the way to go. Not only is this approach anachronistic in a world that seeks balance and sustainable prosperity, but it is also unrealistic, given the infrastructural limitations and socio-economic gaps marring the continent (Fioramonti, 2014).

As expected, the 'rising' mantra has fuelled more exploitation, an excessive focus on foreign direct investment (thus exacerbating the dependency on commodity cycles) and a growing emphasis on large-scale fossil-fuel based industrialisation, rather than a more strategic focus on renewable energy, small, medium and micro-enterprises and human capital development. Such a short-sighted development discourse has focused almost exclusively on the quantity rather than the quality of growth. It has provided policy makers with the wrong incentives, leading to more social tensions, rampant inequality and ultimately laying the macro-economic conditions for more downturns. Indeed, over the past few years, we have seen many African countries plagued by recessions, in particular Nigeria, a country that had become the symbol of the 'rising' continent. We have also seen millions of African people, especially the youth, rising against extractive institutions and their corporate and political masters (Branch and Mampilly, 2015).

Even if the reader is unconvinced by my critique of the growth economy, s/he may agree that we are unlikely to see the high growth rates of the past, especially in a country like South Africa. Even China has slipped to its lowest growth rates in three decades (Allen, 2017). The International Monetary Fund believes that the 21st century may experience a 'secular stagnation' (IMF, 2016) and the World Bank announced in 2018 that global economic growth had 'peaked' (Donnan, 2018). The likelihood of a low growth future compels us to find alternative avenues to develop sustainably and equitably, especially if we want to create a decent livelihood for millions of young people.

Assuming that African policymakers have finally learned the lesson (a dubious assumption), then the real question is: how can Africa develop in the 21st century if traditional recipes for growth are no longer suitable or even desirable?

In this chapter I present some ideas for the transformation of the mineral mining sector as part of a transition to a wellbeing economy in South Africa and in the continent at large. In particular, I show how a development model that is centred on wellbeing would provide enormous opportunities for the sector to reinvent itself and become a champion of job creation as well as sustainable and equitable development. This can be achieved not only by shifting from underground extraction to on-the-ground recycling of minerals, metals and other rare earths (as some pioneering mining companies globally and in South Africa have begun to do), but also through the development of an integrated network of local mining 'artisans' operating like local-level collectors of electronic waste (e-waste), in line with the principles of decentralisation and empowerment underlying the wellbeing economy.

BEYOND GROWTH: THE DAWN OF THE WELLBEING ECONOMY

The dominant approach to development based on economic growth only counts formal market economic activities and emphasises the positive impacts of large transactions (economies of scale), disregarding the negative impacts thereof. It thereby privileges the 'big' economic actors at the expense of small ones (see, for example, Kniivila, 2008). Following such an approach, mainstream politicians, economists and the media have celebrated Africa's growth 'miracle' at least since 2010, yet the development reality is quite different. Disenchanted with an approach to progress that has deteriorated the environment and benefited a few, often at the expense of the majority, many young Africans are voicing their dissatisfaction, whether through service delivery protests (as is the case in South Africa) or through permanent mobilisation (as we have seen in North Africa) (Branch and Mampilly, 2015). There is, of course, no clear trajectory of resistance, but such widespread dissatisfaction coupled with the demographic boom could prompt new generations of Africans to repudiate an economic approach that is losing traction also in the West and in the Far East, as indicated by the international agreements on climate change and sustainable development. African economies must break their dependence on commodities and foreign direct investment (whose real financial gains ultimately leave the continent through licit or illicit financial flows) with a view to focusing on a form of industrialisation which takes into account local needs, which is sensitive to environmental and social dynamics, and which is labour intensive, given the fast-increasing multitude of young people seeking decent employment. Alternatively, these economies will suffer the most in an age in which conventional economic growth is decreasing massively, automation is replacing industrial workers and environmental and social factors are becoming ever-more crucial to achieve sustainable prosperity.

As opposed to the growth economy, what is an economy that pursues wellbeing? A wellbeing economy is centred on social relations and natural ecosystems as the pillars of value creation (Fioramonti, 2017b). The wellbeing economy is a highly integrated system driven by locally based and community-oriented businesses, which are more in tune with local needs, more labour intensive and better suited to customise the production of goods and services in order to avoid over-production, problems of scale and the resulting massive amount of pollution and waste (Fioramonti, 2017b).

Unlike the conventional approach to growth, according to which 'more is better' regardless of its impact on society and nature, the wellbeing economy does not strive for scale at all cost and takes full account of externalities, both positive and negative. Indeed, economies of scale can be efficient for profit maximisation, but they result in separation between producers and consumers and over-production and waste. According to a study conducted in partnership with the United Nations (UN), the 20 largest industrial sectors globally produce more negative consequences on the environment than revenues (Trucost, 2013). This means that, if they were to pay for their impacts on natural capital, they would simply not have enough capital to do so. These companies may look productive from a profit maximisation point of view, but only because society is then saddled with the losses, which take the form of increasing taxes to deal with environmental degradation and lower quality of life due to pollution and climate change. In many cases, the current model of growth trades private profits for social losses, also due to a model of macro-economic accounting that attributes no value to social and environmental impacts (Fioramonti, 2013; Fioramonti, 2017a).

In a wellbeing economy, producing in excess of what is needed does not count as a plus, as is the case in the growth economy, but as a minus. Optimisation rather than maximisation is the key objective. All negative externalities, from the overuse of natural resources to environmental pollution and social stress, must be fully integrated into the measurement of economic success. As a consequence, an economy that fuels inequality, that weakens the social fabric and that consumes more resources than it regenerates, must be seen as underdeveloped and underperforming. To the contrary, an economy that produces positive externalities such as sustainable use of resources, regenerative practices, social integration and empowerment and creation of decent and fulfilling work opportunities (not just jobs) should be seen as more advanced.

The growth economy suffers from a productivity paradox (Fioramonti, 2017b; Jackson, 2009). On the one hand, companies must compete to reduce production costs, both in terms of time and labour, which allegedly makes them more competitive. This is why the market celebrates corporations that cut down costs, generally seen as a sign of 'efficiency' and competitiveness. On the other hand, however, unless more stuff is produced and consumed, an increasing number of people lose their employment, given that the same levels of production can be achieved with fewer workers. If robots are brought into the growth game, then it is clear that many jobs will simply disappear in conventional large industries (as is argued, for example, by West, 2015).

Rethinking work is crucial not only for industrialised economies, which are trapped in a vicious cycle of low growth and structural unemployment. It is indispensable, also for developing economies, where population (especially the young segments thereof) will skyrocket during this century. A record 2.8 billion people by 2060 will generate a massive 'employment bomb' (World Bank, 2015). The 'Fourth Industrial Revolution' will possibly challenge conventional industrialisation models (Rifkin, 2014; Schwab, 2016). But the quality of this disruption may vary enormously, depending on the overarching development approach a society endorses. In a growth economy, machines will arguably replace workers due to their cutting edge at driving economies of scale and mass production; in a wellbeing economy, by contrast, machines will be used by artisans to customise production in accordance with local needs, repairing and fixing goods for durability and reducing excess waste. In one case, we may have systematic joblessness; in the other we may be able to generate millions of livelihoods by replacing top-down industrial systems with bottom-up production processes aided by technology (Schwab, 2016).

In the traditional growth economy, small businesses have long felt isolated and marginalised, especially in Africa. But now we have technologies that support decentralised production while guaranteeing integration through networks. Mobile connections are rather widespread across the continent. Many services are available to millions of users via the internet. Obviously, ideas can travel easily and so can the business strategies of small and micro-enterprises operating in the 'light' economy. But what about manufacturing, which has traditionally required economies of scale, massive upfront investment and complex value chains? Recent developments in information and communications technology (ICT) and new manufacturing technologies, from the blockchain to 3D printing and energy production through small grids that are powered by renewable resources, may very well provide the type of opportunities that we need (see, for example, UNIDO, 2017). By reducing the capital costs involved in high-tech production, new technologies are potentially creating the conditions for a 'rise of the artisans', thus providing opportunities for millions of micro-enterprises to be developed in millions of local communities rather than corporate giants dominating global value chains (Anderson, 2012; Rifkin, 2014).

The wellbeing economy requires a conceptual U-turn regarding the nature of work, insofar as it moves away from the simple quantification of the production–consumption cycle (and the maximisation thereof) with a view to placing emphasis on the quality of the relations underpinning the economic system. While the pursuit of growth induces policies to encourage mass consumption, mostly through an impersonal relationship between producers and consumers (which can be more efficiently performed by robots), the pursuit of wellbeing requires economic policies that pay significant attention to the quality of the transaction, in which the human interaction is essential to determine the value of goods and services. Rather than a top-down 'supermarket economy', the wellbeing economy is a personalised economy. In this case, instead of replacing workers, decentralised machines will empower mechanics, carpenters, plumbers and the like to become high-tech small business leaders (Fioramonti, 2017b).

In the wellbeing economy, the skills that will drive development will concern sustainable farming (e.g., agro-ecology), green energy, integrated production systems, smart transport and goods distribution, new approaches to healthcare and education, as well as resource maintenance (as opposed to resources extraction) and a variety of activities in personal services industry (Fioramonti, 2017b). Technology will be used to improve the quality of jobs and achieve decent work for as many people as possible rather than simply pursue productivity gains at the cost of workers, the environment and society at large.

At the core of the wellbeing economy are small businesses, not large corporations. Small businesses are not only more labour intensive, but they are also more efficient at increasing local wellbeing. While the overall profit of a small-business-based economy may be lower than that generated by large corporations, the social benefits are much higher. Small businesses are more likely to be in touch with customers, to maintain a healthy proximity between employers and employees, to hire local staff and to reinvest in the economies they serve (see, for example, Jackson, 2009). Often driven by families, small businesses' social responsibility is implicit in their own operations, which means these companies are less likely to be driven by legalistic principles of shareholder value. Finally, small businesses tend to be more flexible and can more easily adapt to changing conditions.

As small businesses usually operate in the very communities in which they reside, they tend to internalise externalities better than large companies. Their (often informal) business plans invariably include how to deal with environmental degradation and social impacts, given that small business leaders and staff tend to live in the very community in which their impacts are felt. This is not to say that small businesses cannot act improperly. Like any form of business, small businesses too may act improperly when they are subjected to incentives, both positive and negative. In the growth economy, the marginalisation of small businesses has often resulted in survivalist strategies, which include dubious operations. But in a wellbeing economy placing small businesses at the core of value creation, their transparency and accountability would be increased not only by the right infrastructure and legislation, but also by the fact that local communities will be better equipped to keep them in check. In the end, small business owners have no high-security headquarters and teams of lawyers to hide behind in case of popular opposition.

Against this backdrop, the Africa Progress Panel highlights the importance of small-scale approaches to sustainable development (Africa Progress Panel, 2015). In particular, they highlight the critical role of smallholding farmers in making Africa food secure while ensuring sustainable livelihoods for people. They emphasise the need to focus on renewable energy systems, which can be more easily decentralised at the local level, as opposed to the large capital-intensive infrastructure required by fossil energy sources. In the panel's analysis, energy production must be 'democratised' so that 'Africa's poorest and most vulnerable people could be reached through renewable energy on terms that drive down energy costs, stimulate small and medium-sized enterprises, generate jobs and reduce pollution-related health risks' (Africa Progress Panel, 2015: 23).

A transition to a wellbeing economy is facilitated by at least three factors. The first factor is the systemic inefficiency of the current growth model, which has generated resource depletion and overproduction crises, has made large companies often irresponsible vis-à-vis society and the environment and has created inequalities and instabilities across the globe, especially now that climate change has become an existential threat.

The second factor has to do with innovation, given that new technologies are lowering the transaction costs for small businesses to become not only efficient at local production, but also able to be at the forefront of a circular economy based on customised production (as opposed to mass production), recycling, reuse and upgrading. While in the past, start-up costs were highly concentrated and thus posed a barrier to the capacity of small businesses to take off and succeed at building their niche markets, new technologies are helping innovators overcome that problem. As emphasised by one of the leaders of the ICT revolution of the late 1990s, cited in a BBC report, the availability of new manufacturing technologies, which customise production and multiple markets for local producers, may very well trigger a massive industrial transition. 'The 20th century was about dozens of markets of millions of consumers. The 21st century is about millions of markets of dozens of consumers' (Day, 2013).

The third factor concerns a shifting paradigm in terms of value creation. Are large corporations that deplete non-renewable resources for the financial benefits of a few really productive? Are industries that destroy the environment, which is a fundamental input to the wealth of a society, really adding value? Are jobs that impoverish the social fabric, weaken family and community ties and alienate millions of people really opportunities for a better life? Being much less industrialised than other continents and being home to the world's largest generation of young people, Africa is a fertile ground to experiment with new forms of industrialisation.

In the next section, I consider how a wellbeing-economy approach might impact one of Africa's most dominant industrial sectors, which has been responsible for much destruction and oppression. In particular, the section focuses on mineral mining because of its symbolic relevance as Africa's most traditional form of extractive industry. Other extractive sectors such as oil, coal and gas are not included, mostly because their future is already challenged by the international agreements to curb climate change and by the rise of renewable energy systems. Of course, fossil fuel companies are powerful globally and in Africa. In South Africa, the mineral-energy complex is still dominant in the economy and many governmental positions regarding the future developments of the sector, such as Operation Phakisa (see, for example, AIDC, 2015), are unclear about the need to embark on a serious transition to a post-carbon economy. Nevertheless, consensus is growing globally about the need to shift away from fossil fuels, which is already evidenced by the burgeoning industry of alternative energy sources and the plummeting price of solar and wind energy generation (AIDC, 2015). Yet, there is much less debate about the future evolutions of mineral mining more broadly and how these can be made compatible with better sustainability not only in environmental terms, but also in terms of human and social wellbeing.

MINERAL MINING IN A WELLBEING ECONOMY

Globally and in South Africa, mining operations face multiple challenges, from dropping global demand, volatile prices, rising costs, reduction in resources to social and labour unrest; all factors that are squeezing mineral mining's profit margins and threatening not only the sector, but the sustainability of entire economies that rely on this industry (Deloitte, 2017). Environmentally, the situation is dire too. In particular, the depletion of resources (especially the lower quality of ore grades) means more energy is required for extraction, more health risks are encountered and greater consumption of reagents and water takes place, resulting in a substantial increase in solid waste, such as tailings and waste rock (Green Economy, 2017: 22). These impacts are effectively 'dumped' on the rest of society, thus forming part of state liabilities and making the land unusable for other activities that support communities and trigger economic development (Green Economy, 2017:

22). As recognised by industry representatives in South Africa, 'many promises are made in terms of clean-up and rehabilitation initiatives, but in practice, there has been very little, if any, follow-through and delivery on these commitments' (Green Economy, 2017: 22).

Against the discussion of the wellbeing economy outlined in the previous sections, the future of mineral mining in South Africa requires two critical shifts: the first shift is from extraction to recycling; the second is from centralised approaches to distributed networks.

From extraction to recycling

The process of mining is normally associated with digging holes in the ground with a view to finding the metals and rare earths that support the global economy and power our technological revolution (including the renewable energy systems needed to build post-carbon economies). But this conventional approach misses a major point: an enormous amount of these precious materials has already been mined and it is now lying in our landfills, inside the transistors of disposed computers, TV sets, washing machines and so on. Urban mines of electronic waste (e-waste) are proven to be far richer than natural deposits. To illustrate, a conventional open-pit mine is likely to yield between 1 and 5 grams of gold per tonne. By contrast, mobile-phone handsets can contain up to 350 grams per tonne of gold and computer circuit boards up to 250 grams (Owens, 2013). Rather than drilling the ground, which continues posing not only a major threat to our environment but also to mine workers (both in terms of safety hazards and as long-term health problems), the future of mining is about sifting through the vast amount of metal and mineral scrap (especially e-waste). This can generate economic value while helping solve the environmental problem of waste management and creating safe jobs for people.

Mineral mining and the production of metals are among the industrial processes most responsible for the generation of waste globally. These processes account for waste generation of about 10 billion tonnes a year, which is roughly half the total production of waste globally (WEF, 2015). Much of this waste has significant value. The World Economic Forum (WEF) suggests that aluminium production could be increased by 20 per cent if the right treatment for bauxite waste was introduced (WEF, 2015). For base and precious metals the recovery amounts would be smaller, but still economically profitable given the scarcity of these minerals (WEF, 2015). Some mining companies in South Africa have begun to invest in the treatment of tailings, that is, the waste generated by the mining industry itself. The Jubilee Tailings Treatment Company, which is a subsidiary of Jubilee Minerals, has launched a new venture to recover chrome ore in 2017 (Kilian, 2017). Sibanye Gold has also launched a tailings retreatment project that aims to recover 11 million ounces of gold, 170 million pounds of uranium as well as sulphuric acid from 1.3 billion tonnes of tailings (Creamer, 2016). The company estimates that this will offer 35 years of mine life and over 500 sustainable jobs, while lowering water risks by taking the sulphite material off the dolomites and turning it into acid to be used for conventional leaching processes. These are obviously only small preliminary steps in the right direction which, however, attest to increasing sensibility in some sectors.

Many so-called advanced economies have systems in place to sort and reprocess e-waste, but better efficiency gains can be achieved through more intensive and systematic metal collection. In the USA, for instance, over 30 per cent of aluminium scrap is not collected (WEF, 2015). In general, however, the largest gains can be made not in industrialised economies, but in so-called developing countries, where solid waste data in both urban and rural areas shows the greatest gaps (WEF, 2015).

Metal 'scrap' can also be reintroduced in conventional production processes. Scrap comes in two forms: it can be either home scrap (generated by the metal production process, such as in the milling process) or new scrap (from plants that manufacture metal products) (Javaid and Essadiqi, 2003). Obviously home scrap presents fewer treatment problems as it can be seamlessly reintroduced in the process that generated it in the first place, thus optimising production and internal recycling. Given that home scrap is generally of high quality, letting it go to waste is a clear indication of inefficient metal production. New scrap, instead, is further removed from the original metal production and can be assimilated to downstream recycling (Javaid and Essadiqi, 2003), insofar as it requires a network of logistics at different levels, including appropriate infrastructure and partnerships with a variety of actors in the value chain, as I will discuss below. Finally, downstream recycling (also known as 'old scrap') is the process of reclaiming metals from products that have reached their end of life or end of use, that is, from final consumers such as households, companies, the state and the like. This type of recycling is the most complex, but also the most crucial to support a transition of the mining sector from growth and extraction to wellbeing and sustainability, with the potential to generate new business opportunities and significant numbers of good quality jobs. It is in this area that most gains are possible for a truly transformative approach to mining in a wellbeing economy.

These different levels of recycling are crucial to understand the critical role that the mining industry (especially thanks to its technology designed to separate minerals and metals) could play in such a circular approach to industrialisation, where the ultimate goal is to achieve zero e-waste and complete reutilisation of precious materials in the production process. How can the mining industry play such a central task? Here are a few examples. As the prime source of metals and minerals, the mining sector could influence conventional approaches to design in the downstream processes of production, thus ensuring that products are manufactured in ways that make it easy to separate metals and minerals after disposal. A public-private partnership connecting mining and manufacturing with state agencies such as the Industrial Development Corporation and the Public Investment Corporation should be created to support a transition towards a 'circular' approach to industrialisation, ensuring that goods are designed and manufactured through approaches that make their re-use and recycling as simple as possible. In this regard, mining companies can work directly with manufacturers to shift towards modular approaches, which can help circularity of materials by making it easier to disassemble products in reusable segments. For instance, a recently developed smartphone called FairPhone is designed through a modular approach to allow for downstream repairing and upgrading, either by users themselves or qualified technicians. It is also composed of recycled or conflict-free minerals and is produced by a small company of young entrepreneurs (Fairphone, undated).

Construction practices can also be affected by the mining industry, for instance, by designing steel rods and beams so that they can be disassembled, collected, revamped and certified for reuse in new buildings, rather than destroyed during demolition (WEF, 2015). The partnership between mining groups and public institutions (at the national and local levels) would also be crucial to guarantee that there is an appropriate infrastructure for effectively recycling old scrap. Indeed, the efficiency of the approach depends not only on the lifespan of metal currently in use, which is mostly dictated by industry's internal processes, but also on functional public infrastructure designed for re-use, recycling and upgrading. Moreover, as discussed below, this partnership must create enabling conditions for closer collaboration between large-scale recycling plants and a network of distributed collection units, thus altering the traditional imbalance in the sector, which has been long dominated by big corporations.

With a view to ensuring a holistic approach throughout the circular value chain, from 're-mining' to product use and back, it is important to take into consideration energy requirements and emissions involved in the whole life cycle, thus factually reducing environment impacts across the value chain (WEF, 2015). Thanks to already existing mining industry expertise and technology in separation of metals and minerals, such a circular process through existing mining infrastructure makes it also possible to maximise reuse, not only in high quality applications, but also in those instances when alloying or impurities require that metals and minerals be used for different purposes. It is also likely to be much more energy efficient than conventional mining. For instance, with the appropriate techniques and through modern sorting practices, recycling aluminium only takes 5 to 10 per cent of the energy needed to produce its primary equivalent. Producing steel only with scrap inputs generates 60 per cent less thermal, electric and upstream energy than steel production relying on primary ore (Owens, 2013). According to a study conducted by the WEF, if the mineral industry focused mostly on end of use recycling, with targets hovering around 80 to 100 per cent, this would result in a considerable reduction in polluting emissions and impact on energy efficiency across the life cycle of products, while eliminating the risks and fatalities caused by conventional extractive practices (WEF, 2015). *Nature* reports on a study that shows that most plants could be turned into closed-loop systems, in which all waste is reused or recycled (Owens, 2013). Lead and other base metals could be refined from the slag, and even sulphuric acid (used to leach the precious metals from the copper) could be generated as a by-product of the gases produced by the furnace. 'Besides being more ecological, these circular plants can have yields close to 100 per cent', argues Owens (2013: S4).

Innovative examples of mining companies that have voluntarily shifted from extraction to recycling already exist. For instance, Korea Zinc is a major zinc and lead producer, with a market capitalisation and annual revenue of over $5 billion. In 2010, it partnered with a zinc recycling specialist to build a major recycling plant capable to treat over 800,000 tons of materials, with an internal rate of return of 30 per cent and annual pre-tax earnings of roughly $29 million (WEF, 2015). In Canada, the mining company Teck has developed an e-waste recycling process maximising metal recovery from – among others – cathode ray tube glass, computer parts and circuit boards. Teck became the first metal supplier in Olympic history to include metals recovered from e-waste in the Olympic medals and has processed over 70,000 tonnes between 2006 and 2015 (WEF, 2015).

Possibly the best-known example of such a transition is the Belgian mining giant Umicore, which has become a multinational conglomerate focusing on materials technology and employing large numbers of workers not only in Belgium, but also in Brazil, Canada and China (Umicore, 2013). Umicore was not always an efficient recycler of minerals. It indeed originated as Union Minière, a large mining company set up over one century ago to mine minerals in the Congolese province of Katanga, with a debatable track record in both social and environmental conduct (Umicore, 2013). 'Over the years, however, Umicore has divested drastically from conventional mining operations to become a global leader in e-waste recycling' (Umicore, 2013). Today, it is a multi-sectoral company active in the production of metal-based chemicals for a variety of applications, such as cobalt materials for rechargeable batteries, germanium semiconductor substrates for satellite solar cells, precious metals containing catalytic converters for cars and the like. Umicore has also become the world's largest recycler of precious and rare metals, as well as zinc (Umicore, 2013). It is 'regularly ranked as one of *The Global 100's Most Sustainable Corporations in the World* (Umicore, 2013: 1) and, in 2013, it was the world's top-ranked company (Umicore, 2013). Extractive mining ceased to be part of the company's business model in 2005, when Umicore discontinued its copper extraction and refining operations. A minority stake in zinc extraction from natural deposits was its last remaining presence in the sector and was sold in April 2008 (Reuters, 2008).

Nowadays, Umicore's core business is the recycling and refining of various precious and other non-ferrous metals, as well as certain non-metals such as selenium (Umicore, 2018). Around two-thirds of all materials included in the refining process are by-products from the production of non-ferrous metals, such as dross, matte and spiess from the zinc smelting industry and anode sludge built up during electrolysis. Other sources of materials used for recycling include slag, spent fuel cells, automotive and industrial catalysts and scrap electronic equipment (Umicore, 2018). Battery recycling is also an important segment of activity, focusing mostly on the recycling of spent rechargeable batteries from laptops, mobile phones and hybrid electric vehicles (Umicore, 2018).

From centralised approaches to distributed networks

Besides partnering with the business sector to facilitate such a transition, what other role can public institutions play to support the shift from mining natural deposits to mineral recycling? The first step is certainly to introduce legislation that requires companies to 'internalise' the costs of all negative externalities of primary metal production, including through systems such as full cost accounting, whose methodology is well developed and widely employed in most fields. In particular, new approaches to inclusive wealth – as they have been applied by, among others, bodies of the United Nations – are a good starting point (UNU-IHDP and UNEP, 2014). New forms of accounting for value creation would help governments define the right mix of sanctions and incentives. It would also build an enabling environment for businesses that are committed to minimising negative social and environmental impacts but are discouraged from doing so because of the dominant market dynamics dependent on profit maximisation rather than resource optimisation, including the risk of being outcompeted by polluting competitors.

As already anticipated in the previous section, public institutions must also guarantee that all fundamentals are in place if the development of an efficient infrastructure for circularity of production materials, 'cradle to cradle', is to happen. While mining companies can take care of the management of the recycling process at the level of central nodes (e.g., inside their treatment plants), it is essential that a distributed network of e-waste collection operators is present throughout the territory, ideally in each local community. This is the context in which artisanal mining can be a key factor to the overall success of the strategy.

According to the few estimates available, 'artisanal and small-scale miners may range between 20 and 30 million across the world' (Buxton, 2013: 1). In the late 1990s, the International Labour Organisation already estimated artisanal miners at roughly 13 million, made up of 50 per cent women and 10 per cent children, with over 100 million people depending on them for livelihoods (ILO, 1999). Over the past two decades, numbers have likely increased, driven by a host of factors including gold prices (rocketing from $290 an ounce in October 2001 to roughly $1,260 an ounce in 2017, down from $1,740 an ounce in 2011), new conflict areas where artisanal mining activity can be a source of income for displaced communities (ILO, 1999) (but also of coerced labour, particularly in the Democratic Republic of Congo) and increased demand for minerals such as tin, tantalum and tungsten (all used by the global personal electronics industries).

While large-scale mining employs approximately 2 to 3 million people around the world, small-scale mining is likely to employ 10 times more, with earnings varying greatly from subsistence amounts to above-market rates (Telmer, 2009). In some countries, artisanal mining's production equals or exceeds that of conventional mining, with Chinese small-scale miners producing roughly 75 per cent of their bauxite, their Indonesian equivalents being on par with large production in tin mining and their Brazilian equivalents being responsible for 84 per cent of all construction and building materials.

In Ecuador and Ghana, small-scale mining accounts for about 65 and 27 per cent of all gold, respectively (MMSD, 2002). As discussed in chapter 9 by Valiani and Ndebele in this volume, artisanal mining also offers an important source of income and entrepreneurial development to many poor people, especially women, for whom it is a much more viable option than conventional mining (see also Eftimie et al., 2012).

Overall, 'artisanal and small-scale mining contributes about 15 to 20 per cent of global minerals and metals' (ILO, 1999). Within this, the sector is responsible for the production of approximately 80 per cent of all sapphires, 20 per cent of all gold and up to 20 per cent of diamonds (Levin, 2012). While small-scale miners account for the extraction of a fraction of the global gold supply chain, they make up over 90 per cent of the labour force in this sector. As discussed in other chapters of this book, automation is reducing the ability of conventional mining to generate jobs, thus making artisanal mining even more crucial to support a transition towards a labour-intensive form of industrialisation.

In general, artisanal mining is associated with a number of social, economic and environmental challenges, including illegal cartels, violent dispute settlements and destruction of ecosystems (such as the Amazon); and it is among the main anthropogenic sources of mercury release into the environment (Buxton, 2013). At the same time, it uses less energy than conventional mining, releases fewer greenhouse gasses and produces less waste rock and tailings per unit of gold (Buxton, 2013). It is also three to five times more lucrative than other small-scale, poverty-driven economic activities with impacts on both household income and contribution to local economies (Siegel and Viega, 2009). Wages range upwards of $2 a day depending on the mineral, the miner's role and geography. Uganda's 200,000 artisanal miners make a financial contribution to the economy around 20 times larger than fishers and farmers (Eftimie et al., 2012).

Against the backdrop of a transition from underground mining to recycling and up-cycling, it is clear that qualified artisans could play a major role in driving innovation in the urban context. Indeed, one of the key challenges facing the efficient reconversion of mining facilities into recycling hubs is not technology, but access and collection. At present, only a fraction of e-scrap is actually disposed correctly, let alone recycled. For instance, the UN Environment Programme estimates that only 15 per cent of the gold in waste electronics is recovered and recycled (UNEP, 2013; Owens, 2013).

In order to boost the circularity of this industrial sector, there is therefore a need for a distributed network of specialised technicians equipped with the right tools to maximise collection. It is precisely in this area that the expertise of artisanal miners should be seen as fundamental to generate the right kind of infrastructure for mineral recycling in each local community, both in industrialised countries and in developing economies. It is particularly in the latter grouping that a distributed form of collection could make the most significant difference, given the lack of pre-existing infrastructure in particular in and around large cities, where landfills have become a major social and environmental hazard. It is also in these countries that most traditional mining operations are based, thus generating a comparative advantage to turn the existing mining industries (both large and artisanal) into local hubs for global mineral recycling, obviously in line with progressive and efficient regulatory frameworks.

Such an approach would also help countries like South Africa, which have been plagued by centuries of migrant work due to mining operations, to revitalise economic activities in a much more balanced fashion. While both conventional and artisanal mining are geospatially limited to a few mineral-rich areas of the national territory, which requires workers to migrate away from their communities and concentrate around a few spots where business operations are in place, the transition to mineral recycling makes it possible to build a nationwide network of small businesses, operating in virtually every community, mostly in urban and semi-urban areas, but also in rural localities. Ideally, collection units should be operating out of every neighbourhood, like car spares and spaza shops.

While mining corporations can be more effectively turned into large recycling plants, the critical service of reaching out to communities, setting up collection mechanisms, disassembling e-waste and channelling it upstream will be performed by artisanal miners turned e-waste treatment technicians (Figure 1). Either through mobile units or fixed stores operating at the grassroots, these artisans will be the front line of action in the 'mining for recycling' approach of the wellbeing economy.

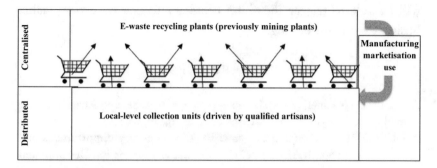

Figure 1: From mining to recycling: e-waste collection and treatment through interaction of mining plants and networks of local artisans

CONCLUSION

The future of mining in a wellbeing economy will not be about mining natural deposits. It will be about mineral reuse and recycling. Such a transition is made imperative by the double goal of minimising environmental impacts and dealing effectively with the ever-growing amount of waste generated by industrial production. It is also a consequence of declining natural deposits and a response to the risk that automation will threaten jobs, in what is already a troubled labour sector.

A country like South Africa, with its dual economy of sophisticated mining plants and a widespread network of artisanal miners, could play a leading role in this new form of mining at the global level. Thanks to its technological cutting edge, it could even expand its reach throughout the southern African region, enabling a transition further afield.

Although there is evidence that mineral recycling is a profitable business, it is unlikely to take root in the short term without a system of full cost accounting for companies as well as a plan to develop the right infrastructure, involving artisanal miners as a network of door-to-door collectors in each and every community. The Sustainable Development

Goals and our commitment to fighting climate change and ecological degradation provide a fertile ground for this kind of change. But only a political vision inspired by the principles of the wellbeing economy will be able to trigger the shifts necessary to achieve a new viable business vision for mining.

REFERENCES

Africa Progress Panel. (2015). 'People, power, planet: Seizing Africa's energy and climate opportunities'. Africa Progress Panel, Geneva.

AIDC. (2015). 'Challenging the state: Renewable energy opportunities and electricity in South Africa in 2015'. Alternative Information Development Centre, Cape Town.

Allen, K. (20 January 2017). 'Chinese growth slips to slowest pace in 26 Years'. *The Guardian*. Available at: https://www.theguardian.com/business/2017/jan/20/chinese-economic-growth-dips-to-67-the-slowest-for-26-years. Accessed 20 March 2018.

Anderson, C. (2012). *Makers*. Random House, London.

Branch, A. and Mampilly, Z. (2015). *Africa Uprising: Popular Protest and Political Change*. Zed Books, London.

Buxton, A. (2013). 'Responding to the challenge of artisanal and small-scale mining: How can knowledge networks help?' International Institute for Environment and Development, London.

Creamer, M. (14 December 2016). 'Sibanye Gold's tailings project launch likely "in next six months". *Mining Weekly*. Available at: http://www.miningweekly.com/article/sibanye-golds-tailings-project-launch-likely-in-next-six-months-2016-12-14. Accessed 20 March 2018.

Day, P. (11 October 2013). 'Imagine a world without shops and factories'. *BBC News Magazine*. Available at: http://www.bbc.com/news/magazine-23990211. Accessed 20 March 2018.

Deloitte. (2017). 'Tracking the trends 2017: Top 10 trends mining companies will face in the coming year'. Available at: https://www2.deloitte.com/global/en/pages/energy-and-resources/articles/tracking-the-trends.html. Accessed 30 September 2017.

Donnan, S. (9 January 2018). 'Global economic growth has peaked, says world bank'. *Financial Times*. Available at: https://www.ft.com/content/4b9e6190-f55e-11e7-88f7-5465a6ce1a00. Accessed 24 January 2018.

Eftimie, A., Heller, K., Strongman, J., Hinton, J., Lahiri-Dutt, K. and Mutemeri, N. (2012). 'Gender dimensions of artisanal and small-scale mining: A rapid assessment toolkit'. World Bank, Washington DC.

FairPhone. (Undated). www.fairphone.com.

Fioramonti, L. (2013). *Gross Domestic Problem: The Politics Behind the World's Most Powerful Number.* Zed Books, London.

Fioramonti, L. (2014). 'Africa rising? Think again'. In Heinrich Böll Stiftung *Perspectives: Political Analyses and Commentary*, Issue 1, pp. 6–9.

Fioramonti, L. (2017a). 'The world after GDP: Economics, politics and international relations in the post-growth era'. Polity, Cambridge.

Fioramonti, L. (2017b). *Wellbeing Economy: Success in a World Without Growth.* MacMillan, Johannesburg.

Green Economy. (2017). 'Tailings and failings'. *Green Economy Journal*, Issue 25, pp. 22.

ILO. (1999). 'Social and labour issues in small-scale mines'. International Labour Organisation, Geneva.

IMF. (2016). 'Too slow for too long. World economic outlook 2016'. International Monetary Fund, Washington DC.

Jackson, T. (2009). *Prosperity Without Growth.* Earthscan, London.

Javaid, A. and Essadiqi, E. (2003). 'Final report on scrap management, sorting and classification of steel'. Government of Canada, Ottawa.

Kilian, A. (5 September 2017). 'Jubilee announces partnership with DCM'. *Engineering News.* Available at: http://www.engineeringnews.co.za/article/jubilee-announces-partnership-agreement-with-dcm-2017-09-05. Accessed 20 March 2018.

Kniivila, M. (2008). 'Industrial development and economic growth: implications for poverty reduction and income inequality'. In O'Connor, D. and Kjollerstrom, M. (eds). *Industrial Development in the 21st Century.* Zed Books, London.

Levin, E. (2012). 'Understanding artisanal and small-scale mining in protected areas and critical ecosystems: A growing global phenomenon'. Presentation for ASM Protected and Critical Ecosystems. London Roundtable.

MMSD. (2002). 'Breaking new ground: Mining, minerals and sustainable development'. IIED, London. Available at: http://pubs.iied.org/pdfs/G00905.pdf. Accessed 30 September 2017.

Owens, B. (2013). 'Mining: Extreme prospects'. *Nature*, Vol. 495, pp. S4–S6.

Reuters. (4 April 2008). 'Update 1-Umicore sells shareholding in Padaeng'. Available at: https://uk.reuters.com/article/umicore-padaeng/update-1-umicore-sells-shareholding-in-padaeng-idUKL0446126420080404. Accessed 20 March 2018.

Rifkin, J. (2014). *The Zero Marginal Cost Society.* Palgrave Macmillan, New York.

Schwab, K. (2016). 'The Fourth Industrial Revolution'. World Economic Forum, Geneva.

Siegel, S. and Viega, M.M. (2009). 'Artisanal and small-scale mining as an extra-legal economy: De Soto and the redefinition of "formalisation"'. *Resources Policy*, Vol. 34, pp. 51–56.

Telmer, K. (2009). 'Artisanal and small-scale gold mining'. Presentation at the CASM Conference.

Trucost. (2013). 'Natural capital at risk: The top 100 externalities of business'. The Economics of Ecosystems and Biodiversity, London.

Umicore. (2013). 'Umicore ranked number one in global 100 Most Sustainable Companies in the World index'. Available at: http://www.umicore.com/pdf/umicore-ranked-number-one-in-global-100-most-sustainable-companies-in-the-world-index.pdf. Accessed 20 March 2018.

UNEP. (2013). 'Global mercury assessment 2013: Sources, emissions, releases, and environmental transport'. United Nations Environment Programme, Nairobi.

UNIDO. (2017). 'Accelerating clean energy through industry 4.0: Manufacturing the next revolution'. A report of the United Nations Industrial Development OrganiSation, Vienna, Austria.

UNU-IHDP and UNEP. (2014). 'Inclusive wealth report 2014: Measuring progress toward sustainability'. Cambridge University Press, Cambridge.

WEF. (2015). 'Mining and metals in a sustainable world 2050'. World Economic Forum, Davos. Available at: http://www3.weforum.org/docs/WEF_MM_Sustainable_World_2050_report_2015.pdf. Accessed 30 September 2017.

West, D.M. (2015). 'What happens if robots take the jobs? The impact of emerging technologies on employment and public policy'. Brookings Institution, Massachusetts. Available at: https://www.brookings.edu/wp-content/uploads/2016/06/robotwork.pdf. Accessed 2 August 2018.

World Bank. (2015). 'Africa's population boom: Will it mean disaster or economic and human development gains?'. Available at: http://www.worldbank.org/en/region/afr/publication/africas-demographic-transition. Accessed 30 September 2017.

CONCLUDING REMARKS
─────

Exhausting the debates

Moving to action?

SALIMAH VALIANI

THIS VOLUME PROVIDES THREE SETS of reflections on the question of the future of mining in South Africa. The first set, in the section 'Transforming Mining for a More Inclusive Future', is forward looking within the scope of the African Union's Africa Mining Vision and notions of 'inclusive growth'. The section offers comprehensive inventories of South Africa's mineral endowments and policy and technology options to develop them. In addition, the section presents a historical understanding of artisanal mining as part of the mining industry.

The second set, in the section, 'The Future Contextualised in the Industry's Continuing Past', places the question of the future of the industry within the context of the industry's continuing past, consisting of arguably extreme forms of super exploitation of mine workers, of land and rural communities. Detailed studies of the nature and constraints of mine workers organising over time, of struggles over land, of mining-induced pollution and of women employed in and affected by mining from the 19th century to the present are used to provide insights into the future of mining through little-known histories, both recent and longstanding.

The third set, in the section 'Beyond Mining: Just Transition and Wellbeing', (re)envisions mining and diversification from mining – the latter a desirable goal acknowledged in the Africa Mining Vision – through critical 21st century lenses of just transition, energy democracy and the wellbeing economy.

In offering insights, reflections and recommendations, the various chapters answer some questions while opening others. Read as a conversation, some chapters construct debates among one another, as do the different sections with their sets of reflections. Within the first section, 'Transforming Mining for a More Inclusive Future', new questions that are opened include: How can the mining industry shift – from public and pension fund investors, to actuaries evaluating investments, to chief executive officers of mining companies – from a focus on short-term gain, to a vision and model of long-term sustainability? What is the time frame for the emergence of the combination of new energies, new ideas and reformulated incentives required to take South Africa from mere mineral extraction to the more inclusive trajectory of mineral-based industrialisation? Is the state, rather than a 'development coalition' or a 'collective of partners in mining', the quicker and more effective vehicle to achieve this? If the state, on behalf of the public, is to become a significant 'owner' in the industry, how will actors of the state erect the necessary mix of intellectual expertise, practical skills and independent oversight? And, if artisanal and small-scale mining is part of making mining inclusive, how to overcome the contradictory relations between artisanal/small-scale miners and large-scale miners that is the current reality not only in South Africa, but throughout the African continent and much of the Global South?

A pressing question that overlaps the first and second sections is: If the mining industry is to switch, in terms of minerals mined, to producing for a low-carbon economy, does the prediction of more than a doubling of metals demand – to supply wind, solar and battery storage technologies – not present a contradiction between saving the climate and saving the earth?

New questions presented in the second section, 'The Future Contextualised in the Industry's Continuing Past' include: What is the appropriate balance of land allocation for agriculture and mining, specifically in the contexts of ongoing inequality rooted in historic land dispossession and unresolved mining-induced environmental damage? Can the mining industry of South Africa exit a formula of profitability based largely on low wages (relative to other industries), underinvestment in the health and safety of both female and male workers and a discounting of needs of mining-affected communities? Conversely, will workers, women and peasants find exit, in the near future, from their current location as the least politically mobilised within the frame of industry and the state?

Big questions to be drawn from the third section, 'Beyond Mining: Just Transition and Wellbeing' (and touching on the first section) include: Are collective and publicly owned models of climate-friendly energy preferable for workers, and society at large, to private models? Within a wellbeing economy, what do downstream linkages based on recycled minerals look like and what is the nature of social negotiation around the types of manufactures produced?

Finally, overarching all three sections of the volume is the question of 'stewardship'. How are the inhabitants of South Africa – or southern Africa, given the interconnectedness of mineral seams, water bodies, air, to mention a few – to define stewardship of the land's vast endowments, both beneath and atop? Does this entail assuring that natural wealth as we know it currently remains for the children of the grandchildren of the next generation? Or does it entail working up renewable and non-renewable resources such that new systems of production reduce poverty, hardship and inequality? If the latter, does comprehensive cost benefit analysis, prior to mineral extraction, have a role to play and if so, who carries the costs?

In closing, presented in this volume is one unifying question, an assembling of reflections and insights and yet more questions. The tasks now are to bring life to the debates, to stage them broadly – far beyond the realm of this book – with many actors in many venues, to reframe current debates, engage in the necessary research to achieve consensus and, from there, work collectively to change the mining status quo.

Who and how many can take the lead? This may initially be unclear, but the words of Theo Sowa (2015: 5) are instructive: 'We need leadership that has a greater purpose – that isn't about being given a title and then telling people what to do.' Alannah Young (2006, 67) stresses respect, responsibility, relationships and reciprocity as key leadership themes. From Hope Chigudu and Rudo Chigudu (2015, 46) comes the emphasis: 'Leadership is about learning to shape the future collectively … decision making is collective … diverse people in diverse positions contributing vitally … enhanced leadership from children to parents to grandparents.' And finally, perhaps most crucially, leadership is about 'the power of conviction to continue innovating' (Chigudu and Chigudu, 2015: 51).

REFERENCES

Chigudu, H. and Chigudu, R. (2015). *Strategies for Building an Organisation with Soul.* AIR for Africa (Graça Machel Trust), Johannesburg.

Sowa, T. (2015). 'Preface'. In Chigudu H. and Chigudu R. (eds). *Strategies for Building an Organisation with Soul.* AIR for Africa (Graça Machel Trust), Johannesburg.

Young, A. (2006). *Elders' Teachings on Indigenous Leadership: Leadership is a Gift.* Master's thesis. University of British Columbia Open Collections. Available at: https://circle.ubc.ca/bitstream/handle/2429/5600/ubc-2006-0350.pdf?sequence. Accessed 31 July 2018.

List of abbreviations

ABC	Alkali-Barium-Calcium
AGA	AngloGold Ashanti
AMCU	Association of Mineworkers and Construction Union
AMD	acid mine drainage
Amplats	Anglo American Platinum
AMV	Africa Mining Vision
AMWU	African Mine Workers' Union
ANC	African National Congress
APP	Africa Progress Panel
ARM	African Rainbow Minerals
ARMI	African Rainbow Mineral Exploration and Investment
ASM	artisanal and small-scale mining
AVI	Anglovaal Industries
B-BBEE	Broad-based Black Economic Empowerment
BEE	Black Economic Empowerment
CALS	Centre for Applied Legal Studies
CCTV	closed-circuit television

CEDT	Centre for Economic Development and Transformation
CEO	chief executive officer
CIGS	copper indium gallium selenide
COP 17	United Nations Conference of the Parties 17
COSATU	Congress of South African Trade Unions
CSI	corporate social investment
CSIR	Council for Scientific and Industrial Research
CSR	China South Rail
CUSA	Council of Unions of South Africa
DBCM	De Beers Consolidated Mines
DEA	Department of Environmental Affairs
DMR	Department of Mineral Resources
DRC	Democratic Republic of Congo
DST	Department of Science and Technology
DTI	Department of Trade and Industry
ECL	environmental critical level
EDZ	export development zone
Eskom	Electricity Supply Commission
ESOPS	employee stock ownership plans
ETF	exchange traded fund
EU	European Union
FAWU	Food and Allied Workers' Union
GDP	gross domestic product
GEPF	Government Employee Pension Fund
HDSAs	historically disadvantaged South Africans
HTS	heat tolerance screening
HySA	Hydrogen South Africa
IBMR	Itireleng Bakgatla Mineral Resources
ICT	information and communications technology
IDC	Industrial Development Corporation
IEA	International Energy Agency
ILR	in line recovery
IMF	International Monetary Fund
Implats	Impala Platinum
INDC	intended nationally determined contributions

IOL	International Labour Organisation
IoT	internet of things
IP	intellectual property
IPAP	industrial policy action plan
IPP	independent power producers
IRP	integrated resources plan
Iscor	Iron and Steel Corporation
ISMO	independent systems and market operator
ISR	in situ recovery
JCI	Johannesburg Consolidated Investment Company
JSE	Johannesburg Stock Exchange
KIO	Kumba Iron Ore
LCOE	levelised cost of electricity
LED	light-emitting diode
LSM	large-scale mining
MEA	Millennium Ecosystems Assessment
MEC	minerals energy complex
MHSA	Mine and Health Safety Act
MIGDETT	Mining Industry Growth Development and Employment Task Team
MISTRA	Mapungubwe Institute for Strategic Reflection
MPRDA	Mineral and Petroleum Resources Development Act
NALEDI	National Labour and Economic Development Institute
NDP	National Development Plan
NEDLAC	National Economic Development and Labour Council
NEMA	National Environmental Management Act
NGO	non-governmental organisation
NUM	National Union of Mineworkers
NUMSA	National Union of Metalworkers of South Africa
NWA	National Water Act
NWSA	National Weather Service Act
OB	overburden
OECD	Organisation for Economic Co-operation and Development

OEMs	original equipment manufacturers
PGI	Platinum Guild International
PGM	platinum group metals
PIC	Public Investment Corporation
PMC	public mining company
PPM	Pilanesberg Platinum Mines
PV	photovoltaic
R&D	research and development
RB Plat	Royal Bafokeng Platinum
RBH	Royal Bafokeng Holdings
RDP	Reconstruction and Development Programme
RDO	rock drill operators
REIPPP	Renewable Energy Independent Power Producer Programme
RMB	Rand Merchant Bank
SACAN	South African Climate Action Network
SACP	South African Communist Party
SAHRC	South African Human Rights Commission
SANT	South African Native Trust
Sasol	South African Coal, Oil and Gas Corporation
SDGs	Sustainable Development Goals
SEDA	Small Enterprise Development Agency
SEZ	special economic zone
SIMS	state intervention in the minerals/mining sector
SIOC	Sishen Iron Ore Company
SLP	social and labour plan
SNT	South African Native Trust
SPV	special purpose vehicles
SWF	sovereign wealth fund
TNC	Trans Natal Coal
UAV	unmanned aerial vehicles
UN	United Nations
UNDP	United Nations Development Plan
UNEP	United Nations Environment Programme
UNGP	United Nations Guiding Principles on Business and Human Rights

WEF	World Economic Forum
Westplats	Western Platinum
WiMSA	Women in Mining South Africa
WITS	University of the Witwatersrand
WWF	World Wide Fund for Nature
ZEPA	zero entry production areas

Index

Printed in the United States
By Bookmasters